Lecture Notes in Electrical Engineering

Volume 620

The book series *Lecture Notes in Electrical Engineering* (LNEE) publishes the latest developments in Electrical Engineering—quickly, informally and in high quality. While original research reported in proceedings and monographs has traditionally formed the core of LNEE, we also encourage authors to submit books devoted to supporting student education and professional training in the various fields and applications areas of electrical engineering. The series cover classical and emerging topics concerning:

- Communication Engineering, Information Theory and Networks
- Electronics Engineering and Microelectronics
- Signal, Image and Speech Processing
- Wireless and Mobile Communication
- Circuits and Systems
- Energy Systems, Power Electronics and Electrical Machines
- Electro-optical Engineering
- Instrumentation Engineering
- Avionics Engineering
- Control Systems
- Internet-of-Things and Cybersecurity
- Biomedical Devices, MEMS and NEMS

For general information about this book series, comments or suggestions, please contact leontina. dicecco@springer.com.

To submit a proposal or request further information, please contact the Publishing Editor in your country:

China

Jasmine Dou, Associate Editor (jasmine.dou@springer.com)

India

Aninda Bose, Senior Editor (aninda.bose@springer.com)

Japan

Takeyuki Yonezawa, Editorial Director (takeyuki.yonezawa@springer.com)

South Korea

Smith (Ahram) Chae, Editor (smith.chae@springer.com)

Southeast Asia

Ramesh Nath Premnath, Editor (ramesh.premnath@springer.com)

USA, Canada:

Michael Luby, Senior Editor (michael.luby@springer.com)

All other Countries:

Leontina Di Cecco, Senior Editor (leontina.dicecco@springer.com)

** Indexing: The books of this series are submitted to ISI Proceedings, EI-Compendex, SCOPUS, MetaPress, Web of Science and Springerlink **

More information about this series at http://www.springer.com/series/7818

Mauro Parodi · Marco Storace

Linear and Nonlinear Circuits: Basic and Advanced Concepts

Volume 2

 Springer

Mauro Parodi
Department of Electric, Electronic,
Telecommunications Engineering and Naval
Architecture (DITEN)
University of Genoa
Genoa, Italy

Marco Storace
Department of Electric, Electronic,
Telecommunications Engineering and Naval
Architecture (DITEN)
University of Genoa
Genoa, Italy

ISSN 1876-1100   ISSN 1876-1119   (electronic)
Lecture Notes in Electrical Engineering
ISBN 978-3-030-35046-8   ISBN 978-3-030-35044-4   (eBook)
https://doi.org/10.1007/978-3-030-35044-4

This Springer imprint is published by the registered company Springer Nature Switzerland AG
The registered company address is: Gewerbestrasse 11, 6330 Cham, Switzerland

Preface

In Volume 1 of *Linear and Nonlinear Circuits: Basic & Advanced Concepts*,[1] after introducing basic concepts, we considered only circuits containing memoryless components, whose equations (both topological and descriptive) are algebraic.

This volume is focused on components with memory and on circuits characterized by time evolution, that is, by *dynamics*. The volume is articulated in three parts, and its structure follows the guidelines described in the preface to the whole book (Volume 1), with each part articulated in two independent lecture levels: basic and advanced. The basic chapters are devoted to linear components and circuits with memory, whereas the advanced chapters focus on nonlinear circuits, whose nonlinearity is provided by memoryless components.

The analysis of dynamical circuits is carried out by making reference to the so-called *state variables*, and the concept of *state* runs through the whole volume, innervating with dendritic structure the language used to describe the concepts and even the equations describing a given circuit. Moreover, owing to their proximity to the concept of energy, the state variables are an effective tool for understanding the behavior of a circuit not only from a mathematical standpoint, but also from a physical perspective.

The formalism adopted in this book to describe the state equations is based on system theory and on its general results, so that each circuit (together with its properties) can be viewed as a particular physical system. To this end, the proposed theoretical results are intermingled with case studies that instantiate general ideas and point out methodological and applicative consequences.

Whenever possible, the circuits are compared to physical systems of different natures (mechanical or biological, for instance) that are governed by equations and properties completely similar and therefore exhibit the same dynamical behaviors. To this end, we note the importance of *normalizations*, which make it possible to analyze, with the same conceptual tools, models of physical systems of any nature.

---

[1]Published by Springer in 2018 (Print ISBN: 978-3-319-61233-1; eBook ISBN: 978-3-319-61234-8).

The reader's comprehension of the proposed concepts can be checked by solving the problems appearing at the end of each chapter (mainly the basic chapters) and comparing the obtained results with the solutions provided at the end of this volume.

Part V shows how to analyze circuits with one state variable, that is, first-order circuits. In the linear case (treated in Chap. 9, basic), the complete analytical solution can be obtained based on well-assessed approaches. In the nonlinear case (Chap. 10, advanced), we usually cannot easily find a closed-form analytical solution, but we can discover some important information about the general properties of the solution(s). To this end, we employ the tools of nonlinear dynamics and bifurcation theory.

Part VI generalizes the above concepts to higher-order circuits, both linear (Chap. 11, basic) and nonlinear (Chap. 12, advanced).

Part VII is focused on the analysis of periodic solutions, that is, on circuits exhibiting persistent oscillations. In the linear case (Chap. 13, basic), we describe how to analyze circuits by working in the so-called *sinusoidal steady state*, induced by a sinusoidal input. In the nonlinear case (Chap. 14, advanced), we show how to analyze circuits (called nonlinear *oscillators*) by working in the periodic steady state also in the absence of a forcing periodic input (autonomous oscillators). The analysis can also be carried out when some circuit parameters are varied, thus inducing qualitative changes (bifurcations) in the circuit dynamics and switching the circuit steady state from periodic to stationary or quasiperiodic or even chaotic. Some examples of oscillators (also of a noncircuit nature) are proposed and studied, isolated or coupled or even networked. Also in these cases, the circuit analysis can be easily related to energetic balances; this allows one to relate the general but often abstract concepts provided by system theory to the physics of the considered examples. In our opinion, this is a great help in understanding not only the specific case under consideration, but also the general laws it obeys. With this perspective, as already stated in the general preface to the whole book (see Volume 1), circuits represent an excellent environment for better understanding the relationships between physics, mathematics, and system theory.

We are indebted to our friend Lorenzo Repetto, who carefully revised the preliminary version of this volume, reporting bugs and providing detailed comments. We also acknowledge our colleagues and friends Giovanni Battista Denegri for his helpful comments on Sect. 13.9.2, Matteo Lodi for his invaluable help with the simulations described in Sect. 14.9, and Alberto Oliveri for constructive discussions and comments.

Genoa, Italy   Mauro Parodi
September 2019   Marco Storace

Contents

Part VI   Second- and Higher-Order Dynamical Circuits

About the Authors

Mauro Parodi was appointed full professor of Basic Circuit Theory by the Engineering Faculty at the University of Genoa, Italy, back in 1985. His scientific and teaching activity has been focusing on nonlinear circuits and systems theory, non-linear modeling, and mathematical methods for treatment of experimental data. He is currently affiliated with the Department of Electrical, Electronic, Telecommunications Engineering and Naval Architecture at the University of Genoa, where he has been teaching *Mathematical Methods for Engineers* and *Applied Mathematical Modeling*.

Marco Storace received a Ph.D. degree in Electrical Engineering from the University of Genoa, Italy, in 1998. He was appointed full professor by the same university in 2011 and is currently affiliated with the Department of Electrical, Electronic, Telecommunications Engineering and Naval Architecture. He was a visiting professor at École Polytechnique Fédérale de Lausanne (EPFL), Lausanne, Switzerland, in 1998 and in 2002. The main focus of his research is on theory and applications of nonlinear circuits, with a special emphasis on circuit models of nonlinear systems, such as systems with hysteresis and biological neurons. He is also concerned with methods for piecewise linear approximation (and circuit synthesis) of nonlinear systems, and bifurcation analysis and nonlinear dynamics alike. He has been teaching *Basic Circuit Theory, Analog and Digital Filters*, and *Nonlinear Dynamics* at the University of Genoa. From 2008 to 2009 he served as an associate editor of the IEEE Transactions on Circuits and Systems. He is serving as a Chair Elect (2019/2021) of the IEEE Technical Committee on Nonlinear Circuits and Systems (TC-NCAS).

Mauro Parodi and Marco Storace are also the authors of *Linear and Nonlinear Circuits: Basic & Advanced Concepts—Volume 1* published by Springer in 2018, ISBN 978-3-319-61233-1 (Hardcover), ISBN 978-3-319-61234-8 (eBook).

Acronyms

AC	Alternating current
CCCS	Current-controlled current source
CCVS	Current-controlled voltage source
DC	Direct current
DP	Driving point
KCL	Kirchhoff's current law
KVL	Kirchhoff's voltage law
l.h.s.	Left-hand side
LTI	Linear and time-invariant
ODE	Ordinary differential equation
PWL	Piecewise-linear
r.h.s.	Right-hand side
RMS	Root mean square
SI	International system of units
SSV	Strict-sense state variable
VCCS	Voltage-controlled current source
VCVS	Voltage-controlled voltage source
WSV	Wide-sense state variable
ZIR	Zero-input response
ZSR	Zero-state response

Part V
Components with Memory
and First-Order Dynamical Circuits

Chapter 9
Basic Concepts: Two-Terminal Linear Elements with Memory and First-Order Linear Circuits

Whenever humanity seems condemned to heaviness, I think I should fly like Perseus into a different space. I don't mean escaping into dreams or the irrational. I mean that I have to change my approach, look at the world from a different perspective, with a different logic and with fresh methods of cognition and verification.
—Italo Calvino, Six Memos for the Next Millennium

Abstract In this chapter, we introduce two-terminal components with memory, whose descriptive equations involve the time derivative of one descriptive variable. This implies generally speaking, that the circuit is described by a system of algebraic differential equations. As a consequence, the generic circuit variable can no longer be expressed as an algebraic function of the inputs, but has its own *dynamics*, which will be analyzed for simple (first-order) circuits. The concept of circuit stability is introduced. Moreover, discontinuous functions are defined, in particular the so-called unit step and impulse functions, which are often used in circuit inputs.

9.1 Two-Terminal Linear Elements with Memory

Here we introduce the descriptive equations for two widely used components with memory.

9.1.1 Capacitor

The capacitor (originally called a condenser) is a *passive* electrical device (some examples are shown in Fig. 9.1a), whose two-terminal model is shown in Fig. 9.1b.

© Springer Nature Switzerland AG 2020

M. Parodi and M. Storace, *Linear and Nonlinear Circuits: Basic and Advanced Concepts*,
Lecture Notes in Electrical Engineering 620,
https://doi.org/10.1007/978-3-030-35044-4_9

Fig. 9.1 Capacitor: **a** six
physical devices of different
natures; **b** model

(a)   **(b)**

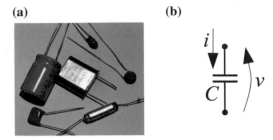

The **capacitor descriptive equation** is

$$i(t) = C \frac{dv}{dt} \qquad (9.1)$$

Here C is a parameter called *capacitance*. The (derived) SI unit of measurement of capacitance is the *farad*,[1] whose symbol is F. According to the descriptive equation, it is evident that $[F] = [A \, s \, V^{-1}]$. Capacitance values of typical capacitors for use in electronics range from picofarad to millifarad (see Table B.1 in Appendix B.1).

It is apparent from the descriptive equation that knowing the specific value of v at a given time t does not provide any information about the corresponding value of $i(t)$. Analogously, knowing the specific value of i at a given time t does not provide any information about the corresponding value of $v(t)$, since (by integrating the descriptive equation and assuming that $v(-\infty) = 0$)

$$v(t) = \frac{1}{C} \int_{-\infty}^{t} i(\tau) d\tau. \qquad (9.2)$$

In both cases, to obtain one of the descriptive variables, we have to know the other one's *history*; in other words, this component *keeps memory* of the past. Then it is linear, time-invariant (assuming that the capacitance is constant), with memory.

The shapes of real devices vary widely (see Fig. 9.1a), but most of them contain at least two electrical conductors (plates) separated by a dielectric; the model in Fig. 9.1b sketches this physical situation. The conductors are typically thin films or metal foils. Materials commonly used as dielectrics include glass, ceramic, plastic film, paper, mica, and oxide layers.

---

[1]The unit is named after Michael Faraday (1791–1867), an English scientist who contributed to the study of electromagnetism and electrochemistry. His main discoveries include the principles underlying electromagnetic induction, diamagnetism, and electrolysis.

The physical principle at the basis of the capacitor's behavior is quite simple. When a voltage v is applied across the conductive plates, a positive charge $+q$ accumulates on one plate and a negative charge q on the other plate. The ratio of the electric charge $q(t)$ to the voltage $v(t)$ is the capacitance C, namely $q = Cv$, where the positive coefficient C depends on the physical (material) and geometric (shape) characteristics of both dielectric and plates. This could be chosen as the capacitor's descriptive equation, but in this book we have from the beginning chosen voltages and currents as descriptive variables, since they can be easily measured (see Sect. 1.3 in Vol. 1). Then, by deriving this equation with respect to t, under the assumption that C does not change with time, we obtain the capacitor's descriptive equation (Eq. 9.2) in terms of the variables v and i.

The term capacitor refers to the property of storing charge. In any physical device, this can be done up to a given limit, after which the dielectric is no more able to insulate the two plates, changes its nature and allows a flow of ohmic current, as if it were a resistor. This limit corresponds to the so-called *breakdown voltage*.

Other properties will be discussed in Sect. 9.2.

Particular case: when $v(t)$ is constant, the descriptive equation reduces to $i(t) = 0$, that is, the capacitor becomes equivalent to an open circuit, apart from the aspects related to the stored energy (that will be treated in Sect. 9.2.1). This is the typical situation of the so-called *DC steady state*, as we will see in Sect. 9.4.

9.1.2 Inductor

The inductor (also called coil or reactor) is a *passive* electrical device (some examples are shown in Fig. 9.2a), whose two-terminal model is shown in Fig. 9.2b.

(a)   **(b)**

Fig. 9.2 Inductor: **a** three physical devices of different nature; **b** model

The **inductor descriptive equation** is

$$v(t) = L \frac{di}{dt} \tag{9.3}$$

L is a parameter called *inductance*. The (derived) SI unit of measurement of inductance is the *henry*,[2] whose symbol is H. According to the descriptive equation, it is evident that $[H] = [V \, s \, A^{-1}]$. Inductance values of typical inductors for use in electronics range from microhenry to henry (see Table B.1 in Appendix B.1, Vol. 1).

As well as the capacitor, the inductor is a component linear, time-invariant (assuming that the inductance is constant) and with memory.

The shapes of real devices vary widely (see Fig. 9.2a), but most of them contain at least a coil of conductive material, whose turns are often wound around a ferromagnetic core (usually made of ferrite); the model in Fig. 9.2b sketches a coil winding.

The physical principle at the basis of the inductor behavior is quite simple. When a current i flows through the coil, it induces a magnetic flux ϕ. The ratio of the magnetic flux $\phi(t)$ to the current $i(t)$ is the inductance L, namely, $\phi = Li$, where the positive coefficient L depends on the physical (material) and geometrical (e.g., shape, number of turns) characteristics of both coil and core. Once more, this could be chosen as the inductor descriptive equation, but we want voltages and currents as descriptive variables. Then, by deriving this equation with respect to t, under the assumption that L does not change with time, we obtain the inductor descriptive equation (Eq. 9.3) in terms of the variables v and i.

In this book, we assume (unless otherwise stated) that the inductance is constant and then that the component is linear and time-invariant. These are assumptions not always satisfied by real inductors; for instance, inductors with ferromagnetic cores are nonlinear, since the inductance changes with the current. Moreover, the model neglects the presence of resistance (due to the resistance of the wire and energy losses in core material) and capacitance in real devices.

Remark: Every wire or other conductor generates a magnetic field when current flows through it, so every portion of wire has some inductance.

Particular case: when $i(t)$ is constant, the descriptive equation reduces to $v(t) = 0$, that is, the inductor becomes equivalent to a short circuit, apart from the aspects related to the stored energy (see the next section). This is the typical situation of the *DC steady state* (see Sect. 9.4).

---

[2]The unit is named after Joseph Henry (1797–1878), an American scientist who discovered the electromagnetic phenomenon of self-inductance. He also discovered mutual inductance, independently of Michael Faraday.

9.2 Capacitor and Inductor Properties

9.2.1 Energetic Behavior

The energy absorbed by a capacitor is

$$w_C(t) = \int_{-\infty}^{t} p_C(\tau)d\tau = \int_{-\infty}^{t} Cv(\tau)\frac{dv}{d\tau}d\tau, \tag{9.4}$$

whereas the energy absorbed by an inductor is

$$w_L(t) = \int_{-\infty}^{t} p_L(\tau)d\tau = \int_{-\infty}^{t} Li(\tau)\frac{di}{d\tau}d\tau. \tag{9.5}$$

It is common to assume that $v(-\infty) = 0$ and $i(-\infty) = 0$. In other words, we assume that there exists a time in the past at which both the capacitor and the inductor were uncharged. Owing to this assumption, we have

$$w_C(t) = \tfrac{1}{2}Cv^2(t) \quad \text{and} \quad w_L(t) = \tfrac{1}{2}Li^2(t). \tag{9.6}$$

Capacitor and inductor are passive components, so we must have $w_C(t) \geq 0$ and $w_L(t) \geq 0$ for all t (see Sect. 3.3.5 in Vol. 1). These inequalities are valid if and only if we have, as anticipated, $C > 0$ and $L > 0$ respectively. Moreover, passivity also means that these components cannot deliver more energy than they absorbed previously. However, in contrast to the resistor, which dissipates the absorbed power, the capacitor and inductor *store* the absorbed energy; in particular, the capacitor stores it through an electric field, the inductor through a magnetic field. In other words, the resistor is passive and dissipative, whereas the capacitor and inductor are passive and *conservative*. This can be easily proved as follows (the proof is given for the capacitor, but can be applied, mutatis mutandis, also to the inductor).

The energy variation between two generic times t_A and $t_B > t_A$ is

$$\Delta w_C = w_C(t_B) - w_C(t_A) = \frac{1}{2}C(v^2(t_B) - v^2(t_A)).$$

This means that if $v(t_B) = v(t_A)$, the energy variation is 0, even if $w_C(t)$ changed in the interval $[t_A, t_B]$ and independently of the shape of $v(t)$ between $v(t_A)$ and $v(t_B)$, that is, of the path followed. Figure 9.3 shows two different voltage paths I and II, which connect the same starting and ending points, thus corresponding to the same energy variation.

In mechanics, a completely analogous property holds when we consider a particle subject to the action of a conservative[3] force: the work done by this force in moving this particle from one point to another depends only on the initial and final positions

---

[3] Two examples of conservative forces are gravitational forces and elastic spring forces.

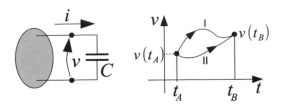

Fig. 9.3 The capacitor as a conservative element

of the particle (with respect to some coordinate system), and it is independent of the path followed.

9.2.2 Gyrator and Two-Terminals with Memory

As stated in Vol. 1, Sect. 5.7.2, the most interesting application of the gyrator is the possibility of converting an inductor into a capacitor and vice versa. Indeed, if a capacitor is connected to the second port of a gyrator, as shown in Fig. 9.4a, we obtain

$$\begin{cases} i = g_m v_C, \\ -C\dfrac{dv_C}{dt} = -g_m v. \end{cases}$$

Then v_C can be obtained from the first equation and substituted in the second one, thus giving us the descriptive equation of the composite two-terminal:

$$v = \frac{C}{g_m^2}\frac{di}{dt}. \tag{9.7}$$

(a)   **(b)**

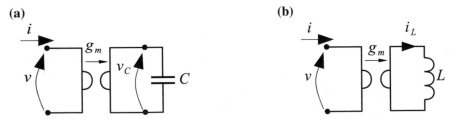

Fig. 9.4 **a** Inductor obtained by connecting a gyrator and a capacitor; **b** capacitor obtained by connecting a gyrator and an inductor

This confirms that when a capacitor of capacitance C is connected to the second port of a gyrator with gyration conductance g_m, the two-terminal "viewed" at the first port is equivalent to an inductor of inductance C/g_m^2.

Similarly, if an inductor of inductance L is connected to the second port of a gyrator with gyration conductance g_m, as shown in Fig. 9.4b, the two-terminal "viewed" at the first port is equivalent to a capacitor of capacitance Lg_m^2.

9.2.3 Series and Parallel Connections

The parallel connection of two capacitors C_A and C_B (see Fig. 9.5a) is equivalent (macromodel) to a single capacitor with capacitance $C_A + C_B$ (see Fig. 9.5b).

This can be shown by considering that $i_A = C_A \dfrac{dv}{dt}$ and $i_B = C_B \dfrac{dv}{dt}$ and that $i = i_A + i_B = (C_A + C_B) \dfrac{dv}{dt}$. This is in line with the physical interpretation of the capacitor as a container of charge: for any voltage v, the overall charge is the sum of the individual contributions $C_A v$ and $C_B v$.

The series connection of two capacitors C_A and C_B (see Fig. 9.6a) is equivalent (macromodel) to a single capacitor with capacitance $\dfrac{C_A C_B}{C_A + C_B}$ (see Fig. 9.6b).

This can be easily shown by considering that $i = C_A \dfrac{dv_A}{dt} = C_B \dfrac{dv_B}{dt}$ and $v = v_A + v_B$. Then $\dfrac{dv}{dt} = \left(\dfrac{1}{C_A} + \dfrac{1}{C_B}\right)i$, that is, $i = \dfrac{C_A C_B}{C_A + C_B} \dfrac{dv}{dt}$.

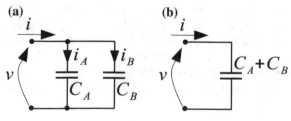

Fig. 9.5 a Parallel connection of capacitors; **b** equivalent macromodel

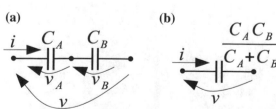

Fig. 9.6 a Series connection of capacitors; **b** equivalent macromodel

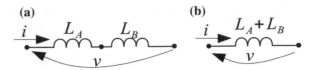

Fig. 9.7 **a** Series connection of inductors; **b** equivalent macromodel

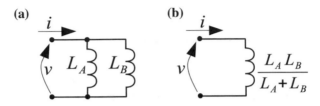

Fig. 9.8 **a** Parallel connection of inductors; **b** equivalent macromodel

Similarly, you can easily check that the series connection of two inductors L_A and L_B (see Fig. 9.7a) is equivalent to a single inductor with inductance $L_A + L_B$ (see Fig. 9.7b), whereas the parallel connection of L_A and L_B (see Fig. 9.8a) is equivalent to a single inductor with inductance $\dfrac{L_A L_B}{L_A + L_B}$ (see Fig. 9.8b).

9.3 State and State Variables

Systems described by a set of algebraic differential equations evolve with time, that is, they are *dynamical systems*. The presence of dynamics is related to memory, and so modeling the evolution of a dynamical system requires some information about its history. The **state** of a dynamical system at a given time t_0 is the information necessary and sufficient to summarize the circuit history before t_0, that is, the information that, together with the knowledge of the system inputs[4] for any $t \geq t_0$, allows one to predict the state evolution for any $t \geq t_0$.

Because the circuits are particular systems, we can define the state of a circuit as follows.

> The **state of a circuit** at a given time t_0 is a set of *independent initial conditions* that are necessary and sufficient to determine the circuit evolution for all $t \geq t_0$, provided that the system inputs are known for all $t \geq t_0$. The word "independent" plays a key role; it means that these initial conditions might be fixed arbitrarily, that is, there are no algebraic constraints involving exclusively one or more of these terms (and possibly one or more circuit inputs).

---

[4]In a circuit, the inputs are voltages/currents impressed by voltage/current sources.

Fig. 9.9 Example of a circuit with state

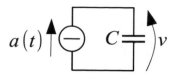

> The **state variables** are the circuit variables corresponding to the chosen set of independent initial conditions.

For instance, as shown in Vol. 1, a memoryless circuit can be solved at any time without need of information about its past history, in which case it is stateless. In contrast, in the very simple circuit shown in Fig. 9.9, the voltage v can be determined for some time $t \geq t_0$ only if we know the circuit's initial condition $v(t_0)$. The capacitor descriptive equation and Kirchhoff's current law (KCL) immediately give $a(t) = C\dfrac{dv}{dt}$: by integrating this equation from t_0 to t and assuming that $a(t)$ is continuous, we obtain $v(t) = v(t_0) + \dfrac{1}{C}\displaystyle\int_{t_0}^{t} a(\tau)d\tau$. Then, given $a(t)$ for $t \geq t_0$, knowing v for $t \geq t_0$ requires the knowledge of the initial condition $v(t_0)$. We remark that v is the only state variable of this circuit.

The above definition of state has two remarkable implications. The first is that in the presence of algebraic constraints, the choice of state variables can be nonunique. The second is that the circuit variables can be divided into two sets: the state variables and the other circuit voltages and currents.

On the one hand, the first implication requires an answer to the following question: *how can we choose the state variables?* On the other hand, the second implication provides a way to simplify the circuit analysis, which is called the *state variables method*.

9.3.1 Wide-Sense and Strict-Sense State Variables

How can we choose the state variables? We can begin by considering the so-called *wide-sense state variables* (WSVs) of the circuit, which are all the voltages across capacitors and all the currents through inductors. Indeed, the capacitor (inductor) is a conservative component, and at time t_0, its voltage (current) is all we need to know about the component's previous history to predict its evolution for $t \geq t_0$, owing to the path-independence proved in Sect. 9.2.1. The voltage of each capacitor and the current of each inductor are *candidates* for being state variables, but they are not necessarily so, since they also have to be *independent*.

For instance, the circuit shown in Fig. 9.10 contains a capacitor, in which case v is a WSV. But in this circuit, KVL imposes that $v(t) = e(t)$, which means that we

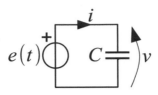

Fig. 9.10 Example of a circuit with one WSV and no SSV

(a) **(b)**

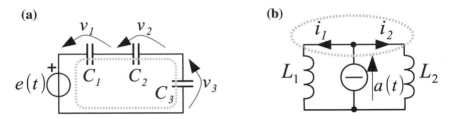

Fig. 9.11 Circuits with number of SSVs less than the number of WSVs

cannot impose arbitrarily an initial condition $v(t_0)$. This implies that $v(t)$ cannot be a state variable. In fact, in this case we do not need an initial condition to find any circuit variable, which can be found algebraically, because $v(t) = e(t)$ and $i(t) = C\dfrac{de}{dt}$. On the other hand, in the "similar" example of Fig. 9.9, there are no algebraic constraints on $v(t)$; thus we can have any initial condition $v(t_0)$ (which is needed to find $v(t)$, as shown above), and $v(t)$ is a state variable.

The real state variables of a circuit are called *strict-sense state variables* (SSVs) and are a subset of independent WSVs, that is, a set of WSVs none of which is related by algebraic constraints to other WSVs and/or to inputs, that is, independent sources. In the example above (Fig. 9.10), v was a WSV, but not an SSV, because of the algebraic constraint $v(t) = e(t)$.

Two further examples are shown in Fig. 9.11. In the circuit of Fig. 9.11a we have three capacitors and then three WSVs. But KVL poses the algebraic constraint $e(t) = v_1 + v_2 + v_3$, meaning that the three WSVs are not independent, because a linear combination of them (their sum) is determined by an input. The same conclusion would hold even in the more general case in which a WSV was a (linear or nonlinear) combination of other WSVs and/or inputs or input time derivatives. In the considered example, the SSVs are two, and we can arbitrarily choose any pair of capacitor voltages.

In the circuit of Fig. 9.11b we have two inductors and then two WSVs. But KCL poses the constraint $a(t) = i_1 + i_2$, meaning that the two WSVs are not independent, for the same reason as before. In this case, there is only one SSV, and we can arbitrarily choose any inductor current.

In the considered examples, the algebraic relationships between WSVs are due to specific topological structures, which are quite simple to detect by inspection:

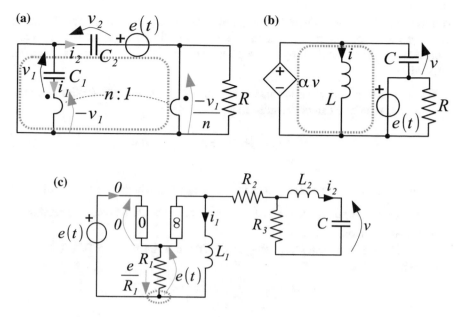

Fig. 9.12 Circuits with number of SSVs less than the number of WSVs

loops containing only capacitors (as in Fig. 9.5a) or capacitors and voltage sources (as in Figs. 9.10 and 9.11a) and cut-sets involving only inductors (as in Fig. 9.7a) or inductors and current sources (as in Fig. 9.11b). In other cases, the constraints are due to the presence of controlled sources or other memoryless two-ports. Three examples are shown in Fig. 9.12. In the circuit of Fig. 9.12a the algebraic constraint between WSVs is $v_2 + e(t) = v_1/n$ (KVL for the dashed loop). In the circuit of Fig. 9.12b, KVL for the dashed loop ($\alpha v = v + e(t)$) implies that v is not an SSV. Finally, in the circuit of Fig. 9.12c, KCL for the dashed nodal cut-set ($i_1 + e(t)/R_1 = 0$) prevents i_1 from being an SSV.

9.3.2 Circuit Models of Algebraic Constraints

Each independent algebraic constraint between a set of WSVs (and possibly inputs) implies the exclusion of one of them from the SSV set of the circuit. After the variable to be excluded is chosen (arbitrarily), the corresponding component (capacitor or inductor) can be represented in the circuit by an equivalent model, entirely formulated in terms of nonstate variables and independent sources. Let $\mathcal{X}$ denote the set of WSVs subject to a given algebraic constraint inside a circuit and consider two rather general forms of linear algebraic constraints: the first is concerned with capacitor voltages, and the second with inductor currents. According to the examples discussed in the previous section, these constraints on WSVs allow the presence of sources, represented in compact form by a voltage source $e(t)$ and by a current source $a(t)$, respectively.

- When the linear constraint is concerned with capacitor voltages (collected in the set $\mathcal{X}$), its structure can be formulated as

$$\sum_{p:v_p\in\mathcal{X}} \alpha_p v_p + e(t) = 0, \qquad \alpha_p \neq 0, \tag{9.8}$$

where v_p denotes the voltage across the capacitor C_p, whose associated current is $i_p = C_p \dfrac{dv_p}{dt}$. As anticipated, $e(t)$ can represent in general the overall contribution of a set of sources.

Let us choose to exclude the voltage $v_k \in \mathcal{X}$ from the set of SSVs. Then from Eq. 9.8, we have

$$v_k = -\frac{1}{\alpha_k} \sum_{p:v_p\in\mathcal{X};p\neq k} \alpha_p v_p - \frac{1}{\alpha_k} e(t).$$

By taking the time derivative of this expression and using the relation $\dfrac{dv_p}{dt} = \dfrac{i_p}{C_p}$ for all p, we easily obtain

$$i_k = -\frac{C_k}{\alpha_k} \sum_{p:v_p\in\mathcal{X};p\neq k} \frac{\alpha_p}{C_p} i_p - \frac{C_k}{\alpha_k} \frac{de(t)}{dt}. \tag{9.9}$$

Note that this expression does *not* involve state variables. The corresponding circuit model for C_k is shown in Fig. 9.13. Each term $-\dfrac{C_k}{\alpha_k} \dfrac{\alpha_p}{C_p} i_p$ of the sum on the right-hand side of Eq. 9.9 can be interpreted as a current contribution supplied by a CCCS. For all $p \neq k$, the driving current i_p is that flowing through the pertinent capacitor C_p. The last term of the right-hand side of Eq. 9.9 is an independent current source.

As an example, in the (previously discussed) circuit of Fig. 9.12a, we can write the algebraic constraint between WSVs as $v_2 = \dfrac{v_1}{n} - e(t)$ and replace the capacitor C_2 with the model shown in Fig. 9.14. After this replacement, the only state variable in the circuit is v_1. A completely analogous result holds, mutatis mutandis, by choosing as SSV the voltage v_2, that is, swapping the roles of the variables v_1 and v_2 inside the algebraic constraint and using the model of the capacitor C_1.

- Similarly, a linear constraint between inductor currents (collected in the set $\mathcal{X}$) can be written as

$$\sum_{p:i_p\in\mathcal{X}} \alpha_p i_p + a(t) = 0, \qquad \alpha_p \neq 0, \tag{9.10}$$

Fig. 9.13 Model for the capacitor C_k, whose voltage v_k is not an SSV

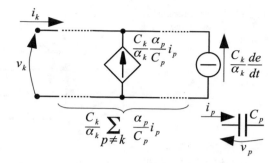

Fig. 9.14 Model for the capacitor C_2 in the circuit of Fig. 9.12a

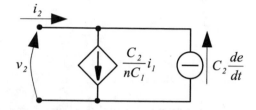

where i_p denotes the current through the inductor L_p, whose associated voltage is $v_p = L_p \dfrac{di_p}{dt}$. The current $a(t)$ can represent the overall contribution of a set of sources.

Following the same line of reasoning as in the previous case, we can decide to exclude $i_k \in \mathcal{X}$ from the SSV set. To this end, we write

$$i_k = -\frac{1}{\alpha_k} \sum_{p:i_p \in \mathcal{X};\, p \neq k} \alpha_p i_p - \frac{1}{\alpha_k} a(t).$$

The time derivative of this expression and the relations $\dfrac{di_p}{dt} = \dfrac{v_p}{L_p}$ give

$$v_k = -\frac{L_k}{\alpha_k} \sum_{p:i_p \in \mathcal{X};\, p \neq k} \frac{\alpha_p}{L_p} v_p - \frac{L_k}{\alpha_k} \frac{da(t)}{dt}. \tag{9.11}$$

The structure of this result is identical, mutatis mutandis, to that found in the previous case. Again, notice that the expression obtained does *not* involve state variables. The corresponding circuit model for L_k contains a set of VCVSs, each driven by a voltage v_p ($p \neq k$) and an independent voltage source, as shown in Fig. 9.15. As an example, the algebraic constraint between WSVs in the circuit of Fig. 9.11b can be written as $i_2 = -i_1 + a(t)$. Therefore, in the circuit we can replace, for instance, the inductor L_2 (whose descriptive variables are i_2 and v_2) with the model shown in Fig. 9.16.

Fig. 9.15 Model for the inductor L_k whose current i_k is not an SSV

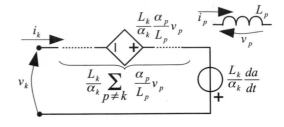

Fig. 9.16 Model for the inductor L_2 in the circuit of Fig. 9.11b

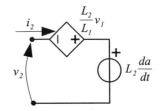

Given a circuit containing N two-terminal linear components with memory (namely, N WSVs) and subject to M ($< N$) independent algebraic constraints, these models allow one to redraw the circuit so that it contains only the $N - M$ memory components associated with the chosen SSVs and to write directly the circuit equations consistent with this setting.

9.3.3   State Variables Method

As will be shown later (see Sect. 9.12), every other circuit variable can be expressed algebraically in terms of state variables and inputs. Owing to this general property, the original system of algebraic differential equations can be solved in two simpler steps:

- first, we solve a system of *differential* equations whose only unknowns are the state variables (i.e., the SSVs);
- then, we solve a system of *algebraic* equations whose only unknowns are all the other circuit variables.

This is the core of the so-called *state variables method*, which will be detailed in the rest of this chapter.

> The **order of a circuit** is the number of its SSVs.

Generally speaking, a linear circuit with N WSVs and $M(< N)$ independent algebraic constraints is said to be of order $N - M$. In the next section, we start analyzing the simplest case, with $N = 1$ and $M = 0$, i.e., first-order circuits with one WSV that is also an SSV.

9.4 Solution of First-Order Linear Circuits with One WSV

A circuit containing only linear and time-invariant (LTI) components and independent sources is called a **linear time-invariant circuit**.

In other words, an LTI circuit can be represented schematically as a linear time-invariant P-port connected to P independent sources, as shown in Fig. 9.17a.

As stated above, in this section we consider LTI circuits with one WSV that is also an SSV and analyze them by defining a common framework.

The generic circuit belonging to this class is a first-order circuit. It can be represented as the parallel connection of the component with memory (say, mc) and a memoryless two-terminal, called the *complementary component*, which contains the rest of the circuit, as shown in Fig. 9.17b, c.

The complementary component can be represented through either its Thévenin equivalent or its Norton equivalent, according to the admitted basis. Generally speaking, the equivalent source in both cases is a linear combination of the P independent sources $\hat{u}_k(t)$ included in the complementary component; in other words, it can be expressed as $\sum_{k=1}^{P} c_k \hat{u}_k(t)$.

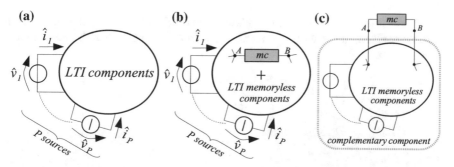

Fig. 9.17 **a** General structure of an LTI circuit; **b** Particular case of a first-order LTI circuit containing only one two-terminal with memory (memory component mc), either an inductor or a capacitor; **c** its equivalent representation: parallel connection between mc and a complementary component

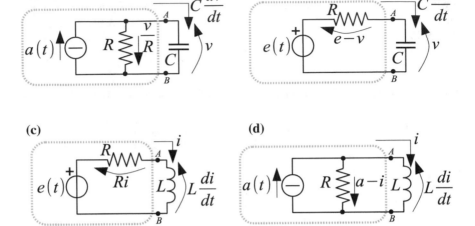

Fig. 9.18 The four possible combinations of two-terminal with memory (L/C) and complementary component (Thévenin/Norton equivalent)

There are four possible combinations, shown in Fig. 9.18: panels a and b show two RC circuits, whereas panels c and d show two RL circuits.

Before analyzing these circuits, we introduce the following concept.

> In a first-order circuit, an **input–output (I/O) relationship** is a differential equation relating a given circuit variable (not necessarily the SSV) considered as the circuit output to the circuit input(s). If the chosen output variable is also the SSV, the I/O relationship is also called a **state equation**.

In the case of LTI circuits, the I/O relationship is a linear differential equation with constant coefficients, as will become apparent from the analysis of the four circuits shown in Fig. 9.18.

> **Case Study a**
> For the RC circuit shown in Fig. 9.18a, we easily obtain (through KCL and multiplying by R) the following differential equation for the state variable v:
>
> $$RC\frac{dv}{dt} + v = Ra(t). \tag{9.12}$$

This is both the state equation and the I/O relationship for v and can also be recast as follows:

$$\frac{dv}{dt} = -\frac{v}{RC} + \frac{a}{C}. \tag{9.13}$$

Case Study b
For the RC circuit shown in Fig. 9.18b, the state equation (I/O relationship) for v is (KVL)

$$RC\frac{dv}{dt} + v = e(t), \tag{9.14}$$

which can also be recast as follows:

$$\frac{dv}{dt} = -\frac{v}{RC} + \frac{e}{RC}. \tag{9.15}$$

Case Study c
For the RL circuit shown in Fig. 9.18c, the state equation (I/O relationship) for i is (KVL)

$$L\frac{di}{dt} + Ri = e(t), \tag{9.16}$$

which can also be recast as follows:

$$\frac{di}{dt} = -\frac{R}{L}i + \frac{e}{L}. \tag{9.17}$$

Case Study d
For the RL circuit shown in Fig. 9.18d, the state equation (I/O relationship) for i can be easily obtained through KCL and multiplying by R:

$$L\frac{di}{dt} + Ri = Ra(t), \tag{9.18}$$

and it can also be recast as follows:

$$\frac{di}{dt} = -\frac{R}{L}i + \frac{R}{L}a. \tag{9.19}$$

In all possible cases, the structure of the state equation (linear and with constant coefficients, as stated above) is the same, and thus we can study the solutions of all these circuits by making reference to the generic equation described below.

> The **state equation** for an LTI circuit with one WSV that is also an SSV (say, x) can be always written either in the form (**I/O relationship**)
>
> $$a_1\frac{dx}{dt} + a_0x = \sum_{k=1}^{P} h_k\hat{u}_k(t), \tag{9.20}$$
>
> or in the following alternative way (**canonical form**):
>
> $$\frac{dx}{dt} = \dot{x} = ax + \sum_{k=1}^{P} b_k\hat{u}_k(t). \tag{9.21}$$

For instance, in Case Study a, we have $x(t) = v(t)$, $a_1 = RC$, $a_0 = 1$, $P = 1$, $h_1 = R$, $\hat{u}_1(t) = a(t)$, $a = -\frac{1}{RC}$, $b_1 = \frac{1}{C}$.

The left-hand side of Eq. 9.20 can be also written as $\mathcal{L}(x)$, to point out that we are applying a linear operator $\mathcal{L} = (a_1\frac{d}{dt} + a_0)$ to the variable $x(t)$. Similarly, Eq. 9.21 can be recast as $\hat{\mathcal{L}}(x) = \sum_{k=1}^{P} b_k\hat{u}_k(t)$, where $\hat{\mathcal{L}} = (\frac{d}{dt} - a)$.

The **homogeneous** differential equation associated with Eq. 9.20 (9.21) is obtained by setting $\hat{u}_k(t) = 0$ for all k, thereby yielding $\mathcal{L}(x) = 0$ ($\hat{\mathcal{L}}(x) = 0$).

In the following sections we provide two possible alternative ways for finding the **unique solution** to Eqs. 9.20 and 9.21 with a given **initial condition** $x(t_0) = x_0$.

9.4.1   General Solution of the State Equation

In this section, we refer to the state equation in canonical form (Eq. 9.21). We introduce a new variable $\tilde{x}$, related to x as follows:

$$x = e^{a(t-t_0)}\tilde{x}. \tag{9.22}$$

By deriving with respect to time, we obtain

$$\dot{x} = a e^{a(t-t_0)} \tilde{x} + e^{a(t-t_0)} \dot{\tilde{x}}.$$

By substituting the above expressions of x and $\dot{x}$ in Eq. 9.21, we obtain

$$e^{a(t-t_0)}(a\tilde{x} + \dot{\tilde{x}}) = a e^{a(t-t_0)} \tilde{x} + \sum_{k=1}^{P} b_k \hat{u}_k(t),$$

i.e.,

$$\dot{\tilde{x}} = e^{-a(t-t_0)} \sum_{k=1}^{P} b_k \hat{u}_k(t).$$

By integrating both sides of this equation between t_0 and t, we now have [1]:

$$\tilde{x}(t) - \tilde{x}(t_0) = \sum_{k=1}^{P} b_k \int_{t_0}^{t} e^{-a(\tau-t_0)} \hat{u}_k(\tau) d\tau.$$

Finally, we can express this solution in terms of the original variable x (recall that $x(t_0) = x_0$ and notice that owing to Eq. 9.22, we have $x(t_0) = e^{a(t_0-t_0)}\tilde{x}(t_0) = \tilde{x}(t_0)$),

$$x(t) = \underbrace{x_0 e^{a(t-t_0)}}_{ZIR} + \underbrace{e^{at} \sum_{k=1}^{P} b_k \int_{t_0}^{t} e^{-a\tau} \hat{u}_k(\tau) d\tau}_{ZSR} \qquad \text{for } t \geq t_0. \qquad (9.23)$$

In the above expression, ZIR is the **zero-input response**, that is, the response that we would obtain in the absence of inputs ($\hat{u}_k(t) = 0$ for all k). It depends only on the circuit parameters (through a) and (linearly) on the initial condition. The term ZSR is the **zero-state response**, that is, the response that we would obtain with initial condition $x_0 = 0$. It depends only on the circuit parameters (through the coefficients a and b_k) and (linearly) on the inputs.

As an example, in the absence of inputs, the state variables of the two physical systems represented in Fig. 9.19 show a similar physical behavior. In case (a), the voltage v of the capacitor is governed by the (homogeneous) equation $\dfrac{dv}{dt} + \dfrac{v}{RC} = 0$; by assuming that there are no inputs and $v(t_0) = V_0$, the solution is the ZIR term: $v(t) = V_0 e^{-\frac{(t-t_0)}{RC}}$. Therefore, the energy initially stored by the capacitor is gradually dissipated by the resistor R. In case (b), a kayak of mass m has an initial velocity $v(t_0) = V_0$. Inasmuch as nobody is paddling and there is no current, the only force acting on the kayak is the viscous resistance Kv oriented in the direction opposite to v. Newton's law $m\dfrac{dv}{dt} = -Kv$ implies that $v(t) = V_0 e^{-\frac{(t-t_0)K}{m}}$, which is a solution similar to that for the capacitor voltage. The role of the viscous force corresponds to that played by the resistance R in the first case, from both a mathematical and a physical (energy dissipation) standpoint.

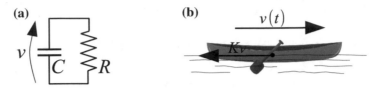

Fig. 9.19 Two physical systems showing a similar physical behavior in the absence of input: **a** an RC circuit; **b** a kayak

9.4.2   Free Response and Forced Response

A second way to find the (unique) solution $x(t)$ to Eq. 9.20 (or 9.21) with initial condition x_0 is based on the superposition principle. In this case, the solution is found by summing different contributions as follows:

$$x(t) = x_{fr}(t) + x_{fo}(t) = x_{fr}(t) + \sum_{k=1}^{P} \hat{x}_k(t), \qquad (9.24)$$

where $x_{fr}(t)$ is called the **free response** (or transient response), $x_{fo}(t)$ is called the **forced response**—since it is the part of the solution induced (forced) by the corresponding inputs—and $\hat{x}_k(t)$ is a **particular integral** due to the kth input $\hat{u}_k(t)$.

The **free response**, or **transient response**, $x_{fr}(t)$ is the solution of the homogeneous differential equation associated with Eq. 9.20 ($\mathcal{L}(x_{fr}) = 0$) or Eq. 9.21 ($\hat{\mathcal{L}}(x_{fr}) = 0$) such that the initial condition holds, that is, such that $x_{fr}(t_0) + x_{fo}(t_0) = x_0$. It has the following structure:

$$x_{fr}(t) = K e^{\lambda(t-t_0)} = \tilde{K} e^{\lambda t}. \qquad (9.25)$$

The term λ is called the **natural frequency** of the circuit. In terms of Eq. 9.20, λ must be such that $\mathcal{L}(\tilde{K}e^{\lambda t}) = 0$, that is, $(a_1\lambda + a_0)e^{\lambda t} = 0$. This requires that

$$a_1\lambda + a_0 = 0, \qquad (9.26)$$

which is called the *characteristic equation* (or *auxiliary equation*) of Eq. 9.20. In practice, the characteristic equation can be directly obtained from the homogeneous differential equation by replacing the term $\dfrac{dx}{dt}$ with λ and $x(= \dfrac{d^0x}{dt^0})$ with $\lambda^0 = 1$.

From the characteristic equation, we obtain $\lambda = -\dfrac{a_0}{a_1}$, provided that $a_1 \neq 0$.

Similarly, the homogeneous differential equation associated with Eq. 9.21 is

$$\dot{x} - ax = 0. \tag{9.27}$$

Then the corresponding characteristic equation is simply $\lambda = a$.

In other words, λ is the **eigenvalue** of Eq. 9.27. It contains information about the way in which a given circuit naturally reacts to any input change, namely, if it tends to contrast this change ($\lambda < 0$), as in the kayak example, or if, on the contrary, it tends to enhance the change ($\lambda > 0$). A deeper discussion about this point will be given in Sect. 9.4.3. For instance, in case studies a and b we find that $\lambda = -\dfrac{1}{RC}$, whereas in case studies c and d, we have $\lambda = -\dfrac{R}{L}$. Notice that both natural frequencies are negative.

We remark that the natural frequency is an intrinsic property of the P-port shown in Fig. 9.17, and then it can be determined by turning off all of the P independent sources. That is why λ is found by analyzing the homogeneous equation associated with the I/O relationship.

The reciprocal of the absolute value of a natural frequency has the physical dimension of time and is called the **time constant** τ of the circuit: $\tau = \dfrac{1}{|\lambda|}$.

A **particular integral** $\hat{x}_k(t)$ is a specific solution of the inhomogeneous differential equation associated with Eq. 9.20 (resp. 9.21) and related to the kth input, that is, $\mathcal{L}(\hat{x}_k(t)) = h_k \hat{u}_k(t)$ (resp. $\hat{\mathcal{L}}(\hat{x}_k(t)) = b_k \hat{u}_k(t)$).

At this point, we can show that solution 9.24 indeed solves Eq. 9.20:

$$\mathcal{L}(x) = \mathcal{L}(x_{fr} + x_{fo}) = \mathcal{L}\left(x_{fr} + \sum_{k=1}^{P} \hat{x}_k\right) = \underbrace{\mathcal{L}(x_{fr})}_{=0} + \sum_{k=1}^{P} \underbrace{\mathcal{L}(\hat{x}_k)}_{=h_k \hat{u}_k} = \sum_{k=1}^{P} h_k \hat{u}_k(t).$$

You can check that solution Eq. 9.24 also satisfies Eq. 9.21.

Remark: $x(t)$ can be written as a sum of terms owing to the linearity of the state equation (Eq. 9.20 or 9.21), which makes it possible to apply the superposition principle (see Sect. 7.3 in Vol. 1).

Once the particular integrals are found (see the examples below), the constant $\tilde{K}$ in Eq. 9.25 can be determined by imposing the initial condition $x(t_0) = x_0$,

$$x(t_0) = x_{fr}(t_0) + \sum_{k=1}^{P} \hat{x}_k(t_0) = \tilde{K} e^{\lambda t_0} + \sum_{k=1}^{P} \hat{x}_k(t_0), \tag{9.28}$$

from which follows

$$\tilde{K} = \left[x_0 - \sum_{k=1}^{P} \hat{x}_k(t_0) \right] e^{-\lambda t_0}. \tag{9.29}$$

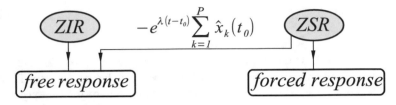

Fig. 9.20 Description of the relationship between solutions Eqs. 9.23 and 9.24

Then

$$x_{fr}(t) = \underbrace{\left(x_0 - \sum_{k=1}^{P} \hat{x}_k(t_0) \right)}_{=K \quad \text{(see Eq. 9.25)}} e^{\lambda(t-t_0)}. \tag{9.30}$$

We remark that solutions Eqs. 9.23 and 9.24 are equal. However, their structures are similar, but not identical: $x_{fr}(t)$ contains the ZIR and part of the ZSR, whereas the remaining part of the ZSR is given by the forced response, as symbolically summarized in Fig. 9.20.

In the simplest cases, each particular integral can be found on the basis of a *similarity criterion*, by assuming that $\hat{u}_k(t)$ is a function of the same form as the corresponding input $\hat{u}_k(t)$. For instance, when $\lambda \neq 0$, the forced response term corresponding to a constant (DC) input U is in turn a constant, say U_0. Indeed, $\mathcal{L}(U_0) = h_k U$, that is, $a_0 U_0 = h_k U$, or equivalently, $U_0 = \dfrac{h_k U}{a_0}$ (recall that we assumed $\lambda \neq 0$, i.e., $a_0 \neq 0$).

Case Study 1: constant (DC) input

Solve the state equation $a_1 \dfrac{dx}{dt} + a_0 x = hU$ for $t \geq t_0$, with initial condition $x(t_0) = x_0$ and $\lambda = -\dfrac{a_0}{a_1} < 0$.

We already know that $x_{fr}(t) = \tilde{K} e^{\lambda t}$ and that $x_{fo}(t) = \hat{x} = \dfrac{hU}{a_0}$. The coefficients h, a_0, and U are given by the problem and are expressed in terms of circuit parameters. The only unknown term is $\tilde{K}$, which can be determined by exploiting the initial condition, according to Eq. 9.28: $x(t_0) = x_{fr}(t_0) + \hat{x}(t_0) = \tilde{K} e^{\lambda t_0} + \dfrac{hU}{a_0} = x_0$, which leads to find $\tilde{K} = \left(x_0 - \dfrac{hU}{a_0} \right) e^{-\lambda t_0}$.

Then the complete solution for $t \geq t_0$ is

$$x(t) = \underbrace{\left(x_0 - \frac{hU}{a_0}\right) e^{\lambda(t-t_0)}}_{x_{fr}(t)} + \underbrace{\frac{hU}{a_0}}_{x_{fo}(t)},$$ (9.31)

which can be also recast as follows:

$$x(t) = \underbrace{x_0 e^{\lambda(t-t_0)}}_{ZIR} + \underbrace{\frac{hU}{a_0}\left(1 - e^{\lambda(t-t_0)}\right)}_{ZSR}.$$

For instance, if we analyze the circuit of Fig. 9.18a for $t \geq t_0$, with $a(t) = A$ and initial condition $v(t_0) = V_0$, the state equation (see Eq. 9.12) is

$$RC\frac{dv}{dt} + v = RA.$$ (9.32)

The circuit natural frequency is $\lambda = -\frac{1}{RC} < 0$ (under the standard assumption $R, C > 0$). The free response is $v_{fr}(t) = Ke^{-\frac{1}{RC}(t-t_0)}$, and the (constant) forced response is $\hat{v} = RA$.

Therefore, the solution for $t \geq t_0$ is

$$v(t) = Ke^{-\frac{1}{RC}(t-t_0)} + RA.$$

To determine K, we impose the initial condition $v(t_0) = K + RA = V_0$, from which follows $K = V_0 - RA$.

Thus, the complete solution is

$$v(t) = \underbrace{(V_0 - RA)e^{-\frac{1}{RC}(t-t_0)}}_{v_{fr}(t)} + \underbrace{RA}_{v_{fo}(t)\equiv\hat{v}} = \underbrace{V_0 e^{-\frac{1}{RC}(t-t_0)}}_{ZIR} + \underbrace{RA(1 - e^{-\frac{1}{RC}(t-t_0)})}_{ZSR}.$$

Similarly, the forced response term corresponding to an input $\hat{u}_k(t) = Ue^{\sigma t}$, with $\sigma \neq \lambda$, is $\hat{x}_k(t) = U_0 e^{\sigma t}$. Indeed, $\mathcal{L}(U_0 e^{\sigma t}) = h_k U e^{\sigma t}$, and therefore, $(a_1\sigma + a_0)U_0 e^{\sigma t} = h_k U e^{\sigma t}$. Then we obtain $U_0 = \dfrac{h_k U}{a_1\sigma + a_0}$; notice that the denominator is nonzero, owing to the assumption $\sigma \neq \lambda$.

As a final remarkable example, we consider a sinusoidal (AC) input $\hat{u}_k(t) = U\cos(\omega t)$. The corresponding forced response term is a sinusoid with the same angular frequency $\hat{x}_k(t) = K_1\cos(\omega t) + K_2\sin(\omega t)$, with K_1 and K_2 such that $\mathcal{L}(K_1\cos(\omega t) + K_2\sin(\omega t)) = h_k U\cos(\omega t)$. A detailed example is provided below.

Case Study 2: sinusoidal (AC) input

Solve the state equation $a_1 \dfrac{dx}{dt} + a_0 x = hU \cos(\omega t)$ *for* $t \geq t_0$, *with initial condition* $x(t_0) = x_0$ *and* $\lambda = -\dfrac{a_0}{a_1} < 0$.

We already know that $x_{fr}(t) = K e^{\lambda(t-t_0)}$. Owing to the similarity criterion, the particular integral is a sinusoidal term with the same frequency and, in general, different amplitude and phase, that is, $\hat{x}(t) = K_1 \cos(\omega t) + K_2 \sin(\omega t)$. Since $\hat{x}(t)$ must be a solution of the state equation, we have

$$a_1 \frac{d\hat{x}}{dt} + a_0 \hat{x} = hU \cos(\omega t)$$

that is,

$$-a_1 K_1 \omega \sin(\omega t) + a_1 K_2 \omega \cos(\omega t) + a_0 K_1 \cos(\omega t) + a_0 K_2 \sin(\omega t) = hU \cos(\omega t).$$

At this point, K_1 and K_2 can be easily identified by solving the following equations:

$$\begin{cases} a_1 K_2 \omega + a_0 K_1 = hU, \\ -a_1 K_1 \omega + a_0 K_2 = 0. \end{cases}$$

Now the only unknown term is K, which can be determined by exploiting the initial condition, according to Eq. 9.28: $x(t_0) = x_{fr}(t_0) + \hat{x}(t_0) = K + K_1 \cos(\omega t_0) + K_2 \sin(\omega t_0) = x_0$, which leads $K = x_0 - K_1 \cos(\omega t_0) - K_2 \sin(\omega t_0)$.

The complete solution for $t \geq t_0$ is then

$$x(t) = \underbrace{(x_0 - K_1 \cos(\omega t_0) - K_2 \sin(\omega t_0)) e^{\lambda(t-t_0)}}_{x_{fr}(t)} + \underbrace{K_1 \cos(\omega t) + K_2 \sin(\omega t)}_{x_{fo}(t)},$$

$$(9.33)$$

which can be also recast as follows:

$$x(t) = \underbrace{x_0 e^{\lambda(t-t_0)}}_{ZIR} + \underbrace{K_1 \cos(\omega t) + K_2 \sin(\omega t) - (K_1 \cos(\omega t_0) + K_2 \sin(\omega t_0)) e^{\lambda(t-t_0)}}_{ZSR}.$$

For instance, if we analyze the circuit of Fig. 9.18c for $t \geq t_0$, with $e(t) = E \cos(\omega t)$ and initial condition $i(t_0) = I_0$, the state equation (see Eq. 9.16) is

$$L \frac{di}{dt} + Ri = E \cos(\omega t).$$

You can check that each addend in the above equation has the physical dimension of volts. The circuit's natural frequency is $\lambda = -\frac{R}{L} < 0$ (under the standard assumption $R, L > 0$). The free response is $i_{fr}(t) = Ke^{-\frac{R}{L}(t-t_0)}$, and the forced response $\hat{i}(t)$ is a sinusoid with the same angular frequency imposed by the input: $\hat{i}(t) = I_c \cos(\omega t) + I_s \sin(\omega t)$. By substituting $\hat{i}(t)$ in the state equation, we obtain

$$L[-I_c \omega \sin(\omega t) + I_s \omega \cos(\omega t)] + R[I_c \cos(\omega t) + I_s \sin(\omega t)] = E \cos(\omega t).$$

By solving the linear system

$$\begin{cases} -LI_c\omega + RI_s = 0, \\ LI_s\omega + RI_c = E, \end{cases}$$

we obtain $I_c = \dfrac{RE}{R^2 + (\omega L)^2}$ and $I_s = \dfrac{\omega LE}{R^2 + (\omega L)^2}$.
The solution for $t \geq t_0$ is

$$i(t) = Ke^{-\frac{R}{L}(t-t_0)} + I_c \cos(\omega t) + I_s \sin(\omega t),$$

whose only unknown is the coefficient K. To find K, we impose the initial condition $i(t_0) = K + I_c \cos(\omega t_0) + I_s \sin(\omega t_0) = I_0$, from which follows $K = I_0 - I_c \cos(\omega t_0) - I_s \sin(\omega t_0)$.
Therefore, the complete solution is

$$i(t) = \underbrace{[I_0 - I_c \cos(\omega t_0) - I_s \sin(\omega t_0)]e^{-\frac{R}{L}(t-t_0)}}_{i_{fr}(t)} + \underbrace{I_c \cos(\omega t) + I_s \sin(\omega t)}_{i_{fo}(t) \equiv \hat{i}(t)} =$$

$$= \underbrace{I_0 e^{-\frac{R}{L}(t-t_0)}}_{ZIR} + \underbrace{I_c \cos(\omega t) + I_s \sin(\omega t) - [I_c \cos(\omega t_0) + I_s \sin(\omega t_0)]e^{-\frac{R}{L}(t-t_0)}}_{ZSR}.$$

In all cases, once the particular integral's form is determined by similarity, $\hat{x}_k(t)$ can be obtained by substitution in the state equation by applying the superposition principle.

Remark: From the complete solution expressed in terms of free response and forced response, it is easy to obtain the specific initial condition that would ensure that there is no transient response and the complete solution coincides with the forced response. In the first case study, it would be $x_0 = \dfrac{hU}{a_0}$ (see Eq. 9.31); in the second case study, it would be $x_0 = K_1 \cos(\omega t_0) + K_2 \sin(\omega t_0)$ (see Eq. 9.33). Similar conditions can be found for the circuit examples.

Fig. 9.21 Case Study 3

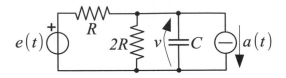

Fig. 9.22 Solution of Case Study 3

(a)

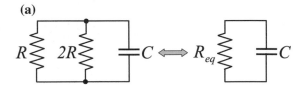

(b)

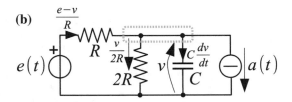

Case Study 3

Solve the circuit shown in Fig. 9.21 for $t \geq t_0$, with initial condition $v(t_0) = V_0$ and with inputs $e(t) = E$ and $a(t) = A \sin(\omega t)$.

We can compute the circuit's natural frequency by turning off all the independent sources (see Fig. 9.22a), thereby obtaining $\lambda = -\dfrac{1}{R_{eq}C} = -\dfrac{3}{2RC} < 0$.

According to Fig. 9.22b, the state equation can be found using the KCL corresponding to the dashed cut-set and recast as follows:

$$2RC\frac{dv}{dt} + 3v = 2E - 2RA\sin(\omega t).$$

Of course, λ is the solution of the characteristic equation.

The solution of the state equation (with the prescribed initial condition) can be expressed as follows:

$$v(t) = v_{fr}(t) + v_{fo}(t) = Ke^{\lambda(t-t_0)} + \hat{v}_{DC}(t) + \hat{v}_{AC}(t).$$

To find the particular integrals $\hat{v}_{DC}(t)$ and $\hat{v}_{AC}(t)$, we apply the superposition principle. By turning off only the current source, we have to solve the equation

$$2RC\frac{d\hat{v}_{DC}}{dt} + 3\hat{v}_{DC} = 2E,$$

whose solution, assuming $\hat{v}_{DC}(t)$ constant, is $\hat{v}_{DC}(t) = \frac{2}{3}E$.

Now, by turning off only the voltage source, we have to solve the equation

$$2RC\frac{d\hat{v}_{AC}}{dt} + 3\hat{v}_{AC} = -2RA\sin(\omega t),$$

from which, by assuming $\hat{v}_{AC}(t) = V_C\cos(\omega t) + V_S\sin(\omega t)$, after some tedious but straightforward manipulations we identify $V_S = -\dfrac{6RA}{9 + (2\omega RC)^2}$ and $V_C = \dfrac{4\omega R^2 C A}{9 + (2\omega RC)^2}$.

Finally, K is obtained by imposing the initial condition

$$v(t_0) = K + \hat{v}_{DC}(t_0) + \hat{v}_{AC}(t_0) = K + \frac{2}{3}E + V_C\cos(\omega t_0) + V_S\sin(\omega t_0) = V_0.$$

Therefore, the complete solution is

$$v(t) = \underbrace{\left(V_0 - \frac{2}{3}E - \frac{4\omega R^2 C A}{9 + (2\omega RC)^2}\cos(\omega t_0) + \frac{6RA}{9 + (2\omega RC)^2}\sin(\omega t_0)\right)e^{-\frac{3}{2RC}(t-t_0)}}_{v_{fr}(t)} +$$

$$+ \underbrace{\frac{2}{3}E + \frac{4\omega R^2 C A}{9 + (2\omega RC)^2}\cos(\omega t) - \frac{6RA}{9 + (2\omega RC)^2}\sin(\omega t)}_{v_{fo}(t)} =$$

$$= \underbrace{V_0 e^{-\frac{3}{2RC}(t-t_0)}}_{ZIR} + ZSR.$$

Remark: If the inputs were $e(t) = E$ and $a(t) = A$, the state equation would be as follows:

$$2RC\frac{dv}{dt} + 3v = 2E - 2RA.$$

In this case, we can choose one of the following alternatives to find the solution:

- consider $2E - 2RA$ as if it were a single constant input; in this case, the forced response coincides with the particular integral $v_{fo} = \hat{v} = \frac{2}{3}E - \frac{2}{3}RA$;
- apply the superposition principle; in this case, we obtain two particular integrals, $\hat{v}_e = \frac{2}{3}E$ and $\hat{v}_a = -\frac{2}{3}RA$.

In both cases, the complete solution is

$$v(t) = \underbrace{\left(V_0 - \frac{2}{3}E + \frac{2}{3}RA \right) e^{-\frac{3}{2RC}(t-t_0)}}_{v_{fr}(t)} + \underbrace{\frac{2}{3}E - \frac{2}{3}RA}_{v_{fo}(t)} =$$

$$= \underbrace{V_0 e^{-\frac{3}{2RC}(t-t_0)}}_{ZIR} + \underbrace{\frac{2}{3}(E - RA)\left[1 - e^{-\frac{3}{2RC}(t-t_0)} \right]}_{ZSR}.$$

Remark: In particular cases, the similarity criterion cannot be applied. For instance, if the state equation is $a_1 \dfrac{dx}{dt} = hU$, we have $a_0 = 0, \lambda = 0$, and $\hat{x}(t)$ cannot be constant. It is easy to check (by substitution) that in this case, the forced response is $\hat{x}(t) = \hat{K} + \dfrac{hU}{a_1}t$, with $\hat{K}$ constant. The whole solution for $t \geq t_0$ can be found by integrating each side from t_0 to t,

$$a_1[x(t) - x(t_0)] = \int_{t_0}^{t} hU dt = hU(t - t_0),$$

i.e., the solution grows linearly with time, starting from a constant value (see also Sect. 9.4.3):

$$x(t) = \underbrace{x_0}_{ZIR} + \underbrace{\frac{hU}{a_1}(t - t_0)}_{ZSR}.$$

In this case, it is impossible to distinguish between $x_{fr}(t) = K$ and $x_{fo}(t) = \hat{x}(t)$. Indeed, we know only that $x(t) = x_0 - \dfrac{hU}{a_1}t_0 + \dfrac{hU}{a_1}t = x_{fr}(t) + \hat{x}(t) = K + \hat{K} + \dfrac{hU}{a_1}t$, and therefore, $x_0 - \dfrac{hU}{a_1}t_0 = K + \hat{K}$. Then K and $\hat{K}$ cannot be determined separately.

The next section is focused on one of the most important concepts related to circuit (or more generally, system) analysis: the concept of *stability*.

9.4.3 Circuit Stability

As stated in the previous section, the natural frequency of a first-order LTI circuit does not depend on the specific input; thus in this section, we initially assume that

Fig. 9.23 Free response for
an absolutely stable circuit
$(\lambda < 0)$

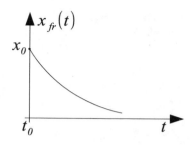

all inputs are turned off. For first-order circuits, the natural frequency is necessarily a real term,[5] whose sign determines the **stability** of the circuit.

If $\lambda < 0$, the free response (coincident with the ZIR, owing to the previous assumption about inputs) generated by an initial condition x_0 at t_0 is a decreasing exponential, as shown in Fig. 9.23. This means that after some time (about 5–10 τ), the free response vanishes and the circuit is said to be **absolutely stable**.

Now we assume that the absolutely stable circuit has one or more inputs turned on. In this case, once the free response vanishes, the circuit reaches a **steady state** imposed by the input(s) through the forced response. This can happen, for instance, once the transient response vanishes in an absolutely stable circuit containing only:

- constant sources, in which case the circuit reaches a *constant (DC) steady state* in which each circuit variable remains constant;
- sinusoidal sources with a unique frequency, in which case the circuit reaches a *sinusoidal (AC) steady state* in which each circuit variable oscillates sinusoidally at the same frequency imposed by the input(s), but in general with different amplitude and phase;
- generic periodic sources (e.g., a square wave or a sawtooth wave), in which case the circuit reaches a *periodic steady state* in which each variable describes a periodic waveform with the period imposed by the input(s), but in general with a different shape.

If $\lambda = 0$, the free response generated by the initial condition x_0 remains constant, as shown in Fig. 9.24. In this case, the circuit is **simply stable**.

As anticipated in Sect. 9.4.2, in the presence of inputs, free response x_{fr} and forced response x_{fo} cannot be fully distinguished.

If $\lambda > 0$, the free response generated by the initial condition x_0 is a diverging exponential, as shown in Fig. 9.25. In this case, the circuit is **unstable**.

In the presence of inputs, the free response x_{fr} is eventually dominant with respect to the forced response x_{fo}. In this perspective, it no longer makes sense to talk about "transient" response in the strict sense; it is more correct simply to state that $x(t)$ diverges.

---

[5]The constants a_0 and a_1 are combinations of the circuit parameters, which are real. Therefore, the solution of the characteristic equation must be real.

Fig. 9.24 Free response for
a simply stable circuit
($\lambda = 0$)

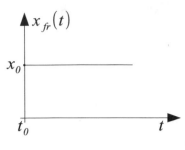

Fig. 9.25 Free response for
an unstable circuit ($\lambda > 0$)

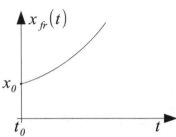

Case Study 4

Find the voltage $v(t)$ for $t \geq t_0$ in the circuit of Fig. 9.18b, by assuming that $e(t) = E$ and $v(t_0) = V_0$.

Equation 9.14 corresponds to the characteristic equation $RC\lambda + 1 = 0$. Then the circuit's natural frequency is $\lambda = -\dfrac{1}{RC}$, and the time constant is $\tau = RC$. Under the standard assumptions ($R, C > 0$), the circuit is absolutely stable, and then, after a transient of some time constants, it reaches a steady state, which is constant, due to the input form.

Then the required solution is $v(t) = Ke^{\lambda(t-t_0)} + \hat{v}$, with $\hat{v}$ constant.

In particular, by substituting $v = \hat{v}$ in the state equation $RC\frac{dv}{dt} + v = E$, we obtain $\hat{v} = E$, since ($\hat{v}$ being constant) $\frac{d\hat{v}}{dt} = 0$.

Now we can obtain the coefficient K by imposing the initial condition $v(t_0)(= V_0) = Ke^{\lambda(t_0-t_0)} + \hat{v} = K + E$; therefore, $K = V_0 - E$.

Then the complete solution is $v(t) = \underbrace{(V_0 - E)e^{\lambda(t-t_0)}}_{v_{fr}(t)} + \underbrace{E}_{v_{fo}(t)\equiv\hat{v}} = \underbrace{V_0 e^{\lambda(t-t_0)}}_{ZIR}$

$+ \underbrace{E(1 - e^{\lambda(t-t_0)})}_{ZSR}$.

Remark: The forced response $\hat{v}$ in the case study above is the solution of the memoryless circuit shown in Fig. 9.26, where the capacitor is replaced by an open circuit. Notice that the forced response is also the DC steady state solution only

Fig. 9.26 Equivalent circuit
for Case Study 4 in DC
steady state

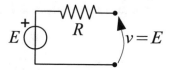

for absolutely stable circuits, as stated in Sect. 9.1.1. Similar considerations can be applied (mutatis mutandis) to absolutely stable circuits containing one inductor, which has to be replaced by a short circuit.

9.5 Forced Response to Sinusoidal Inputs

This section provides an alternative method for finding the particular integral forced by a sinusoidal input of a given frequency. For a given circuit variable, this is the solution of a linear differential equation with constant (real) coefficients, and in the presence of multiple inputs, this is part of a more complex forced response, which can be found by applying the superposition principle, as usual.

Provided that the frequency of the sinusoidal input does not coincide with a natural frequency, this alternative solution strategy is once more based on the *similarity criterion*, but it obtains the particular integral by solving an algebraic equation in the complex domain instead of solving a differential equation in the real domain. It is based on the concept of *phasor*.

For a brief summary about complex numbers, the reader is referred to Appendix A.

9.5.1 Sinusoids and Phasors

A sinusoid can be expressed in general as

$$u(t) = U \cos(\omega t + \phi), \tag{9.34}$$

where U is the *amplitude*, ω is the *angular frequency*, and ϕ is the *phase* of the sinusoid.

Remark: A constant $u(t)$ can be seen as a particular case of a sinusoid with $\omega = 0$ and $\phi = 0$.

A **phasor** $\dot{U}$ is a complex number associated with a sinusoid with fixed angular frequency (say, ω) and with amplitude U and phase ϕ:

$$\dot{U} = U e^{j\phi}. \tag{9.35}$$

For instance, a pure cosine term is

$$u(t) = U \cos(\omega t) = U \cos(\omega t + 0);$$

thus its representing phasor is

$$\dot{U} = U \underbrace{e^{j0}}_{= 1} = U. \tag{9.36}$$

Similarly, a pure sine term is

$$u(t) = U \sin(\omega t) = U \cos\left(\omega t - \frac{\pi}{2}\right);$$

thus its representing phasor is

$$\dot{U} = U \underbrace{e^{-j\frac{\pi}{2}}}_{= -j} = -jU. \tag{9.37}$$

Based on the phasor definition (Eq. 9.35), the **sinusoid** of Eq. 9.34 can be expressed also as

$$u(t) = \Re\{\dot{U}e^{j\omega t}\}. \tag{9.38}$$

Indeed,

$$\Re\{\dot{U}e^{j\omega t}\} = \Re\{Ue^{j\phi}e^{j\omega t}\} = U\Re\{e^{j(\omega t + \phi)}\} = U \cos(\omega t + \phi).$$

We already know (on the basis of the similarity criterion; see Case Study 2 in Sect. 9.4.2) that in an LTI circuit, the forced response induced by a sinusoidal input of angular frequency ω is a sinusoid with the same angular frequency and, in general, different amplitude and phase. This implies that the only features that allow one to distinguish a circuit variable from the others in AC steady state are its amplitude and its phase, that is, the information carried by the corresponding phasor.

This suggests that we can substitute the real circuit variables (voltages and currents) with the corresponding phasors, solve the assigned problem with respect to phasors, and finally obtain the sinusoid corresponding to the phasor solution using Eq. 9.38.

Fig. 9.27 Geometric interpretation of the relationship between a sinusoid and the corresponding phasor

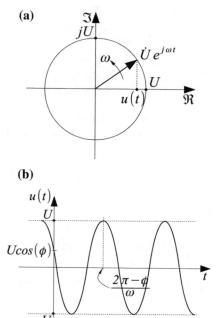

9.5.1.1 Geometric Interpretation

On the one hand, a phasor is a complex parameter, and as such can be represented as a vector with amplitude U and phase ϕ in the complex plane (see Appendix A).

On the other hand, $e^{j\omega t}$ is a time-varying phasor with unitary amplitude and phase ωt. Therefore, it is a vector that rotates with angular velocity ω in the complex plane.

From a geometric standpoint, Eq. 9.38 means that the vector $\dot{U}e^{j\omega t}$ has amplitude U and phase $\omega t + \phi$; thus it rotates in the complex plane. Its projection on the real axis is the sinusoid $u(t)$, as shown in Fig. 9.27.

We remark that in a given circuit forced by a sinusoid with angular frequency ω, the phasors of all the electric variables rotate in the complex plane with the same angular speed. Therefore, two given phasors maintain their fixed phase difference.

9.5.1.2 Phasor of the Derivative of a Sinusoid

Given that a sinusoid $u(t) = U \cos(\omega t + \phi) = \Re\{\dot{U}e^{j\omega t}\}$ is associated with a phasor $\dot{U} = Ue^{j\phi}$, what can we say about the phasor associated with $\dfrac{du}{dt}$?

The phasor of $\dfrac{du}{dt}$ is $j\omega \dot{U}$.

In general, the phasor of $\dfrac{d^n u}{dt^n}$ is $(j\omega)^n \dot{U}$.

This property can be simply proved by taking the derivative with respect to time of either of the two alternative expressions of $u(t)$:

$$\frac{du}{dt} = \frac{d}{dt}\{\Re\{\dot{U}e^{j\omega t}\}\} = \Re\left\{\dot{U}\frac{d}{dt}\{e^{j\omega t}\}\right\} = \Re\{(j\omega\dot{U})e^{j\omega t}\}$$

or

$$\frac{du}{dt} = \frac{d}{dt}\{U\cos(\omega t + \phi)\} = -U\omega\sin(\omega t + \phi) = -U\omega\cos\left(\omega t + \phi - \frac{\pi}{2}\right).$$

In the first case, the phasor is the coefficient of the term $e^{j\omega t}$ in parentheses. In the second case, the phasor is

$$-U\omega e^{j(\phi - \frac{\pi}{2})} = -U\omega e^{j\phi}\underbrace{e^{-j\frac{\pi}{2}}}_{=-j} = j\omega\underbrace{U e^{j\phi}}_{=\dot{U}} = j\omega\dot{U}.$$

You can check that the phasor of $\dfrac{d^2 u}{dt^2}$ is $-\omega^2\dot{U}$.

By replacing each term of a given I/O relationship with the corresponding phasor, we apply the so-called *phasor transform* and map a differential equation with real coefficients in the time domain into an algebraic equation with complex coefficients in the so-called phasor domain.

9.5.2 Phasor-Based Method for Finding a Particular Integral

The particular integral in response to a sinusoidal input with angular frequency ω can be obtained by following a four-step method based on phasors, as an alternative to the method described in Sect. 9.4.2:

Step 1 (time domain). We find the circuit I/O relationship and keep only the sinusoidal input(s) with a given angular frequency ω.

Step 2 (mapping from time domain to phasor domain). By exploiting the relationships between sinusoids and phasors defined in the previous section, we can easily associate the equation found in Step 1 with a corresponding algebraic equation in the complex domain: each variable is replaced by its phasor, and the operator $\dfrac{d^n}{dt^n}$ is replaced by $(j\omega)^n$.

Step 3 (phasor domain). We find the expression of the phasor of the unknown variable in the complex domain.

Step 4 (inverse mapping from phasor domain to time domain). By exploiting Eq. 9.38, we obtain the sinusoid associated with the phasor found in Step 3, which is the forced response.

Case Study 1

Find the forced response $i_{fo}(t)$ *for the circuit analyzed in Case Study 2 in Sect. 9.4.2 using the phasors method.*

Step 1. The I/O relationship is

$$L\frac{di}{dt} + Ri = E\cos(\omega t).$$

Step 2. The I/O relationship in the phasor domain becomes (recall that the phasor of a pure cosine function is simply its amplitude)

$$j\omega L\dot{I} + R\dot{I} = E.$$

Step 3. We algebraically obtain $\dot{I}$:

$$\dot{I} = \frac{E}{R + j\omega L} = \frac{E}{R^2 + (\omega L)^2}(R - j\omega L).$$

Step 4. We apply Eq. 9.38:

$$i_{fo}(t) = \Re\{\dot{I}e^{j\omega t}\} = \Re\{\frac{E}{R^2 + (\omega L)^2}(R - j\omega L)[\cos(\omega t) + j\sin(\omega t)]\} =$$
$$= \frac{E}{R^2 + (\omega L)^2}[R\cos(\omega t) + \omega L\sin(\omega t)].$$

Of course, this solution coincides with the one found in Sect. 9.4.2.

Case Study 2

Find the sinusoidal part of the forced response for the circuit analyzed in Case Study 3 in Sect. 9.4.2 using the phasors method.

Step 1. The complete I/O relationship is

$$2RC\frac{dv}{dt} + 3v = 2E - 2RA\sin(\omega t);$$

thus the forced response contains a constant part $v_{DC}(t)$ due to the DC input and a sinusoidal part $v_{AC}(t)$ due to the sinusoidal input. By applying the super-position principle and removing the inputs that are not sinusoids with angular frequency ω, we obtain the differential equation to be satisfied by $v_{AC}(t)$:

$$2RC\frac{dv_{AC}}{dt} + 3v_{AC} = -2RA\sin(\omega t). \tag{9.39}$$

Step 2. Equation 9.39 in the phasor domain becomes (recall that the phasor of a pure sine is its amplitude multiplied by $-j$)

$$2RCj\omega\dot{V} + 3\dot{V} = -2RA(-j) = 2jRA.$$

Step 3. We algebraically obtain $\dot{V}$:

$$\dot{V}(3 + 2j\omega RC) = 2jRA,$$

that is,

$$\dot{V} = \frac{2jRA}{3 + 2j\omega RC} = \frac{2RA}{9 + (2\omega RC)^2}j(3 - 2j\omega RC) = \frac{2RA}{9 + (2\omega RC)^2}(2\omega RC + 3j).$$

Step 4. We apply Eq. 9.38 to find the AC part of the forced response:

$$v_{AC}(t) = \Re\{\dot{V}e^{j\omega t}\} = \Re\left\{\frac{2RA}{9 + (2\omega RC)^2}(2\omega RC + 3j)[\cos(\omega t) + j\sin(\omega t)]\right\} =$$
$$= \frac{2RA}{9 + (2\omega RC)^2}[2\omega RC\cos(\omega t) - 3\sin(\omega t)].$$

You can check that in this case as well, this part of the solution coincides with the one found in Sect. 9.4.2.

9.5.3 Multiple Periodic Inputs: Periodic and Quasiperiodic Waveforms

In the presence of multiple sinusoidal inputs whose frequencies differ from any natural frequency of the circuit, the forced response is the sum of the forced responses corresponding to each input term. Though each of the sinusoidal terms (both in the input and in the forced response) has its own periodic behavior, the sum of these terms is not necessarily periodic. To point this out, it is sufficient to consider the sum of two sinusoidal terms operating at different angular frequencies:

$$f(t) = A \cos{(\omega_A t)} + B \cos{(\omega_B t)}.$$

In the simplest case, $\dfrac{\omega_B}{\omega_A}$ is an integer p, that is, ω_B is a multiple of ω_A. In this case, $f(t)$ is periodic with period $\dfrac{2\pi}{\omega_A}$, as you can easily check. Moreover, it oscillates p times per period.

For instance, by choosing $\omega_A = \omega_0$, $\omega_B = 3\omega_0$, $A = 1$, and $B = \dfrac{1}{2}$, we obtain the function $f_1(t)$ plotted in Fig. 9.28a.

In general, $\dfrac{\omega_B}{\omega_A}$ is a rational number $\dfrac{p}{q}$, with p and q integers. In this case, the resulting function $f(t)$ is still periodic, with period $T = q \dfrac{2\pi}{\omega_A}$, and it makes p oscillations per period. This can be proved as follows.

Consider $\omega_A = \omega_0$ and $\omega_B = \dfrac{p}{q} \omega_0$ and take $p < q$ without loss of generality. Assuming that the sum is periodic with period T, the terms $\omega_0 T$ and $\dfrac{p}{q} \omega_0 T$ must differ by a *minimum* integer (say k) multiple of 2π:

$$\omega_0 T \left(1 - \frac{p}{q} \right) = k\, 2\pi.$$

Therefore,

$$\omega_0 T = \frac{kq}{q - p}\, 2\pi,$$

which implies $k = q - p$ and $\omega_0 T = q\, 2\pi$.

For instance, by choosing $\omega_A = \omega_0$, $\omega_B = \dfrac{7}{10} \omega_0$, $A = \dfrac{1}{2}$, and $B = 1$, we obtain the function $f_2(t)$ plotted in Fig. 9.28b, where $p = 7$, $q = 10$, $k = 3$, and $T = 10 \dfrac{2\pi}{\omega_0}$.

In the most general case, $\dfrac{\omega_B}{\omega_A}$ is an irrational number. In this case, the resulting function $f(t)$ is *quasiperiodic*, that is, it never repeats itself exactly, despite the periodicity of each of its components. As an example, by choosing $\omega_A = \omega_0$, $\omega_B = \dfrac{\omega_0}{\sqrt{2}}$, $A = \dfrac{1}{2}$, and $B = 1$, we obtain the function $f_3(t)$ plotted in Fig. 9.28c.

(a)

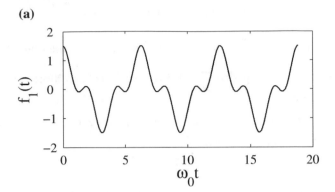

(b)

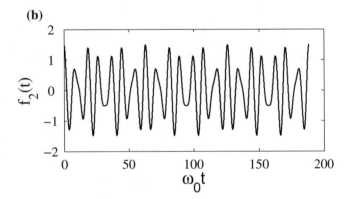

(c)

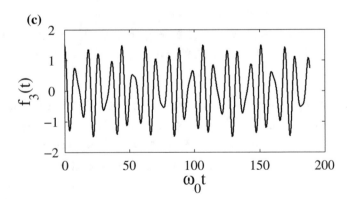

Fig. 9.28 Three examples of a function $f(t)$ with **a** $\dfrac{\omega_B}{\omega_A} = 3$ (periodic); **b** $\dfrac{\omega_B}{\omega_A} = \dfrac{7}{10}$ (periodic); **c** $\dfrac{\omega_B}{\omega_A} = \dfrac{1}{\sqrt{2}}$ (quasiperiodic)

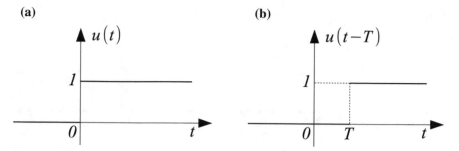

(a)   **(b)**

Fig. 9.29 **a** The unit step function; **b** the same function delayed by T

9.6 Generalized Functions (Basic Elements)

In this section we introduce some discontinuous functions largely used in circuit theory. Inasmuch as they are not continuous, their classical derivative is not defined at the points of discontinuity. This requires that we extend the definition of derivative, and that is the reason for the attribute ("generalized") commonly associated with these functions.

The *unit step* (or Heaviside[6] step) function $u(t)$ is defined as follows:

$$u(t) = \begin{cases} 0 & \text{for } t < 0, \\ 1 & \text{for } t > 0 \end{cases}$$

it is dimensionless and has a so-called *first-order discontinuity* at $t = 0$, as shown in Fig. 9.29a.

In some contexts, this function is arbitrarily defined at $t = 0$, for instance, as $u(0) = 0$ or (more commonly) $u(0) = 1$.

The function $u(t - T)$ corresponds to a unit step delayed by T, as shown in Fig. 9.29b.

The unit rectangular pulse (or impulse) p_T shown in Fig. 9.30a can be analytically expressed as the following combination of two unit step functions:

$$p_T(t) = \frac{1}{T}\left[u\left(t + \frac{T}{2}\right) - u\left(t - \frac{T}{2}\right) \right]. \tag{9.40}$$

---

[6]Oliver Heaviside (1850–1925) was an English electrical engineer, mathematician, and physicist who adapted complex numbers to the study of electrical circuits, invented mathematical techniques for the solution of differential equations, reformulated Maxwell's field equations in terms of electric and magnetic forces and energy flux, and independently coformulated vector analysis, thus making fundamental contributions to telecommunications, mathematics, and science.

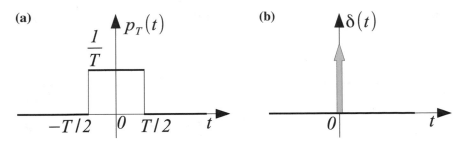

Fig. 9.30 **a** Unit rectangular pulse; **b** Dirac δ-function

This waveform has first-order discontinuities at $t = -\dfrac{T}{2}$ and $t = \dfrac{T}{2}$. The term T plays the role of a parameter. Irrespective of the value of T, the (dimensionless) area of the rectangular pulse is 1, as shown in Fig. 9.30a.

When $T \to 0$, the rectangular pulse height becomes infinite, and its basis infinitely narrow and positioned at $t = 0$. The result, not a function in the conventional sense, is the so-called *Dirac δ-function*,[7] which has a *second-order discontinuity* at $t = 0$:

$$\delta(t) = \begin{cases} 0 & \text{for } t \neq 0, \\ +\infty & \text{for } t = 0, \end{cases} \tag{9.41}$$

whose symbol is shown in Fig. 9.30b. The unit value of the rectangular pulse area is not affected by the above limit process. Thus we have, for all $\alpha > 0$,

$$\int_{-\alpha}^{\alpha} \delta(t) dt = 1. \tag{9.42}$$

Remark: It may happen that the δ-function argument is multiplied by a coefficient. In this case, we have

$$\int_{-\alpha}^{\alpha} \delta(\beta t) dt = \frac{1}{\beta} \int_{-\alpha}^{\alpha} \delta(\beta t) \beta dt. \tag{9.43}$$

By changing the integration variable to $\tau = \beta t$ and assuming $\beta > 0$ for the sake of simplicity, we obtain

$$\int_{-\alpha}^{\alpha} \delta(\beta t) dt = \frac{1}{\beta} \int_{-\alpha\beta}^{\alpha\beta} \delta(\tau) d\tau = \frac{1}{\beta}. \tag{9.44}$$

---

[7]It is named after Paul Dirac (1902–1984), an English theoretical physicist who made fundamental contributions to quantum mechanics. He was awarded the 1933 Nobel Prize in Physics jointly with Erwin Schrödinger.

When $\beta < 0$ (that is, $\beta = -|\beta|$), the value of the previous integral becomes $\dfrac{1}{|\beta|}$. As a general result, we therefore have

$$\delta(\beta t) = \frac{1}{|\beta|}\delta(t). \tag{9.45}$$

The following two relationships between the step function $u(t)$ and the δ-function are a consequence of the definitions:

$$u(t) = \int_{-\infty}^{t} \delta(\tau)d\tau \qquad \text{(for any } t \neq 0\text{)},$$

$$\delta(t) = \frac{du(t)}{dt}. \tag{9.46}$$

The above definition of $\delta(t)$ (as well as any other we can give) is meaningless in the context of ordinary functions, but it finds its meaning in the context of the so-called *generalized functions*, or *distributions* [2].

Remark: The physical dimension of $\delta(t)$ is $[\text{s}^{-1}]$. Then a *voltage pulse* is defined as $v(t) = \Phi\delta(t)$ (with $[\Phi] = [\text{V s}]$), and a *current pulse* is defined as $i(t) = Q\delta(t)$ (with $[Q] = [\text{A s}]$).

Generally speaking, a family of functions can be obtained by integrating or taking the derivative of $\delta(t)$ one or more times with respect to t. The number n of integrations or derivations of $\delta(t)$ is indicated by a superscript $(-n)$ or $(+n)$, respectively. For example, from Eq. 9.46 we can call $u(t)$ the $\delta^{(-1)}(t)$-element of the family.

The *unit ramp* function $t \cdot u(t)$, shown in Fig. 9.31a, is obtained by integrating the unit step function $u(t)$:

$$\int_{-\infty}^{t} u(\tau)d\tau = t \cdot u(t).$$

Then we can write $t \cdot u(t) = \delta^{(-2)}(t)$.

The time derivative of the unit rectangular pulse p_T is the key to introducing another element of the family. From Eq. 9.40 we directly obtain

$$\frac{d}{dt}p_T(t) = \frac{1}{T}\left[\delta\left(t + \frac{T}{2}\right) - \delta\left(t - \frac{T}{2}\right)\right],$$

that is, a couple of δ-functions having opposite signs and whose distance is T. For $T \to 0$, the above expression becomes

$$\frac{d}{dt}\delta(t) = \delta^{(1)}(t),$$

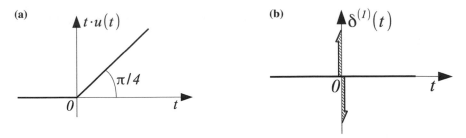

Fig. 9.31 **a** Unit ramp $\delta^{(-2)}(t)$; **b** unit doublet $\delta^{(1)}$

where $\delta^{(1)}(t)$ is the so-called *unit doublet*, which has a *third-order discontinuity* at $t = 0$ and whose symbol is shown in Fig. 9.31b. In full analogy with the integral of Eqs. 9.46, we have (for all $t \neq 0$)

$$\delta(t) = \int_{-\infty}^{t} \delta^{(1)}(\tau)d\tau.$$

9.7 Discontinuity Balance

In this section, a discussion about the continuity of the solution in response to discontinuous inputs is provided. The I/O relationship Eq. 9.20 is reported here for ease of reference:

$$a_1 \frac{dx}{dt} + a_0 x = \sum_{k=1}^{P} h_k \hat{u}_k(t).$$

Due to this equivalence, it is evident that the highest-order discontinuity on the right-hand side—that is, the input side—at a given time must be balanced, at the same time, by the same order of discontinuity on the left-hand side, that is, the output side. At this side, the highest-order discontinuity is generated by the highest (here, the first) derivative term.

The **discontinuity balance** is obtained by equating the highest-order discontinuities at the two sides of the I/O relationship at a given time t_0. On the input side, the information about the highest-order discontinuity is provided by the knowledge of the inputs. On the output side, the highest-order discontinuity is generated by the highest output derivative term and must equal the highest-order input discontinuity. This allows us to obtain the discontinuity order of the output variable at t_0, inasmuch as each derivative increases the discontinuity order by one.

For instance, if the input highest-order discontinuity at a given time t_0 is a step, this means that at t_0:

- $\dfrac{dx}{dt}$ must exhibit a step discontinuity as well;
- $x(t)$ is continuous across t_0.

In this case, we can conclude that $x(t_0^+) = x(t_0^-)$, where $x(t_0^\pm) = \lim\limits_{t \to t_0^\pm} x(t)$.

In contrast, if the input highest-order discontinuity at t_0 is a Dirac impulse $\delta(t - t_0)$ with coefficient h_1, at t_0, then:

- $\dfrac{dx}{dt}$ must exhibit a δ-discontinuity as well;
- $x(t)$ has a corresponding step discontinuity at t_0.

In this case, we can find $x(t_0^+)$ by integrating the I/O relationship between t_0^- and t_0^+:

$$
\underbrace{\int_{t_0^-}^{t_0^+} a_1 \frac{dx}{dt} dt}_{= a_1(x(t_0^+) - x(t_0^-))} + \underbrace{\int_{t_0^-}^{t_0^+} a_0 x \, dt}_{= 0} = \underbrace{\sum_{k=1}^{P} \int_{t_0^-}^{t_0^+} h_k \hat{u}_k(t) dt}_{= h_1 \int_{t_0^-}^{t_0^+} \delta(t - t_0) dt = h_1} \quad ,
$$

that is,

$$
x(t_0^+) = \frac{h_1}{a_1} + x(t_0^-).
$$

In other words, the height of the step jump exhibited by $x(t)$ across t_0 is $\dfrac{h_1}{a_1}$.

We remark that the integral across a step discontinuity is 0, since we are computing the area of a rectangle with finite height and infinitesimal basis.

In any case, owing to the structure of the considered I/O relationship, at every time, *the state variable has a degree of discontinuity less than that exhibited by the most discontinuous input.*

9.8 Response of Linear Circuits with One WSV and One SSV to Discontinuous Inputs

Now we have all the tools necessary to analyze the response to an input with discontinuity at $t = t_0$. Without loss of generality, we can refer to the case of a single input, since with multiple inputs we just consider one input at a time and apply the superposition principle to obtain either the forced response or the ZSR.

Notice that the initial condition after a discontinuity at $t = t_0$, if not specified, must be determined by computing the variable x *before* t_0 (following the guidelines described in Sect. 9.4.2) and by applying the discontinuity balance across t_0

(following the guidelines described in Sect. 9.7). Once $x(t_0^+)$ has been found, the analysis after t_0 is carried out separately, following again the guidelines described in Sect. 9.4.2.

In summary, the solution across a discontinuity at $t = t_0$ can be found in three steps:

Step 1:   analysis for $t < t_0$; we find $x(t)$ by solving the I/O relationship for $t < t_0$ and finally compute $x(t_0^-)$; for this analysis, we need an initial condition at a time before t_0, which might be assigned, or else must be computed. Otherwise, some assumptions must be provided by the problem to bypass the need for an initial condition. For instance, in some cases it is assumed that the circuit (if it is absolutely stable) is in its DC or AC steady state, which means that the free response has vanished and we can compute $x(t) = x_{fo}(t)$ without an initial condition.

Step 2:   analysis across t_0, by applying the discontinuity balance, in order to find $x(t_0^+)$;

Step 3:   analysis for $t > t_0$, by solving the I/O relationship for $t > t_0$, with the initial condition $x(t_0^+)$ found at step 2.

Now we analyze two particularly significant cases.

9.8.1 Step Response

In this case, the I/O relationship can be written as follows:

$$a_1 \frac{dx}{dt} + a_0 x = h u(t - t_0). \tag{9.47}$$

We preliminarily compute the natural frequency, which in this case is $\lambda = -\dfrac{a_0}{a_1}$. The following analysis is based on the assumption that $\lambda < 0$.

Step 1:   for $t_1 \leq t < t_0$, the I/O relationship can be recast as follows, keeping in mind that $u(t - t_0) = 0$ for $t < t_0$ (see Fig. 9.29b):

$$a_1 \frac{dx}{dt} + a_0 x = 0. \tag{9.48}$$

Let $x(t_1) = x_1$ be known/assigned. Then for $t_1 \leq t < t_0$, we have $x(t) = x_{fr}(t) = ZIR = K_1 e^{\lambda(t - t_1)}$, with $\lambda = -\dfrac{a_0}{a_1}$.

By imposing $x(t_1)(= K_1) = x_1$, we obtain $x(t) = x_1 e^{-\frac{a_0}{a_1}(t - t_1)}$, then $x(t_0^-) = x_1 e^{-\frac{a_0}{a_1}(t_0 - t_1)}$.

Step 2: from the discontinuity balance, it follows that $x(t)$ is continuous across t_0, that is, $x(t_0^+) = x(t_0^-)$.

Step 3: for $t > t_0$, the I/O relationship can be recast as follows:

$$a_1 \frac{dx}{dt} + a_0 x = h. \tag{9.49}$$

Then the solution is

$$x(t) = x_{fr}(t) + x_{fo}(t) = K_2 e^{\lambda(t-t_0)} + x_{fo}(t). \tag{9.50}$$

According to the similarity criterion, $x_{fo}(t)$ is a constant. By substituting in Eq. 9.49, it immediately follows that this constant is $\dfrac{h}{a_0}$. Now using the initial condition $x(t_0^+)$, we obtain

$$x(t_0^+) = K_2 + \frac{h}{a_0}, \tag{9.51}$$

that is,

$$K_2 = x(t_0^+) - \frac{h}{a_0}. \tag{9.52}$$

Thus, we finally obtain

$$x(t) = \underbrace{\left(x(t_0^+) - \frac{h}{a_0}\right)e^{\lambda(t-t_0)}}_{x_{fr}(t)} + \underbrace{\frac{h}{a_0}}_{x_{fo}(t)} = \underbrace{x(t_0^+)e^{\lambda(t-t_0)}}_{ZIR} + \underbrace{\frac{h}{a_0}(1 - e^{\lambda(t-t_0)})}_{ZSR}. \tag{9.53}$$

Figure 9.32 shows the complete response.

For instance, let us analyze the circuit of Fig. 9.18a for $t \geq t_1$, with $a(t) = Au(t - t_0)$ and $v(t_1) = V_0$ (see Case Study 1 in Sect. 9.4.2); for $t_1 \leq t < t_0$ (step 1) we have $v(t) = V_0 e^{-\frac{1}{RC}(t-t_1)}$ and $v(t_0^-) = V_0 e^{-\frac{1}{RC}(t_0-t_1)}$. Across the discontinuity (step

Fig. 9.32 Step response $x(t)$ for an absolutely stable circuit

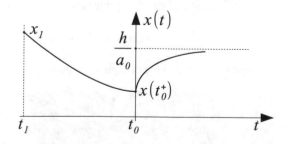

2), by applying the discontinuity balance we obtain $v(t_0^+) = v(t_0^-)$. Finally, for $t > t_0$ (step 3), the I/O equation is Eq. 9.32, and its solution is

$$v(t) = \underbrace{(v(t_0^+) - RA)e^{-\frac{1}{RC}(t-t_0)}}_{v_{fr}(t)} + \underbrace{RA}_{v_{fo}(t) \equiv \hat{v}} .$$

9.8.2 Impulse Response

In this case, the I/O relationship can be written as follows:

$$a_1 \frac{dx}{dt} + a_0 x = h\delta(t - t_0). \tag{9.54}$$

The natural frequency $\lambda = -\frac{a_0}{a_1}$ is assumed to be < 0.

Step 1: for $t_1 \leq t < t_0$, the I/O relationship can be recast as follows:

$$a_1 \frac{dx}{dt} + a_0 x = 0. \tag{9.55}$$

Then also in this case, $x_{fo}(t) = 0$ and $x(t) = x_{fr}(t) = ZIR = x(t_1)e^{\lambda(t-t_1)}$.

Step 2: from the discontinuity balance, it follows that $x(t)$ is discontinuous across t_0. Then we integrate the I/O relationship between t_0^- and t_0^+, thus obtaining (see Sect. 9.7)

$$x(t_0^+) = \frac{h}{a_1} + x(t_0^-). \tag{9.56}$$

Step 3: for $t > t_0$, the I/O relationship can be recast as follows:

$$a_1 \frac{dx}{dt} + a_0 x = 0. \tag{9.57}$$

Then there is no forced response, and the solution is

$$x(t) = x_{fr}(t) = ZIR = x(t_0^+)e^{\lambda(t-t_0)}. \tag{9.58}$$

Figure 9.33 shows the complete response.

Case Study 1
Find the state variable $v(t)$ in the circuit shown in Fig. 9.34, which works at DC steady state for $t < 0$, with $a(t) = Q_0\delta(t)$ and $e(t) = Eu(t)$. Determine the

Fig. 9.33 Impulse response $x(t)$ for an absolutely stable circuit

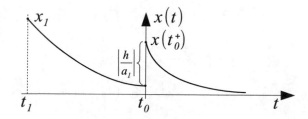

Fig. 9.34 Case Study 1

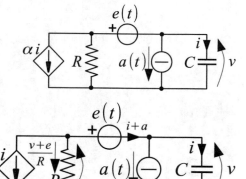

Fig. 9.35 Solution of Case Study 1

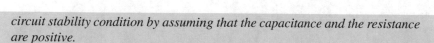

circuit stability condition by assuming that the capacitance and the resistance are positive.

First of all, it is evident that $i = C\frac{dv}{dt}$. Following Fig. 9.35, from KCL at the dashed nodal cut-set we obtain

$$\alpha i + \frac{v + e(t)}{R} + i + a(t) = 0.$$

By combining these two equations, the state equation follows:

$$(\alpha + 1)RC\frac{dv}{dt} + v = -e(t) - Ra(t). \tag{9.59}$$

The natural frequency is $\lambda = -\dfrac{1}{(\alpha + 1)RC}$; then the (absolute) stability condition is $\alpha > -1$. Henceforth, we will assume that this condition is fulfilled. For $t < 0$ (**step 1**), the state equation is

$$(\alpha + 1)RC\frac{dv}{dt} + v = 0.$$

The solution is $v(t) = 0 = v(0^-)$.

Owing to the discontinuity balance (**step 2**), $v(t)$ has a step discontinuity across 0. By integrating Eq. 9.59 between 0^- and 0^+, we obtain the initial condition necessary for the third step:

$$(\alpha + 1)RC[v(0^+) - v(0^-)] + 0 = 0 - RQ_0,$$

that is,

$$v(0^+) = \frac{-Q_0}{(\alpha + 1)C}.$$

For $t > 0$ (**step 3**), the state equation is

$$(\alpha + 1)RC\frac{dv}{dt} + v = -E.$$

The solution is

$$v(t) = v_{fr}(t) + v_{fo}(t) = Ke^{\lambda t} + v_{fo}(t). \qquad (9.60)$$

The forced response is constant and can be obtained by substitution, thereby yielding $v_{fo} = -E$. The coefficient K is obtained by imposing the initial condition $v(0^+)$, thus yielding $K = v(0^+) - v_{fo} = v(0^+) + E$. Since the circuit is absolutely stable, once the free response vanishes we have a constant steady state $(-E)$.

Case Study 2
Find the state variable $v(t)$ in the circuit shown in Fig. 9.36, which works at DC steady state for $t < 0$, with $a(t) = A_1u(-t) + Q_0\delta(t) + A_2u(t)$. Determine the circuit stability condition by assuming that the capacitance and the resistances are positive.

Fig. 9.36 Case Study 2

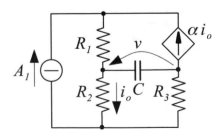

Fig. 9.37 Solution of Case
Study 2

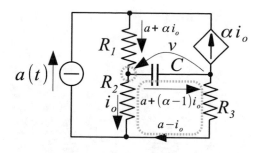

Following Fig. 9.37, from KCL at the dashed nodal cut-set we obtain

$$C\frac{dv}{dt} = a(t) + (\alpha - 1)i_o,$$

and from KVL for the dashed mesh we have $v = R_2 i_o - R_3(a - i_o)$.
By combining these two equations, the state equation follows:

$$(R_2 + R_3)C\frac{dv}{dt} + (1 - \alpha)v = (R_2 + \alpha R_3)a(t). \qquad (9.61)$$

The natural frequency is $\lambda = \dfrac{\alpha - 1}{(R_2 + R_3)C}$; then the (absolute) stability
condition is $\alpha < 1$. Owing to the problem statement (constant steady state for
$t < 0$), we have to assume that this condition is fulfilled.
    For $t < 0$ (**step 1**), the state equation is

$$(R_2 + R_3)C\frac{dv}{dt} + (1 - \alpha)v = (R_2 + \alpha R_3)A_1.$$

The solution is $v(t) = \dfrac{R_2 + \alpha R_3}{1 - \alpha}A_1 = v(0^-)$.
    Owing to the discontinuity balance (**step 2**), $v(t)$ has a step discontinuity
across 0. By integrating Eq. 9.61 between 0^- and 0^+, we obtain the initial
condition necessary for the third step:

$$(R_2 + R_3)C[v(0^+) - v(0^-)] + 0 = (R_2 + \alpha R_3)Q_0,$$

that is,

$$v(0^+) = \frac{(R_2 + \alpha R_3)}{(R_2 + R_3)C}Q_0 + v(0^-).$$

For $t > 0$ (**step 3**), the state equation is

$$(R_2 + R_3)C\frac{dv}{dt} + (1 - \alpha)v = (R_2 + \alpha R_3)A_2.$$

The solution is

$$v(t) = v_{fr}(t) + v_{fo}(t) = Ke^{\lambda t} + v_{fo} \tag{9.62}$$

with $v_{fo} = \dfrac{R_2 + \alpha R_3}{1 - \alpha}A_2$ (see step 1) and $K = v(0^+) - v_{fo}$. Also in this case, since the circuit is absolutely stable, once the free response vanishes, we have a constant steady state (v_{fo}).

9.9 Convolution Integral

In an LTI circuit, the knowledge of the zero-state response to a pulse $\delta(t)$ enables one to obtain the ZSR for a generic input $u(t)$. The mathematical tool used to obtain this result is the so-called *convolution integral*.

To get the convolution integral structure and show its properties, we assume that the $u(t)$ input to the circuit is zero up to an instant t_0. Also, for definiteness, we consider as output the state variable $x(t)$ (however, the treatment is valid for *any* circuit variable). Obviously, we assume $x(t_0^-) = 0$. The first step of the procedure consists in defining a proper function that approximates the input in the interval $[t_0, t]$; then, since t represents the upper limit of a time interval, in the following we will denote the time variable by $\hat{t}$. This is shown in Fig. 9.38a, where the function $u(\hat{t})$ is represented together with a piecewise-constant approximation $u_\Delta(\hat{t})$ obtained by first dividing the segment $t - t_0$ into n segments having equal width Δ; then, over the interval $I_k = [t_k, t_{k+1})$, with $t_k = t_0 + k\Delta$; $k = 0, \ldots, n - 1$ shown in Fig. 9.38b, the constant value chosen for the approximation is $u(t_k)$. Inasmuch as $t - t_0 = n\Delta$, when the number n of intervals increases in $[t_0, t]$, their width Δ tends to zero, and the function $u_\Delta(\hat{t})$ becomes closer and closer to $u(\hat{t})$. We assume that

$$\lim_{\Delta \to 0} u_\Delta(\hat{t}) = u(\hat{t}).$$

The zero-state response to the piecewise-constant input $u_\Delta(\hat{t})$ can be found by appealing to the concepts of linearity and time-invariance. To this end, we first define the unit rectangular pulse $p_\Delta(\hat{t})$:

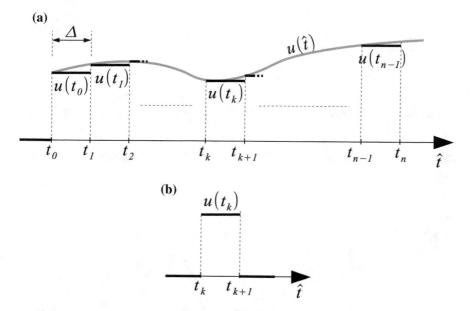

Fig. 9.38 **a** A circuit input $u(\hat{t})$ (gray line) and its piecewise-constant approximation $u_\Delta(\hat{t})$; **b** detail of $u_\Delta(\hat{t})$ over the interval I_k

$$p_\Delta(\hat{t}) = \begin{cases} 0 & \text{for } \hat{t} \le 0; \\[2mm] \dfrac{1}{\Delta} & \text{for } 0 < \hat{t} < \Delta; \\[2mm] 0 & \text{for } \hat{t} \ge 0, \end{cases}$$

which is shown in Fig. 9.39a. By taking a right t_k-shift of $p_\Delta(\hat{t})$, the function $u_\Delta(\hat{t})$ over the interval I_k can be expressed through the product

$$\underbrace{p_\Delta(\hat{t} - t_k)\Delta\, u(t_k)}_{=1 \text{ over } I_k}, \tag{9.63}$$

which is null outside of the range I_k, as evidenced in Fig. 9.40a. This amounts to saying that the expression of $u_\Delta(\hat{t})$ over the whole interval $[t_0, t]$ can be written as a sum of products 9.63 for $k = 0, \ldots, n-1$:

$$u_\Delta(\hat{t}) = \sum_{k=0}^{n-1} p_\Delta(\hat{t} - t_k)\Delta u(t_k). \tag{9.64}$$

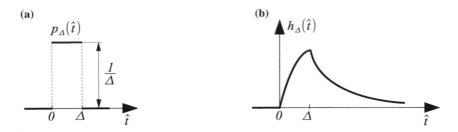

Fig. 9.39 **a** The rectangular pulse function $p_\Delta(\hat{t})$ and **b** the corresponding zero-state response $h_\Delta(\hat{t})$

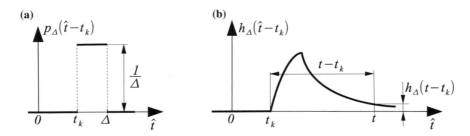

Fig. 9.40 **a** The shifted input $p_\Delta(\hat{t} - t_k)$ and **b** the corresponding shifted response

Let us denote by $h_\Delta(\hat{t})$ the zero-state response of the circuit to the rectangular pulse $p_\Delta(\hat{t})$, as (qualitatively) shown in Fig. 9.39b. Inasmuch as the response to the shifted pulse $p_\Delta(\hat{t} - t_k)$ is $h_\Delta(\hat{t} - t_k)$ as shown Fig. 9.40, the ZSR to the input term 9.63 is $h_\Delta(\hat{t} - t_k)\Delta u(t_k)$. Then the response $x_\Delta(\hat{t})$ to $u_\Delta(\hat{t})$ is given by the sum

$$\sum_{k=0}^{n-1} h_\Delta(\hat{t} - t_k)\Delta u(t_k),$$

and at the upper end t of the interval (namely, for $\hat{t} = t$), we have

$$x_\Delta(t) = \sum_{k=0}^{n-1} h_\Delta(t - t_k)\Delta u(t_k). \tag{9.65}$$

We can now consider what happens when Δ tends to zero. First we observe that the rectangular pulse $p_\Delta(\hat{t})$, whose area remains unitary, becomes a Dirac pulse $\delta(\hat{t})$; correspondingly, the term $h_\Delta(\hat{t})$ becomes the response $h(\hat{t})$ to that impulse. In the sum 9.65, the t_k points are replaced by a continuous variable τ ranging over the interval $[t_0, t]$, and the sum is replaced by an integral:

$$\lim_{\Delta \to 0} x_\Delta(t) = \underbrace{\int_{t_0}^{t} h(t - \tau)u(\tau)d\tau}_{x(t)}.$$

The integral

$$x(t) = \int_{t_0}^{t} h(t - \tau)u(\tau)d\tau, \qquad t \geq t_0, \tag{9.66}$$

is the convolution integral that gives the zero-state response to an input function $u(t)$ null before t_0. This result is a direct consequence of the linearity and time-invariance properties holding for the circuit. Notice that the linear dependence of the integral with respect to the input is immediately evident. Once the impulse response $h(t)$ is known, the convolution integral gives the zero-state response to any input $u(t)$. Moreover, inasmuch as the initial reference to a state variable does not play any specific role in the developed reasoning, we conclude that the convolution integral can be applied to obtain the ZSR of any circuit variable, provided that its specific $h(t)$ is known. The convolution integral is an important conceptual tool in the field of circuit theory as well as in other fields of mathematical physics, such as electromagnetism, mechanics, heat diffusion, and so on.

Case Study
In the circuit shown in Fig. 9.41, the input voltage $e(t)$ is null before t_0 and equals $E\cos(\omega t)$ for $t \geq t_0$. Apply the convolution integral to find the zero-state response of the variable $i(t)$ for $t \geq t_0$.

The differential equation for $i(t)$ is

$$e(t) = Ri + L\frac{di}{dt}.$$

Fig. 9.41 Case Study

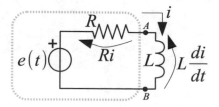

Assuming $i(0^-) = 0$ and setting $e(t) = \delta(t)$, the impulse response $h(t)$ is the solution of the equation

$$\delta(t) = Rh + L\frac{dh}{dt}.$$

By integrating the previous equation between 0^- and 0^+ and since $h(0^-) = 0$, we have

$$\underbrace{\int_{0^-}^{0^+} \delta(t)dt}_{=1} = R\underbrace{\int_{0^-}^{0^+} h(t)dt}_{=0} + L\int_{0^-}^{0^+} \frac{dh}{dt}dt \quad \Rightarrow \quad 1 = Lh(0^+) \quad \Rightarrow \quad h(0^+) = \frac{1}{L}.$$

Therefore, for $t \geq 0$ we have

$$h(t) = \frac{1}{L}e^{\lambda t} \quad \text{with} \quad \lambda = -\frac{R}{L}.$$

Notice that since we have set $e(t) = \delta(t)$, the physical dimension of $h(t)$ is not that of a current. Therefore, for the assigned $e(t)$, the convolution integral Eq. 9.66 gives

$$i(t) = \int_{t_0}^{t} E\cos(\omega\tau)\frac{1}{L}e^{\lambda(t-\tau)}d\tau = \frac{E}{L}e^{\lambda t}\int_{t_0}^{t} \cos(\omega\tau)e^{-\lambda\tau}d\tau, \quad t \geq t_0.$$

Now, setting $\omega\tau = z$, the previous expression can be reformulated as

$$i(t) = \frac{E}{\omega L}e^{-\frac{R}{L}t}\int_{\omega t_0}^{\omega t} \cos(z)e^{\frac{R}{L}z}dz = \frac{1}{\left(\frac{R}{\omega L}\right)^2 + 1}\left[e^{\frac{R}{L}z}\left(\frac{R}{\omega L}\cos(z) + \sin(z)\right)\right]\Big|_{\omega t_0}^{\omega t}.$$

After calculating the expression into square brackets at ωt and at ωt_0, and defining the terms

$$I_C = \frac{ER}{R^2 + \omega^2 L^2}; \quad I_S = \frac{E\omega L}{R^2 + \omega^2 L^2},$$

the above expression can be rewritten in the compact form

$$i(t) = I_C\cos(\omega t) + I_S\sin(\omega t) - (I_C\cos(\omega t_0) + I_S\sin(\omega t_0))e^{-\frac{R}{L}(t-t_0)}, \quad t \geq t_0,$$

which is just the ZSR term of the $i(t)$ expression found within Case Study 2, Sect. 9.4.2 for the same circuit.

9.10 Circuit Response to More Complex Inputs

In this section we consider a generic absolutely stable circuit governed by the state equation

$$T\frac{dv}{dt} + \alpha v = \beta e(t)$$

with two more complex inputs.

9.10.1 Multi-input Example

In the first example, the input is

$$e(t) = V_0 u(-t + t_1) - \Phi_0 \delta(t - t_2) + \Phi_0 \delta(t - t_3) + V_0 \cos[2\pi f(t - t_3)]u(t - t_3),$$

and it is shown in Fig. 9.42a. We assume that at the beginning we have a DC steady state generated by a constant input of amplitude $V_0 = 1$ V, then at $t = t_1 = 2$ s, the input is turned off; at $t = t_2 = 4$ s, the input generates a negative impulse with area $\Phi_0 = 0.1$ Vs; finally, at $t = t_3 = 6$ s there is a positive impulse with area Φ_0, and an AC source with amplitude V_0 and frequency $f = 10$ Hz is turned on.

Even without analyzing in detail the response, one can easily guess its qualitative behavior for given parameters. Let us assume that $T = 1$ s, $\alpha = 5$, and $\beta = 4$. Then the natural frequency is $\lambda = -5$ s^{-1}, corresponding to a time constant $\tau = 0.2$ s. This means that after about $5\tau = 1$ s the free response is negligible with respect to its initial value, given that $e^{-\frac{5\tau}{\tau}} = e^{-5} \ll e^0$.

This is confirmed by the plot of the solution $v(t)$, shown in Fig. 9.42b. After the DC steady state (where $v(t) = \frac{\beta V_0}{\alpha} = 0.8$ V), at $t_1 = 2$ s, a free response starts, which vanishes at about $t = 3$ s. Then the state remains at zero until the next input variation occurs. At $t_2 = 4$ s the input impulse causes a step discontinuity in the state variable, after which a new free response starts, which vanishes at about $t = 5$ s. Finally, the new input impulse at $t_3 = 6$ s causes another step discontinuity, after which $v(t)$ is given by the sum of free and forced responses. The transient vanishes at about $t = 7$ s, and then the circuit reaches an AC steady state.

If we set $\alpha = 1$ (instead of 5), the natural frequency is $\lambda = -1$ s^{-1}, corresponding to a time constant $\tau = 1$ s. This means that the free response vanishes after about $5\tau = 5$ s. This is confirmed by the plot of the solution $v(t)$, shown in Fig. 9.42c. After the DC steady state $v(t) = \frac{\beta V_0}{\alpha} = 4$ V (different from what it was before due to the different value of α), at $t_1 = 2$ s a free response starts, but it does not vanish before the next input variation. At $t_2 = 4$ s the input impulse causes a step discontinuity in the state variable, after which a new free response starts. Finally, the new input impulse at $t_3 = 6$ s causes another step discontinuity, after which $v(t)$ is given by the

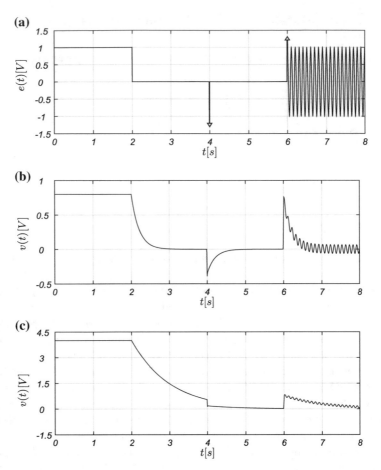

Fig. 9.42 **a** Input $e(t)$; **b** output $v(t)$ for $\lambda = -5\,\mathrm{s}^{-1}$; **c** output for $\lambda = -1\,\mathrm{s}^{-1}$

sum of the free and forced responses. The transient vanishes after $t = 11$ s, and then the circuit reaches an AC steady state.

Remark: This example makes it evident that the closer λ is to 0, the slower the circuit free response. In other words, a high value of $|\lambda|$ denotes a highly reactive circuit, whose free response decays very quickly. This is similar to what happens in a simple mechanical system like the kayak considered in Sect. 9.4.1; in that case, $|\lambda| = \frac{K}{m}$, and we have different behaviors if the kayak (of fixed mass m) is paddled in a swamp (high K) or on a lake (intermediate K), or slides on the icy surface of a frozen lake (low K). For the same initial velocity V_0, the larger the value of K, that is, $|\lambda|$, the shorter the stopping time, that is, the free response and thus the distance covered by the kayak. In other words, the swamp is highly reactive (through high viscous resistance) with respect to the kayak's motion, corresponding to a very fast free response, whereas the frozen lake is scarcely reactive (through a low friction), corresponding to a slow free response.

9.10.1.1 Nonsinusoidal Periodic Input Example

The second example is concerned with the same state equation, but in this case with the square wave with amplitude $V_0 = 1$ V and period $T_0 = 2$ s shown in Fig. 9.43a as input.

We first assume that $T = 1$ s, $\alpha = 5$, and $\beta = 2$. Then the natural frequency is $\lambda = -5$ s^{-1}, corresponding to a time constant $\tau = 0.2$ s. This means that after about $5\tau = 1$ s, the free response vanishes.

This is confirmed by the plot of the solution $v(t)$, shown in Fig. 9.43b (the input square wave $e(t)$ is shown in gray for the sake of comparison). At $t = 0$ s, a free response starts from the assigned initial condition $v(0) = 1.5$ V. After about 1 s, as foreseen, the solution approaches its DC steady state value $\frac{\beta V_0}{\alpha} = 0.4$ V (upper horizontal dashed line). At $t = 1$ s, because of the input discontinuity, a new free response starts, which vanishes after about 1 s, making the solution approach its new DC steady-state value $\frac{-\beta V_0}{\alpha} = -0.4$ V (lower horizontal dashed line). Then the process is repeated, producing the eventually periodic output shown in Fig. 9.43b (black solid line).

If we set $\alpha = 2$ (instead of 5), the natural frequency is $\lambda = -2$ s^{-1}, corresponding to a time constant $\tau = 0.5$ s. This means that the free response vanishes after about $5\tau = 2.5$ s. This is confirmed by the plot of the solution $v(t)$, shown in Fig. 9.43c. At $t = 0$ s, a free response starts from the assigned initial condition $v(0) = 1.5$ V. After 1 s, the input jumps at its lower value when the solution has not yet reached its DC steady state value $\frac{\beta V_0}{\alpha} = 1$ V (upper horizontal dashed line). Then a new transient starts from the reached state $v(1 \text{ s})$. Also in this case, the input jumps at its upper value when the solution is still decreasing toward its DC steady-state value $\frac{-\beta V_0}{\alpha} = -1$ V (lower horizontal dashed line). Then the process is repeated, producing the eventually periodic output shown in Fig. 9.43c (black solid line).

> **Case Study**
>
> *For the RC circuit shown in Fig. 9.44a and subject to the rectangular input waveform $e(t) = \hat{e}(t)$ represented in Fig. 9.44b, find the forced response term $v_{fo}(t)$ of the output $v(t)$ (Hint: the forced response term must be periodic, with the same time period as the input term).*
>
> The response $v_{fo}(t)$ must be periodic with the same period T as that of the input waveform $\hat{e}(t)$. Then, considering, for instance, the time interval $[0, T]$, we must have $v_{fo}(0) = v_{fo}(T) = V_0$. The value V_0 assumed by $v_{fo}(t)$ at the extremes of the time interval is unknown a priori. We notice that owing to the

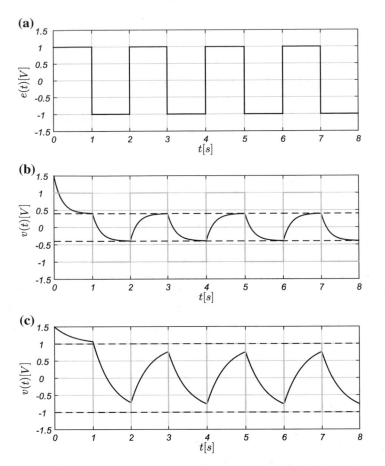

Fig. 9.43 **a** Input $e(t)$; **b** output $v(t)$ for $\lambda = -5\,\mathrm{s}^{-1}$; **c** output for $\lambda = -2\,\mathrm{s}^{-1}$. In panels **b** and **c**, $e(t)$ is shown in gray for the sake of comparison

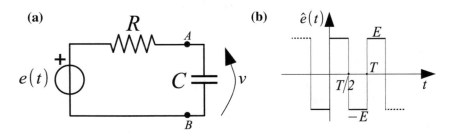

Fig. 9.44 **a** RC circuit and **b** periodic input wave $e(t) = \hat{e}(t)$

discontinuity balance applied to the state equation of the circuit

$$RC\frac{dv}{dt} + v = \hat{e}(t), \tag{9.67}$$

$v(t)$ is always continuous (therefore, in particular, continuity applies to its forced term $v_{fo}(t)$).

Within the interval $[0, T]$, $v_{fo}(t)$ can be found by observing that for $t \in (0, T/2)$, we have $\hat{e}(t) = E$, while for $t \in (T/2, T)$, we have $\hat{e}(t) = -E$. At $t = T/2$, $v_{fo}(t)$ must be continuous despite the first-order discontinuity of $\hat{e}(t)$.

Based on these premises, we can find $v_{fo}(t)$ in the time interval by adopting the following procedure:

- Provisionally assuming the initial value V_0 as known, set $\hat{e}(t) = E$ and find the corresponding capacitor voltage

$$(V_0 - E)e^{-\frac{t}{RC}} + E,$$

which represents $v_{fo}(t)$ for $t \in [0, T/2]$. Denote by V_1 the value $v_{fo}(T/2)$.
- Defining for simplicity the new variable $t' = t - T/2$; $t' \in [0, T/2]$, assume V_1 as the initial value and set $\hat{e}(t') = -E$. The capacitor voltage

$$(V_1 + E)e^{-\frac{t'}{RC}} - E$$

represents $v_{fo}(t')$ over the interval.
- The periodicity assumption implies that $v_{fo}(t')\big|_{T/2} = V_0$, that is,

$$(V_1 + E)e^{-\frac{T/2}{RC}} - E = V_0 \text{ with } V_1 = (V_0 - E)e^{-\frac{T/2}{RC}} + E,$$

which can be immediately reduced to an equation in the only unknown V_0.

For the sake of compactness, we define $\alpha = e^{-\frac{T}{2RC}}$ (notice that $0 < \alpha < 1$). After some tedious but straightforward manipulations, we obtain

$$V_0 = E\frac{\alpha - 1}{\alpha + 1}.$$

For $E > 0$, we obviously have $V_0 < 0$. It is easy to verify that owing to the particular symmetries of $e(t)$, we have $V_1 = -V_0$. For a generic value of α (say α_0, corresponding to $RC = \tau_0$), the waveform $v_{fo}(t)$ is represented in Fig. 9.45a. Now we use this case as a benchmark and analyze two limit cases, in which we hold T fixed and vary α by changing the circuit time constant RC, that is, the natural frequency $\lambda = -1/(RC)$.

- If $RC \ll \tau_0$, the circuit is extremely "reactive" ($|\lambda| \gg \frac{1}{\tau_0}$), $\alpha \ll \alpha_0$, $V_0 \to -E$, and $V_1 \to +E$; in other words, the circuit tends to follow the input, thus producing a wave close to the rectangular wave $\hat{e}(t)$, as shown in Fig. 9.45b;
- if $RC \gg \tau_0$, the circuit is extremely "slow" in reacting to input variations ($|\lambda| \ll \frac{1}{\tau_0}$), $\alpha \to 1$, whereas V_0 and V_1 tend to 0; therefore, for $t \in [0, T/2]$, $v_{fo}(t) = (V_0 - E)e^{-\frac{t}{RC}} + E \approx (V_0 - E)\left(1 - \frac{t}{RC}\right) + E$; a completely similar result holds for $t \in [T/2, T]$, thus producing an almost triangular wave, as shown in Fig. 9.45c.

Let us conclude this example by pointing out that once $v_{fo}(t)$ is known, it is easy to get the voltage $v(t)$ in cases in which the input voltage is zero for $t < 0$ and is $\hat{e}(t)$ for $t > 0$, that is, for $e(t) = \hat{e}(t) \cdot u(t)$. The general structure of the solution for $t \geq 0$ is

$$v(t) = \underbrace{Ke^{-\frac{t}{RC}}}_{v_{fr}(t)} + v_{fo}(t) \text{ for } t \geq 0,$$

where $v_{fr}(t)$ denotes the free-response term and the constant K must be such as to satisfy the (assigned) initial condition. For instance, when C is initially uncharged ($v(0) = 0$), we must have

$$K + v_{fo}(0) = 0, \text{ which implies } K = -V_0.$$

9.11 Normalizations

As seen in many examples considered up to now, by assigning values to the physical parameters of the circuit, an analytic or numerical procedure provides the results for the desired circuit variable in any finite time interval and for the specified initial conditions. This calculation process, however, can be set in such a way as to give the numerical results a more general meaning. The method for doing so consists in dealing with *normalized versions* of the physical circuit variables. Through normalization, these variables (voltages, currents, time) are expressed as new normalized variables, which are physically dimensionless, multiplied by proper coefficients.

For example, in Case Study b in Sect. 9.4 (see Fig. 9.46a), we have parameters R [V/A], C [As/V], e [V], and main variables v [V] and t [s]. The circuit state equation is (Eq. 9.14)

$$RC\frac{dv}{dt} + v = e(t).$$

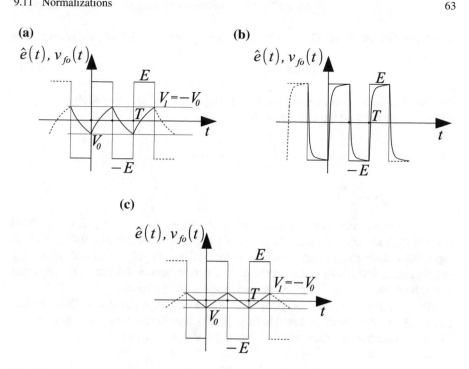

Fig. 9.45 Periodic output terms $v_{fo}(t)$: **a** general case $RC = \tau_0$, **b** almost rectangular wave output obtained for $RC \ll \tau_0$, **c** almost triangular wave output obtained for $RC \gg \tau_0$

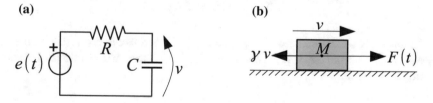

Fig. 9.46 **a** First-order electric circuit and **b** mechanical system

The choice of coefficients for normalization is not unique, but some choices can lead to more compact and effective formulations for the normalized equations. In the considered case, we define a reference voltage V_0 [V] and use as reference time the circuit time constant RC [s]. Therefore, we can define two new (normalized) variables

$$\tilde{t} = \frac{t}{RC} \qquad \text{and} \qquad \tilde{x} = \frac{v}{V_0},$$

which are dimensionless. Conversely, the physical circuit variables can be expressed as

$$t = RC\tilde{t} \qquad \text{and} \qquad v = V_0\tilde{x}. \tag{9.68}$$

By substituting these expressions in the state equation and dividing each term by V_0, we obtain the normalized state equation

$$\frac{d\tilde{x}}{d\tilde{t}} + \tilde{x} = \tilde{u}(\tilde{t}),$$

where $\tilde{u}(\tilde{t}) = \dfrac{e(RC\tilde{t})}{V_0}$.

The same equation could describe physical systems with completely different natures. Consider, for instance, the mechanical system shown in Fig. 9.46b: it could represent a mass moving on a surface with friction and subject to a time-varying external force $F(t)$ or even the kayak considered in Sect. 9.4.1 but in the presence of a driving force produced by an inboard engine or by a rower.

We assume that the physical system is characterized through mass M [Kg], applied force F [Kg m/s^2], friction coefficient γ [Kg/s], which are the system parameters. The main variables are mass velocity v [m/s] and time t [s].

Newton's law

$$M\frac{dv}{dt} = F(t) - \gamma v$$

provides the state equation

$$M\frac{dv}{dt} + \gamma v = F(t).$$

In this case, we define a reference speed v_0 [m/s] and use as a reference time the system time constant M/γ [s]. Therefore, also in this case we can define two new (normalized) variables

$$\tilde{t} = \frac{\gamma t}{M} \qquad \text{and} \qquad \tilde{x} = \frac{v}{v_0},$$

which are dimensionless. Conversely, the physical system variables can be expressed as

$$t = \frac{M\tilde{t}}{\gamma} \qquad \text{and} \qquad v = v_0\tilde{x}. \tag{9.69}$$

By substituting these expressions in the state equation and dividing each term by γv_0, we obtain once more the normalized state equation

$$\frac{d\tilde{x}}{d\tilde{t}} + \tilde{x} = \tilde{u}(\tilde{t}),$$

where $\tilde{u}(\tilde{t}) = \dfrac{F(M\tilde{t}/\gamma)}{\gamma v_0}$.

Therefore, the same state equation can be used to describe two physical systems that at first glance seem to be completely different. The actual variables can be obtained by applying the conversion formulas defined in Eqs. 9.68 and 9.69.

In both examples, the time variable t is normalized to the system time constant. Therefore, the dimensionless natural frequency for the normalized state equation is $\tilde{\lambda} = -1$. To obtain the actual natural frequency for the circuit, we have to divide $\tilde{\lambda}$ by the reference time, thus obtaining $\lambda_{circ} = \frac{-1}{RC}$ [s^{-1}]. As an alternative, we can analyze the circuit characteristic equation

$$\underbrace{\lambda_{circ} RC}_{\tilde{\lambda}} + 1 = 0,$$

thereby obtaining $\lambda_{circ} = \frac{\tilde{\lambda}}{RC}$.

Similarly, the natural frequency for the mechanical system is $\lambda_{mech} = \frac{-\gamma}{M} = \frac{\tilde{\lambda}\gamma}{M}$ [s^{-1}].

In this way, the same normalized equation can be used to represent a large variety of cases. Moreover, proper normalizations can improve the accuracy of numerical solutions, mainly in nonlinear cases.

9.12 Solution for Nonstate Output Variables

Once the solution is found for the state variable $x(t)$, we can apply the substitution theorem to the component with memory (the capacitor is replaced by a voltage source and the inductor by a current source), thereby obtaining a memoryless circuit.

In all cases, we have a linear resistive circuit, whose solution can be found algebraically. In other words, any nonstate output variable $y(t)$ can be found as

$$y(t) = kx(t) + \sum_{k=1}^{P} w_k \hat{u}_k(t). \tag{9.70}$$

This is called the output equation and is solved algebraically, according to the state variable method.

By applying the discontinuity balance to this equation, it is evident that (in contrast to the state variable) $y(t)$ *has the same maximum degree of discontinuity as the most discontinuous input.*

Fig. 9.47 Linear resistive
circuit to be solved to find i_o

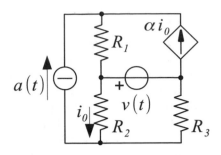

Case Study

Find the nonstate variable $i_o(t)$ in the circuit shown in Fig. 9.36, with $\alpha < 1$ and
$a(t) = A_1 u(-t) + Q_0 \delta(t) + A_2 u(t)$.

By applying the substitution theorem as described above, we have to solve
the linear resistive circuit shown in Fig. 9.47, where $v(t)$ has the analytical
expression found in Sect. 9.8.2 (Case Study 2). Since $v = R_2 i_o - R_3(a -
i_o) = (R_2 + R_3)i_o - R_3 a$ (see the solution of the cited Case Study 2), we have
$i_o = \dfrac{v(t) + R_3 a(t)}{R_2 + R_3}$. From this solution it is apparent that i_o has the same
degree of discontinuity as the input $a(t)$.

As an alternative, i_o can be found by solving the corresponding I/O relationship, which can be found by substituting $v = (R_2 + R_3)i_o - R_3 a$ in Eq. 9.61,
thereby obtaining

$$(R_2 + R_3)C\frac{di_o}{dt} + (1 - \alpha)i_o = a(t) + R_3 C\frac{da}{dt}.$$

Also from this equation it is apparent that i_o has the same degree of discontinuity as the input $a(t)$.

You can check that the solution found in this way is the same as before.

9.13 Thévenin and Norton Equivalent Representations of a Charged Capacitor/Inductor

In this section, we provide equivalent representations of a charged capacitor or inductor. This is done by viewing the corresponding descriptive equation as an I/O relationship, following the line of reasoning discussed in Sect. 9.7. By associating the
input variable role to the current in the capacitor and to the voltage in the inductor,
an equivalent representation (and subsequently, its dual formulation) is obtained in
each case. These representations are built up starting from a certain instant t_0. The

information about the history before t_0 is summarized through an initial value at t_0^- and represented by a proper voltage or current source. Each equivalent representation is formulated in such a way that it takes into account the possible discontinuities of the input variable over time.

9.13.1 Charged Capacitor

The current flowing through a capacitor determines an energy exchange with the rest of the circuit. Suppose you know the capacitor voltage $v(t_0^-)$ at $t = t_0^-$. For all $t > t_0$, the descriptive equation gives

$$v(t) = v(t_0^-) + \frac{1}{C} \int_{t_0^-}^{t} i(\tau)d\tau \qquad \text{for all } t > t_0.$$

The term $v(t_0^-)$ summarizes the capacitor's history before t_0. The integral represents the voltage contribution of an equivalent, initially uncharged, capacitor. This voltage is entirely due to the "input" current flowing through the capacitor from $t = t_0^-$ on.

It is useful to include in the above expression the information concerning its range of validity, $t > t_0$. To do this, we use the delayed unit step function $u(t - t_0)$ shown in Fig. 9.48 and discussed in Sect. 9.6. For $t > t_0$, the value of this step function is 1, while it is 0 before t_0. Then we can write

$$v(t) = v(t_0^-)u(t - t_0) + \underbrace{\frac{1}{C} \int_{t_0^-}^{t} i(\tau)d\tau}_{\hat{v}(t)} \qquad \text{for all } t > t_0. \qquad (9.71)$$

Equation 9.71 can be interpreted as a KVL, where the first term on the right-hand side represents a voltage source with impressed voltage $v(t_0^-)u(t - t_0)$, and the second term represents the initially uncharged capacitor subject to the "input" current $i(t)$. Of course, these two components are connected in series. As a result, the charged capacitor can be represented by the equivalent circuit shown in Fig. 9.49, which recalls the Thévenin equivalent met for memoryless two-terminals.

Fig. 9.48 Delayed unit step function

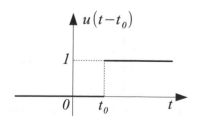

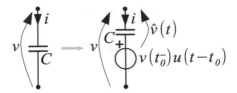

Fig. 9.49 Thévenin equivalent of a charged capacitor

Remark: The equivalent representation is nonlinear, since it contains a voltage source. This is apparent in Eq. 9.71, due to the presence of the constant term $v(t_0^-)$. This is the price to be paid, since this model is valid only for $t > t_0$. In other words, our model becomes nonlinear when we decide (and through this decision we apply a logic function, which is nonlinear) to observe it from t_0 on, by summarizing its previous history by means of $v(t_0^-)$. The same holds also for the other equivalent models proposed in this section.

Owing to the discontinuity balance applied to the capacitor's descriptive equation, $\hat{v}(t)$ is continuous at $t = t_0$ (that is, $\hat{v}(t_0^+) = \hat{v}(t_0^-) = 0$), provided that $i(t)$ has at most a first-order (step) discontinuity at t_0. Under this assumption, by differentiating Eq. 9.71 with respect to time, we obtain

$$\frac{dv}{dt} = v(t_0^-)\delta(t - t_0) + \underbrace{\frac{d\hat{v}}{dt}}_{\frac{i(t)}{C}},$$

that is,

$$i(t) = -Cv(t_0^-)\delta(t - t_0) + C\frac{dv}{dt}, \tag{9.72}$$

which models the charged capacitor for $t > t_0$ in a dual way with respect to Eq. 9.71. The impulse function can be represented by a current source, and the coefficient $Cv(t_0^-)$ is the initial amount of charge on the capacitor. The last term on the right-hand side of Eq. 9.72 is the current flowing through an initially uncharged capacitor C. The corresponding circuit is shown in Fig. 9.50. The parallel of the current source

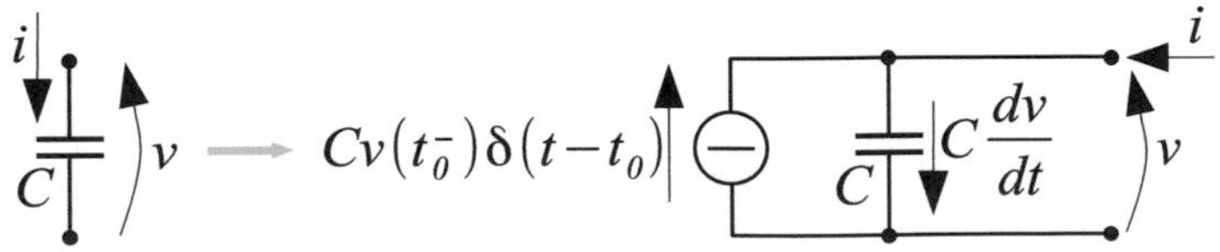

Fig. 9.50 Norton equivalent of a charged capacitor

with the uncharged capacitor recalls the Norton equivalent met for memoryless two-terminals.

Remark: If $i(t)$ contains a Dirac δ-function at t_0, we can write

$$i(t) = Q\delta(t - t_0) + f(t),$$

where the coefficient Q is the area of the current impulse, and $f(t)$ denotes the remaining terms in the expression of $i(t)$, containing at most first-order discontinuities at t_0. In this case, $\hat{v}(t_0^+)$ can be found by integrating the descriptive equation between t_0^- and t_0^+:

$$\hat{v}(t_0^+) - \hat{v}(t_0^-) = \frac{1}{C} \int_{t_0^-}^{t_0^+} [Q\delta(\tau - t_0) + f(\tau)]\, d\tau = \frac{Q}{C}.$$

Therefore, given that $\hat{v}(t_0^-) = 0$, we have $\hat{v}(t_0^+) = \dfrac{Q}{C}$, and $\hat{v}$ exhibits a first-order discontinuity at t_0.

9.13.2 Charged Inductor

Consider now a charged inductor whose current at $t = t_0^-$ is $i(t_0^-)$. Its equivalent representations can be obtained, mutatis mutandis, following the same line of reasoning used in the case of the capacitor.

The inductor's descriptive equation leads directly to the following expression for $i(t)$:

$$i(t) = i(t_0^-)u(t - t_0) + \underbrace{\frac{1}{L} \int_{t_0^-}^{t} v(\tau)d\tau}_{\hat{i}(t)} \quad \text{for all } t > t_0, \tag{9.73}$$

which is completely analogous to Eq. 9.71 and can be interpreted as a KCL. The term $\hat{i}(t)$ is continuous at $t = t_0$, provided that $v(t)$ contains at most a step discontinuity at t_0. The corresponding Norton equivalent representation is shown in Fig. 9.51. The time differentiation of Eq. 9.73, with the substitution $\dfrac{d\hat{i}}{dt} = \dfrac{v(t)}{L}$, gives

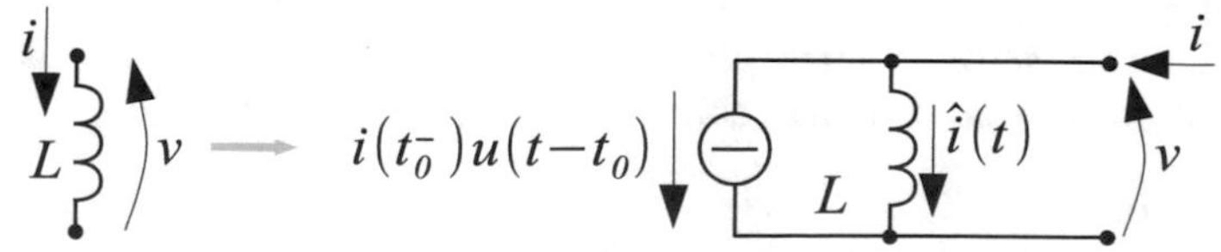

Fig. 9.51 Norton equivalent of a charged inductor

Fig. 9.52 Thévenin
equivalent of a charged
inductor

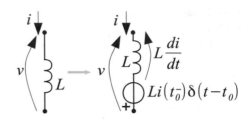

$$v(t) = -Li(t_0^-)\delta(t - t_0) + L\frac{di}{dt}, \tag{9.74}$$

where the coefficient $Li(t_0^-)$ represents the initial amount of flux on the inductor. The Thévenin representation of Eq. 9.74 is shown in Fig. 9.52.

Remark: If $v(t)$ contains a Dirac δ-function at t_0, we can write

$$v(t) = \Phi\delta(t - t_0) + f(t),$$

where the coefficient Φ is the area of the voltage impulse and $f(t)$ denotes the remaining terms in the expression of $v(t)$, containing at most first-order discontinuities at t_0. In this case, $\hat{i}(t_0^+)$ can be found by integrating the descriptive equation between t_0^- and t_0^+:

$$\hat{i}(t_0^+) - \hat{i}(t_0^-) = \frac{1}{L}\int_{t_0^-}^{t_0^+} [\Phi\delta(\tau - t_0) + f(\tau)]\,d\tau = \frac{\Phi}{L}.$$

Therefore, given that $\hat{i}(t_0^-) = 0$, we have $\hat{i}(t_0^+) = \dfrac{\Phi}{L}$, and $\hat{i}$ exhibits a first-order discontinuity at t_0.

9.14   First-Order Linear Circuits with Several WSVs

Now we consider the more general case of first-order circuits with N WSVs and $M = N - 1$ independent constraints. Of course, if there is only one SSV, then the state equation will be again of first order.

The only difference with respect to the case with $N = 1$ is that the constraints can introduce derivatives with respect to time of one or more inputs; then the state equation can always be written in either the form (**I/O relationship**)

$$a_1\frac{dx}{dt} + a_0 x = \sum_{k=1}^{P} h_k\hat{u}_k(t) + \sum_{j=1}^{N-1}\sum_{k=1}^{P} \tilde{h}_k^{(j)}\frac{d^j\hat{u}_k(t)}{dt^j} \tag{9.75}$$

Fig. 9.53 Case Study 1

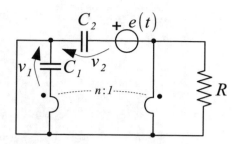

Fig. 9.54 Solution of Case Study 1

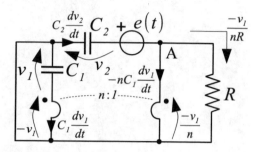

or the following alternative way (**canonical form**):

$$\frac{dx}{dt} = \dot{x} = ax + \sum_{k=1}^{P} b_k \hat{u}_k(t) + \sum_{j=1}^{N-1} \sum_{k=1}^{P} \tilde{b}_k^{(j)} \frac{d^j \hat{u}_k(t)}{dt^j}. \tag{9.76}$$

By applying the discontinuity balance to the state equation in either form, it is evident that (in contrast to the case with $N = 1$) in general (it depends on the coefficients $\tilde{h}_k^{(j)}$ or $\tilde{b}_k^{(j)}$, some of which might be null), $x(t)$ can have the same degree of discontinuity as that for the most discontinuous input, yet can even be more discontinuous.

Case Study 1

Find the state variable $v_1(t)$ in the circuit shown in Fig. 9.53, which works at DC steady state for $t < 0$, with $e(t) = Eu(t)$. Determine the circuit stability condition by assuming that the capacitances and the resistance are positive.

Following Fig. 9.54, from KVL for the outer loop we obtain the constraint $v_2 + e(t) = v_1/n$. Then $N = 2$ and $M = 1$, and thus only one of the two WSVs (v_1 and v_2) is an SSV. From KCL at node A, we have

$$C_2 \frac{dv_2}{dt} = -nC_1 \frac{dv_1}{dt} - \frac{v_1}{nR}.$$

By combining these two equations, the I/O relationship follows:

$$(n^2 C_1 + C_2) R \frac{dv_1}{dt} + v_1 = nRC_2 \frac{de}{dt}. \tag{9.77}$$

By comparing this equation to Eq. 9.75, we have $a_1 = (n^2 C_1 + C_2)R$, $a_0 = 1$, $P = 1, h_1 = 0, \tilde{h}_1^{(1)} = nRC_2$. The natural frequency is $\lambda = -\dfrac{1}{R(n^2 C_1 + C_2)}$, and then the circuit is absolutely stable. We point out the presence of $\frac{de}{dt}$ in the I/O relationship.

For $t < 0$ (**step 1**), the I/O relationship is (since $\frac{de}{dt} = 0$ for $t < 0$)

$$(n^2 C_1 + C_2) R \frac{dv_1}{dt} + v_1 = 0.$$

Thus the DC steady-state solution is $v_1(t) = 0 = v_1(0^-)$.

Owing to the discontinuity balance (**step 2**), $v_1(t)$ has a step discontinuity across 0. By integrating Eq. 9.77 between 0^- and 0^+, we obtain the initial condition necessary for the third step:

$$(n^2 C_1 + C_2) R [v_1(0^+) - \underbrace{v_1(0^-)}_{=0}] + 0 = nRC_2[e(0^+) - e(0^-)] = nRC_2 E,$$

that is,

$$v_1(0^+) = \frac{nC_2}{n^2 C_1 + C_2} E.$$

For $t > 0$ (**step 3**), the I/O relationship is

$$(n^2 C_1 + C_2) R \frac{dv_1}{dt} + v_1 = 0.$$

The solution is

$$v_1(t) = v_{fr}(t) = K e^{\lambda t} \tag{9.78}$$

with $K = v_1(0^+)$.

We remark that this solution can be expected by simply inspecting the circuit. Indeed, the series connection of C_2 and a step voltage source can be viewed as the Thévenin equivalent of a charged capacitor. Thus, it all goes as if the circuit did not have an effective input for $t > 0$ and dissipates (through the resistor) the energy initially stored by the capacitors.

A second remark is concerned with the choice of the state variable. In this case, v_2 could also have been chosen as the SSV. In fact, the constraint $v_2 + e(t) = v_1/n$ just tells us that v_1 and v_2 cannot both be SSVs, but we can freely choose the SSV between them.

Fig. 9.55 Case Study 1: Equivalent formulation for the circuit of Fig. 9.53. The capacitor C_2 is replaced by the model, taking into account the algebraic constraint

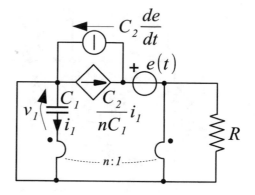

Fig. 9.56 Case Study 2

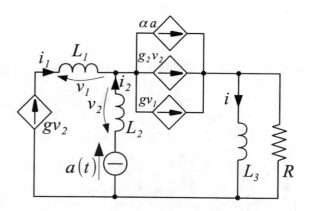

As a final remark, we notice that by taking v_1 as an SSV, C_2 can be replaced by the model discussed in Sect. 9.3.2. The resulting circuit is shown in Fig. 9.55. You are invited to check the equivalence with the original circuit by comparing the two sets of circuit equations and to verify that the circuit of Fig. 9.55 can be viewed as the parallel connection between the only *"mc component"* C_1 and a memoryless *complementary component*, as discussed in Sect. 9.4 and exemplified in Fig. 9.17c.

Case Study 2

Find the state variable $i(t)$ in the circuit shown in Fig. 9.56, which works at DC steady state for $t < 0$, with $a(t) = Au(t)$.

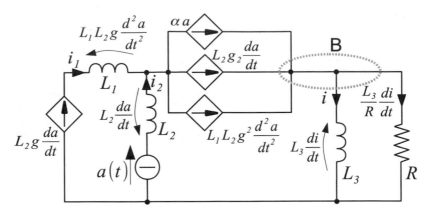

Fig. 9.57 Solution of Case Study 2

According to Fig. 9.57, we find two independent constraints ($M = 2$) $i_2 = a(t)$ and $i_1 = L_2 g \dfrac{da}{dt}$. Then only one of the $N = 3$ WSVs (i_1, i_2, i) is an SSV, namely i. (In this case, there are no alternatives!) From KCL at nodal cut-set B, we obtain

$$\frac{L_3}{R}\frac{di}{dt} + i = \alpha a + L_2 g_2 \frac{da}{dt} + L_1 L_2 g^2 \frac{d^2 a}{dt^2},$$

or (giving to the addends the physical dimension of volts)

$$L_3 \frac{di}{dt} + Ri = \alpha R a + L_2 R g_2 \frac{da}{dt} + L_1 L_2 R g^2 \frac{d^2 a}{dt^2}. \qquad (9.79)$$

By comparing this equation to Eq. 9.75, we have $a_1 = L_3$, $a_0 = R$, $P = 1$, $h_1 = \alpha R$, $\tilde{h}_1^{(1)} = R L_2 g_2$, $\tilde{h}_1^{(2)} = L_1 L_2 R g^2$.

The natural frequency is $\lambda = -\dfrac{R}{L_3} < 0$, and then the circuit is absolutely stable. We could also find it by turning off the independent current sources; in this case, the controlled current sources would also be turned off, and the circuit would reduce to the parallel connection between L_3 and R.

For $t < 0$ (**step 1**), the input is turned off, and the circuit is in its DC steady state by assumption; then $i(t) = 0 = i(0^-)$.

Owing to the discontinuity balance (**step 2**), $i(t)$ has a second-order discontinuity (Dirac δ-function) across 0 (since $a(t)$ has a step discontinuity, $\frac{da}{dt}$ an impulse, and $\frac{d^2 a}{dt^2}$ a doublet, as well as $\frac{di}{dt}$). Then in general, we can expect that $i(t)$ across $t = 0$ is of the kind $I(t)u(t) + Q\delta(t)$, with $I(t)$ continuous and smooth. Since

$$\frac{di}{dt} = \frac{dI}{dt}u(t) + \underbrace{I(t)\delta(t)}_{=I(0)\delta(t)} + Q\delta^{(1)}(t) ,$$

from the I/O relationship we obtain

$$L_3\left[\frac{dI}{dt}u(t) + I(0)\delta(t) + Q\delta^{(1)}(t)\right] + R\left[I(t)u(t) + Q\delta(t)\right] = \\ = \alpha RAu(t) + L_2 Rg_2 A\delta(t) + L_1 L_2 Rg^2 A\delta^{(1)}(t). \tag{9.80}$$

By applying the discontinuity balance to this equation, we obtain the area Q of the Dirac δ-function by equating the $\delta^{(1)}(t)$ coefficients

$$Q = \frac{L_1 L_2}{L_3} Rg^2 A,$$

and the step height $I(0)$ by equating the $\delta(t)$ coefficients:

$$L_3 I(0) + RQ = L_2 Rg_2 A,$$

from which we obtain

$$I(0) = \frac{L_2}{L_3} RA\left(g_2 - \frac{L_1}{L_3} Rg^2\right) .$$

By equating the remaining terms, we obtain the I/O relationship valid for $t > 0$ (notice that $i(t) = I(t)$ for $t > 0$, and thus $i(0^+) = I(0)$).

Indeed, for $t > 0$ (**step 3**), the I/O relationship is

$$L_3 \frac{di}{dt} + Ri = \alpha RA,$$

whose solution is

$$i(t) = i_{fr}(t) + i_{fo}(t) = Ke^{\lambda t} + K_1 \tag{9.81}$$

with $K_1 = \alpha A$ and $i(0^+) = K + \alpha A$, from which $K = i(0^+) - \alpha A = I(0) - \alpha A$. Thus the solution for $t > 0$ is

$$i(t) = \underbrace{(I(0) - \alpha A)e^{-\frac{R}{L_3}t}}_{i_{fr}(t)} + \underbrace{\alpha A}_{i_{fo}(t)} = \underbrace{I(0)e^{-\frac{R}{L_3}t}}_{ZIR} + \underbrace{\alpha A(1 - e^{-\frac{R}{L_3}t})}_{ZSR}. \tag{9.82}$$

The complete solution is sketched in Fig. 9.58.

As an alternative procedure, we can replace L_1 and L_2 by the models that take into account the algebraic constraints, according to Sect. 9.3.2. The resulting circuit is shown in Fig. 9.59. You are invited to check the equivalence with

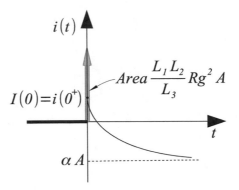

Fig. 9.58 Sketch of $i(t)$ for Case Study 2, assuming $A > 0$ and $\alpha < 0$

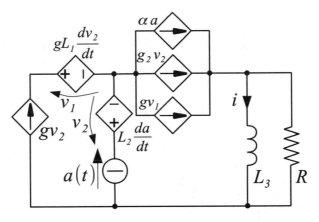

Fig. 9.59 Case Study 2: equivalent formulation for the circuit of Fig. 9.56. Here L_1 and L_2 are replaced by models taking into account the algebraic constraint

the original circuit by comparing the two sets of circuit equations. You can also verify that the circuit of Fig. 9.59 can be viewed as the parallel connection between the only *"mc component"* L_3 and a memoryless *complementary component*, according to the general scheme shown in Fig. 9.17c.

Case Study 3

For the circuit shown in Fig. 9.60, where $e(t) = E$ and $a(t) = A\cos(\omega t)u(t)$, find:

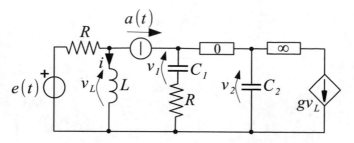

Fig. 9.60 Case Study 3

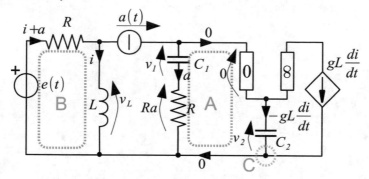

Fig. 9.61 Solution of Case Study 3

1. *the numbers of WSVs and SSVs;*
2. *the I/O relationship for $i(t)$;*
3. *the I/O relationship for $v_2(t)$;*
4. *the circuit's natural frequency;*
5. *$v_2(0^+)$ by assuming that $v_2(0^-)$ is known;*
6. *$v_2(t)$ for $t > 0$.*

The circuit contains one inductor and two capacitors, and it therefore has three WSVs. But KVL for the inner loop A in Fig. 9.61 provides the algebraic constraint $v_2 = v_1 + Ra$.

The current flowing in C_2 is imposed by the VCCS (KCL at node C)

$$C_2\frac{dv_2}{dt} = -gL\frac{di}{dt}.\tag{9.83}$$

By deriving the algebraic constraint, we obtain

$$\frac{dv_2}{dt} = \underbrace{\frac{dv_1}{dt}}_{=\frac{a}{C_1}} + R\frac{da}{dt}.\tag{9.84}$$

Moreover, from the inner loop B in Fig. 9.61, we obtain

$$e = L\frac{di}{dt} + R(i + a),$$

from which

$$\frac{di}{dt} = -\frac{R}{L}i - \frac{Ra}{L} + \frac{e}{L}. \tag{9.85}$$

This may seem to be the state equation for $i(t)$. However, by substituting Eqs. 9.84 and 9.85 into Eq. 9.83, after a few manipulations we obtain a second algebraic constraint,

$$Rgi = \left(\frac{C_2}{C_1} - Rg\right)a + RC_2\frac{da}{dt} + ge,$$

which is also the I/O relationship for $i(t)$. In other words, $i(t)$ cannot be an SSV. Therefore, the circuit has only one SSV, which can be chosen between v_1 and v_2. Notice the key role played by the VCCS, which is a two-port and whose driving voltage is $L\frac{di}{dt}$.

If we choose v_1 as the SSV, the corresponding state equation is

$$C_1\frac{dv_1}{dt} = a(t), \tag{9.86}$$

which provides the circuit natural frequency $\lambda = 0$. Notice that the state equation is an algebraic constraint between the derivative of the SSV and the input.

On the other hand, if we choose v_2 as the SSV, the corresponding state equation is

$$C_1\frac{dv_2}{dt} = a(t) + RC_1\frac{da}{dt}, \tag{9.87}$$

which (of course) provides again the circuit's natural frequency $\lambda = 0$ and is also the required I/O relationship for v_2.

By applying the discontinuity balance to Eq. 9.87, it follows that $v_2(t)$ is not continuous across $t = 0$. We can obtain $v_2(0^+)$ (by assuming that $v_2(0^-)$ is known) by integrating Eq. 9.87 between 0^- and 0^+:

$$C_1[v_2(0^+) - v_2(0^-)] = 0 + RC_1[a(0^+) - a(0^-)] = RC_1A,$$

from which we obtain

$$v_2(0^+) = RA + v_2(0^-).$$

Finally, we can obtain $v_2(t)$ for $t > 0$:

$$v_2(t) = v_{2fr}(t) + v_{2fo}(t) = K + V_C \cos(\omega t) + V_S \sin(\omega t).$$

You can check that $V_C = RA$ and $V_S = \frac{A}{\omega C_1}$. Finally, K is obtained by imposing the initial condition

$$v_2(0^+) = K + V_C,$$

or

$$K = v_2(0^+) - V_C.$$

Thus K is also equal to $v_2(0^-)$.

Therefore, the complete solution for $t > 0$ is

$$v_2(t) = v_2(0^+) - V_C + V_C \cos(\omega t) + V_S \sin(\omega t).$$

9.15 Problems

9.1 Find the descriptive equation of the composite two-terminal shown in Fig. 9.62a.

9.2 For the circuit shown in Fig. 9.62b, where the two-port is described by the resistance matrix $[R] = \begin{pmatrix} R_{11} & R_{12} \\ R_{21} & R_{22} \end{pmatrix}$ and $a(t) = Au(t)$, find:

1. the I/O relationship (state equation) for the state variable $v_2(t)$;
2. the circuit's natural frequency;
3. the absolute stability condition (henceforth, assume that it is satisfied);
4. $v_2(t)$;
5. $v_1(t)$;

by assuming that the circuit works in DC steady state for $t < 0$.

(a)   **(b)**

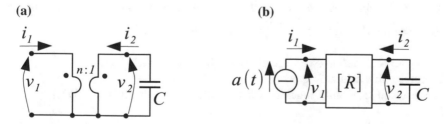

Fig. 9.62 a Problem 9.1; **b** Problem 9.2

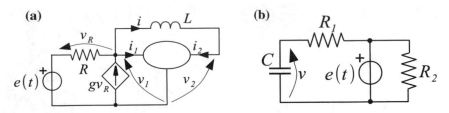

Fig. 9.63 a Problem 9.3; **b** Problem 9.4

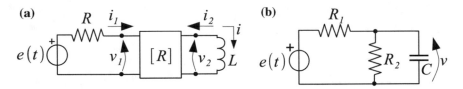

Fig. 9.64 a Problem 9.5; **b** Problem 9.6

9.3 For the circuit shown in Fig. 9.63a, where $e(t) = \Phi\delta(t)$ and the three-terminal is described by the equations $i_1 = gv_2$ and $i_2 = -gv_1$ ($g \neq 0$), find:

1. the I/O relationship (state equation) for the state variable $i(t)$;
2. the circuit's natural frequency;
3. the absolute stability condition;
4. $i(t)$;

by assuming that the circuit works in DC steady state for $t < 0$.

9.4 For the circuit shown in Fig. 9.63b, where $e(t) = E_0u(-t) + E_1u(t)$, find:

1. the I/O relationship (state equation) for the state variable $v(t)$;
2. the circuit's natural frequency;
3. the absolute stability condition;
4. $v(t)$;

by assuming that the circuit works in DC steady state for $t < 0$.

9.5 For the circuit shown in Fig. 9.64a, where the two-port is described by the resistance matrix $[R] = R \begin{pmatrix} 2 & 1 \\ 1 & 2 \end{pmatrix}$ and $e(t) = Eu(t)$, find:

1. the I/O relationship (state equation) for the state variable $i(t)$;
2. the circuit's natural frequency;
3. the absolute stability condition (henceforth, assume that it is satisfied);
4. $i(t)$;

by assuming that the circuit works in DC steady state for $t < 0$.

9.6 For the circuit shown in Fig. 9.64b, where $e(t) = \Phi\delta(t)$, find:

(a)

(b)

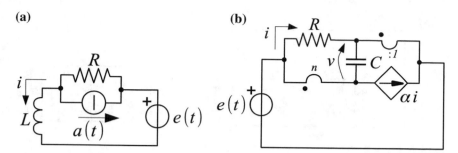

Fig. 9.65 **a** Problem 9.7; **b** Problem 9.8

1. the I/O relationship (state equation) for the state variable $v(t)$;
2. the circuit's natural frequency;
3. the absolute stability condition;
4. $v(t)$;

by assuming that the circuit works in DC steady state for $t < 0$.

9.7 For the circuit shown in Fig. 9.65a, where $e(t) = Eu(t)$ and $a(t) = Q\delta(t)$, find:

1. the I/O relationship (state equation) for the state variable $i(t)$;
2. the circuit's natural frequency;
3. $i(t)$;

by assuming that the circuit works in DC steady state for $t < 0$.

9.8 For the circuit shown in Fig. 9.65b, where $e(t) = Eu(t)$, $n \neq -1$ and $\alpha \neq 1$, find:

1. the I/O relationship (state equation) for the state variable $v(t)$;
2. the circuit's natural frequency;
3. the absolute stability condition;
4. $v(t)$;
5. the power p absorbed by the controlled source;

by assuming that the circuit works in DC steady state for $t < 0$.

9.9 For the circuit shown in Fig. 9.66a, where $e_1(t) = Eu(t)$, $e_2(t) = \Phi\delta(t)$, and the four-terminal is described by the equations

$$\begin{pmatrix} i_1 \\ i_2 \\ i_3 \end{pmatrix} = \begin{pmatrix} g_1 & G & G \\ G & g_2 & G \\ G & G & g_3 \end{pmatrix} \begin{pmatrix} v_1 \\ v_2 \\ v_3 \end{pmatrix},$$

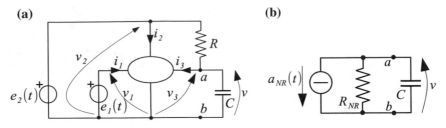

Fig. 9.66 Problem 9.9

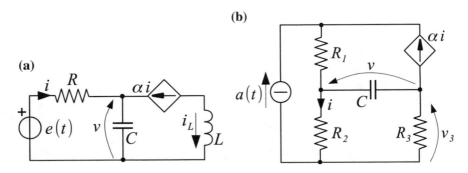

Fig. 9.67 a Problem 9.10; **b** 9.11

find:

1. the Norton equivalent (shown in Fig. 9.66b) of the complementary component;
2. the I/O relationship (state equation) for the state variable $v(t)$;
3. the circuit's natural frequency;
4. the absolute stability condition;
5. $v(t)$;

by assuming that the circuit works in DC steady state for $t < 0$.

9.10 For the circuit shown in Fig. 9.67a, where $e(t) = Eu(-t) + \Phi\delta(t)$ and $\alpha \neq -1$, find:

1. the number of state variables;
2. the I/O relationship for $v(t)$;
3. the circuit's natural frequency;
4. the absolute stability condition;
5. $v(t)$ for $t < 0$;
6. $v(0^-)$;
7. $v(0^+)$;
8. $v(t)$ for $t > 0$;

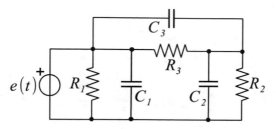

Fig. 9.68 Problem 9.12

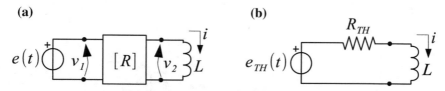

Fig. 9.69 Problem 9.13

by assuming that the circuit works in DC steady state for $t < 0$.

9.11 For the circuit shown in Fig. 9.67b, where $a(t) = Au(t)$ and $\alpha < 1$, find:

1. the state equation;
2. the circuit's natural frequency;
3. $v(t)$ for $t < 0$;
4. $v(0^-)$;
5. $v(0^+)$;
6. $v(t)$ for $t > 0$;
7. $v_3(t)$ for $t > 0$;

by assuming that the circuit works in DC steady state for $t < 0$.

9.12 For the circuit shown in Fig. 9.68, find the number of WSVs and the number of SSVs.

9.13 For the circuit shown in Fig. 9.69a, where $e(t) = Eu(t)$ and $[R] = R \begin{pmatrix} 2 & 3 \\ 1 & 2 \end{pmatrix}$ (with $R > 0$), find:

1. the Thévenin equivalent of the memoryless complementary component (see Fig. 9.69b);
2. the circuit's natural frequency;
3. the value of $i(t)$ for $t \to \infty$.

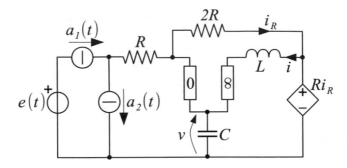

Fig. 9.70 Problem 9.14

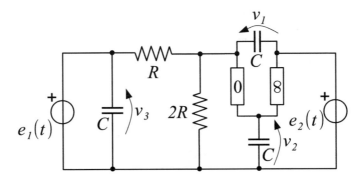

Fig. 9.71 Problem 9.15

9.14 For the circuit shown in Fig. 9.70, where $a_1(t) = A_1$, $a_2(t) = A_2$, and $e(t) = E$, working in DC steady state, find:

1. the numbers of WSVs and SSVs;
2. $v(t)$;
3. $i(t)$.

9.15 For the circuit shown in Fig. 9.71, where $e_1(t) = \Phi\delta(t)$ and $e_2(t) = Eu(-t)$, find:

1. the numbers of WSVs and of SSVs;
2. the I/O relationship for $v_1(t)$;
3. the degree of discontinuity of $v_2(t)$ at $t = 0$;
4. the circuit's natural frequency;
5. the absolute stability condition;
6. $v_1(t)$ for $t < 0$;
7. $v_1(0^-)$;

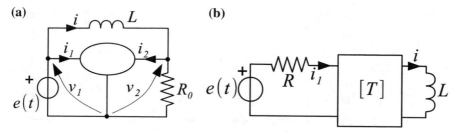

(a)

(b)

Fig. 9.72 a Problem 9.16; **b** Problem 9.17

8. $v_1(0^+)$;
9. $v_1(t)$ for $t > 0$;

by assuming that the circuit works in DC steady state for $t < 0$.

9.16 For the circuit shown in Fig. 9.72a, where the three-terminal descriptive equations are $v_1 = R(i_1 - i_2)$ and $v_2 = R(i_1 + i_2)$, with $R > 0$, find:

1. the state equation;
2. the circuit's natural frequency;
3. $i(t)$ (complete response) for $e(t) = Eu(t)$;
4. $i(t)$ (complete response) for $e(t) = \Phi\delta(t)$;

by assuming that the circuit works in DC steady state for $t < 0$.

9.17 For the circuit shown in Fig. 9.72b, where the two-port is described by the transmission matrix $T = \begin{pmatrix} \alpha & R \\ g & \beta \end{pmatrix}$, with $\alpha, \beta, g \neq 0$ and $R > 0$, and the input is $e(t) = \Phi\delta(t) + Eu(-t)$, find:

1. the state equation;
2. the circuit's natural frequency;
3. the absolute stability condition (henceforth, assume that it is satisfied);
4. $i(t)$ for $t < 0$ and $i(0^-)$;
5. $i(0^+)$;
6. $i(t)$ for $t > 0$;
7. the energy w dissipated by the circuit for $t > 0$;

by assuming that the circuit works in DC steady state for $t < 0$.

9.18 For the circuit shown in Fig. 9.73, find the number of WSVs and the number of SSVs.

9.19 For the circuit shown in Fig. 9.74, where $a(t) = Q\delta(t) + Au(t)$, find:

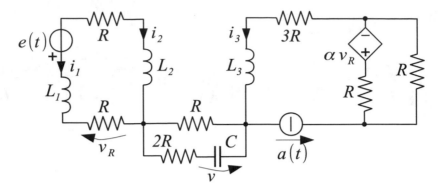

Fig. 9.73 Problem 9.18

Fig. 9.74 Problem 9.19

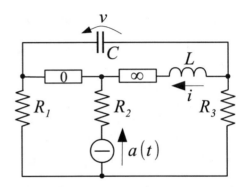

1. the numbers of WSVs and of SSVs;
2. the state equation;
3. the circuit's natural frequency;
4. $v(t)$ (complete response);

by assuming that the circuit works in DC steady state for $t < 0$.

9.20 For the circuit shown in Fig. 9.75, where $e(t) = Eu(-t)$, $a(t) = Au(t)$, and $r > 0$, find:

1. the numbers of WSVs and SSVs;
2. the I/O relationship for $v(t)$;
3. the circuit's natural frequencies;
4. $v(t)$ (complete response);

by assuming that the circuit works in DC steady state for $t < 0$.

9.21 For the circuit shown in Fig. 9.76, where $e(t) = Eu(-t) + \Phi\delta(t)$, find:

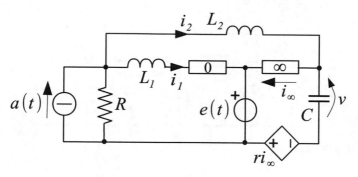

Fig. 9.75 Problem 9.20

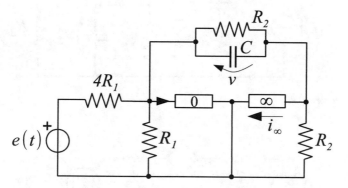

Fig. 9.76 Problem 9.21

1. the state equation;
2. the circuit's natural frequency;
3. $v(t)$ (complete response);

by assuming that the circuit works in DC steady state for $t < 0$.

9.22 For the circuit shown in Fig. 9.77, where $a(t) = Au(t)$, find:

1. the numbers of WSVs and SSVs;
2. the I/O relationship for $v(t)$;
3. the circuit's natural frequencies;
4. $v(0^-)$;
5. $v(0^+)$;
6. $v(t)$ for $t > 0$;
7. an expression for the output variable i_x in terms of $v(t)$ and the inputs;

by assuming that the circuit works in DC steady state for $t < 0$.

9.23 For the circuit shown in Fig. 9.78, where $a(t) = Au(t)$ and $v_1(0^-) = V_0$, find:

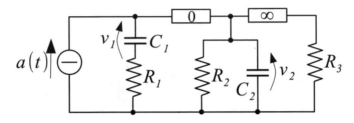

Fig. 9.77 Problem 9.22

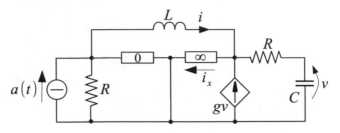

Fig. 9.78 Problem 9.23

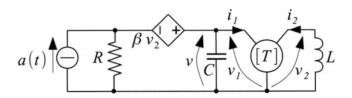

Fig. 9.79 Problem 9.24

1. the numbers of WSVs and SSVs;
2. the I/O relationship for $v_2(t)$;
3. the circuit's natural frequencies;
4. $v_2(0^-)$;
5. $v_2(0^+)$;
6. $v_2(t)$ for $t > 0$;

by assuming that the circuit works in DC steady state for $t < 0$.

9.24 For the circuit shown in Fig. 9.79, where $a(t) = Au(-t) + Q\delta(t)$, $\beta < 1$, and $[T] = \begin{pmatrix} 0 & R \\ 1/R & 0 \end{pmatrix}$, find:

1. the I/O relationship for $v(t)$;

2. the circuit's natural frequencies;
3. $v(0^-)$;
4. $v(0^+)$;
5. $v(t)$ for $t > 0$;

by assuming that the circuit works in DC steady state for $t < 0$.

References

1. Tom A (1967) Calculus, vol 1. Wiley, New York
2. Athanasios P (1960) The Fourier integral and its applications. McGraw-Hill Book Co., Inc, Interscience Publishers, New York

Chapter 10
Advanced Concepts: First-Order Nonlinear Circuits

> *To create consists precisely in not making useless combinations,*
> *and in making those which are useful and which are only in a*
> *small minority ... Among chosen combinations, the most fertile*
> *will often be those formed of elements drawn from domains*
> *which are far apart.*
> —Jules Henri Poincaré

Abstract In this chapter, some of the concepts introduced in Chap. 9 are extended to nonlinear first-order circuits, also making reference to potential functions for resistive circuits. In particular, some circuits with piecewise-linear nonlinearities (either original or obtained by approximation) are studied, showing the advantages of this kind of formulation, which makes it possible to find the solution analytically, whereas in the general case we usually have to resort to numerical solutions.

10.1 Asymptotic Solution of a Particular Class of First-Order Nonlinear Circuits

Like the linear circuits described in Chap. 9, nonlinear circuits containing memory elements and at least one SSV are also usually studied using the state variables method. However, in contrast to the linear case, in dealing with nonlinear circuits it is no longer possible to find the solution (assuming that it exists and is unique) by applying the superposition principle, which implies that the solution can no longer be viewed as the sum of ZIR and ZSR or of free response and forced response. Moreover, an analytical solution exists only in particular cases, for instance circuits whose nonlinear elements are memoryless components with piecewise-linear (PWL) descriptive equations (see Sect. 10.3). In such cases, the solution can be found analytically by composing the "locally valid" solutions of linear circuits.

In general, we can instead either find a *numerical solution* through integration algorithms or obtain some information about the *qualitative behavior* of the solution and about the presence and properties of specific kinds of asymptotic solutions (such

© Springer Nature Switzerland AG 2020

M. Parodi and M. Storace, *Linear and Nonlinear Circuits: Basic and Advanced Concepts*,
Lecture Notes in Electrical Engineering 620,
https://doi.org/10.1007/978-3-030-35044-4_10

as equilibrium points or periodic behaviors) through the analytical and geometric tools provided by the nonlinear dynamics approach [1].

In this section we will restrict our analysis to a specific class of first-order nonlinear circuits, where:

(1) there are no algebraic constraints involving one or more WSVs (and possibly one or more circuit inputs) only, and thus we can impose any initial condition compatible with the circuit nonlinearities, that is, *we assume that there is only one memory element, whose WSV is also the SSV for the circuit*;
(2) all the nonlinear components are memoryless;
(3) all components are time-invariant, including the inputs, which can only be constant sources;
(4) the solution exists and is unique for an assigned initial condition.

Our main goal is finding their DC solutions.

Generally speaking, these circuits can be sketched as shown in Fig. 10.1.

Also in this case, as for first-order linear circuits, we can represent the circuit as the connection of the memory element and a memoryless complementary component that contains the rest of the circuit. In the nonlinear case, of course, the complementary component cannot be represented in general by a Thévenin/Norton equivalent, and will be generically sketched as a nonlinear resistor (active, in general), as shown in Fig. 10.2.

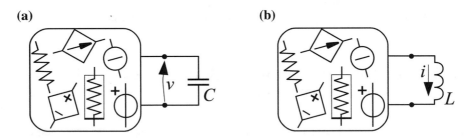

Fig. 10.1 The two possible kinds of nonlinear circuits that will be analyzed in this section

Fig. 10.2 The two possible complementary-component representations for the circuits sketched in Fig. 10.1

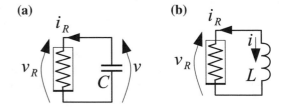

In the case shown in Fig. 10.2a, it is required that the nonlinear resistor admit at least the voltage basis. Thus, it can be described by a driving-point (DP) characteristic $i_R = \hat{i}_R(v_R)$, and its state equation can be written as

$$C\frac{dv}{dt} = -\hat{i}_R(v).$$ (10.1)

Analogously, in the case of Fig. 10.2b, the nonlinear resistor must admit at least the current basis. Thus, it can be described by a DP characteristic $v_R = \hat{v}_R(i_R)$, and its state equation can be written as

$$L\frac{di}{dt} = \hat{v}_R(-i).$$ (10.2)

In both cases, the structure of the state equation is the same, and then we can formulate a unique canonical form to study their solutions.

The **state equation** for a first-order nonlinear circuit belonging to the specified class can always be written in the following **canonical form**:

$$\frac{dx}{dt} = \dot{x} = f(x),$$ (10.3)

where x is the state variable and f is called a **vector field** and defines the law of the state's evolution in time.

The **state space** is the set of all possible states x of the considered circuit, that is, the domain of the function f.

If all the circuit components (including the independent sources) are time-invariant, the vector field is in turn time-invariant and the overall circuit is said to be **autonomous**. Otherwise, it also depends explicitly on time; we should write $f(x, t)$, and the overall circuit is said to be nonautonomous.

According to our assumptions,

- inasmuch as we have only one SSV (assumption (1)), the state space is one-dimensional: typically the real axis or a real interval;
- we will consider only autonomous circuits (assumption (3)); their state evolution is described by Eq. 10.3;
- f is a *smooth enough* function, that is, a Lipschitz-continuous function. This ensures that the solution of the *ordinary differential equation* (ODE) Eq. 10.3 for a given initial condition $x(t_0)$ exists and is unique (assumption (4)) [1].

The solution of Eq. 10.3 for a given initial condition $x(t_0)$ can be represented either as a function of time (as $x(t)$) or by an ordered sequence of points (parameterized by time) in the state space, called its **trajectory**.

Every DC solution of Eq. 10.3 corresponds to the condition $\dot{x} = 0$ and is also called an **equilibrium point** (or simply *equilibrium*) of the circuit (or system, in general) described by Eq. 10.3.

The equilibrium points of a first-order dynamical system $\dot{x} = f(x)$ can be found algebraically by imposing

$$f(x) = 0, \tag{10.4}$$

which is called an **equilibrium condition**.

Thus the DC solutions of the circuit can be found either by analytically solving Eq. 10.4 (*analytical method*) or by graphically finding the zero-crossings of $f(x)$ in the plane $(x, \dot{x})$ (*graphical method*).

Moreover, the graphical method tells us that where $f(x)$ is positive, we also have $\dot{x} > 0$, and then x is increasing, whereas where $f(x)$ is negative, x is decreasing. This point will be better illustrated in the following two case studies.

Case Study 1
Consider the (linear) circuit shown in Fig. 10.3 (Case Study 2 in Sect. 9.8.2 for $t < 0$) and find its equilibrium solutions, both analytically and graphically. Assume that $0 < \alpha < 1$.

The I/O relationship Eq. 9.6.1 can be recast as follows:

$$\frac{dv}{dt} = \frac{\alpha - 1}{(R_2 + R_3)C}v + \frac{R_2 + \alpha R_3}{(R_2 + R_3)C}A_1 = f(v). \tag{10.5}$$

By solving the equilibrium condition $f(v) = 0$, we obtain $v = \dfrac{R_2 + \alpha R_3}{1 - \alpha}A_1$, which coincides with the solution found analytically (Case Study 2 in Sect. 9.8.2) for $t < 0$.

By assuming $A_1 > 0$, the graphical method allows one to easily detect the equilibrium point (zero-crossing point) shown as a dot in Fig. 10.4a. The arrows on the v-axis denote the **flow** induced by the vector field. For a given initial condition, the state trajectory follows the flow. The figure suggests that for values of v far from the equilibrium point, the vector field action is very strong ($|\dot{v}| \gg 0$), whereas it becomes weaker and weaker as v approaches the equilibrium point. For instance, if we consider voltages v_1 and v_2 in Fig. 10.4b, it is evident that $|\dot{v}_1| (= |f(v_1)|)$ is larger than $|\dot{v}_2| (= |f(v_2)|)$; this implies that the absolute value of the time derivative of the state variable v at v_1 is greater than at v_2. Similarly, $|\dot{v}|$ decreases as v approaches its equilibrium value.

This perfectly agrees with the analytical solution found in Sect. 9.8.2, but it also adds some qualitative information about the transient behavior starting from any initial condition. Moreover, the figure suggests that starting from any initial condition, the circuit dynamics drives the state variable toward the equilibrium point; in other words, all possible trajectories converge to the equilibrium point.

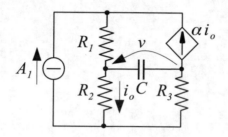

Fig. 10.3 Case Study 1

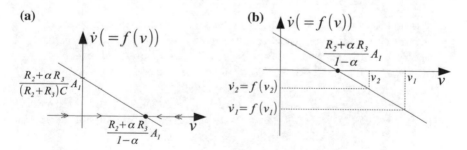

Fig. 10.4 Graphical method applied to Case Study 1

Case Study 2

Consider the circuit shown in Fig. 10.5a, containing a particular type of diode (called a tunnel diode), whose DP characteristic is displayed in Fig. 10.5b. By assuming that $i_d(v_d) = A_0 + \gamma(v_d - E_0)^3 - \alpha(v_d - E_0)$, with positive dimensional coefficients γ [A/V^3] and α [Ω^{-1}], find the equilibrium solutions, both analytically and graphically.

It is easy to see that the descriptive equation of the complementary component within the dashed box in Fig. 10.5a is $i = \hat{i}(v) = \gamma v^3 - \alpha v$. Then the state equation in canonical form is

$$\dot{v} = -\frac{\gamma}{C}v^3 + \frac{\alpha}{C}v = f(v). \tag{10.6}$$

By solving the equilibrium condition $f(v) = 0$, we obtain three solutions: $v = 0$ and $v = \pm\sqrt{\dfrac{\alpha}{\gamma}}$. The equilibria obtained by applying the graphical method are marked by dots and labeled by letters in Fig. 10.6, and they of course coincide with the analytical results.

We remark that the three equilibrium points are distinct in nature, as evidenced by the arrows in Fig. 10.6:

- At the left of A and between B and C, f is positive; thus the time derivative of the state variable is positive and v increases.
- Between A and B and at the right of C, f is negative; thus the time derivative of the state variable is negative and v decreases.

Thus on the whole, the vector field induces a flow directed toward A and C, whereas starting from any initial condition close to (but not coincident with) B, the flow makes the state variable move away from the central equilibrium point. Then in this case, the trajectories converge to one of the two outer equilibrium points (A or C), according to the initial condition.

Figure 10.7 shows the result of numerical integrations (by setting $\gamma/C = 1$ [$V^{-2}s^{-1}$] and $\alpha/C = 1$ [s^{-1}]) starting from several initial conditions uniformly distributed in the range $[-2, 2]$ V. It is evident that solutions $v(t)$ starting from any equilibrium point remain constant, whereas the other solutions converge to one of the equilibrium points, A or C. The corresponding trajectories are the projections of these solutions on the v-axis.

The last example points out the following two important concepts.

In the nonlinear case, the **stability** is no longer a **property** of the whole circuit, as in the linear case, but of **each single equilibrium point**.

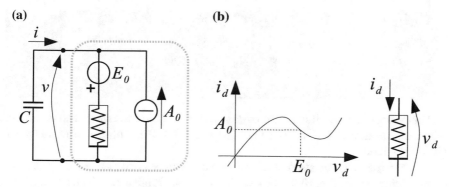

Fig. 10.5 Case Study 2: **a** circuit to be analyzed and **b** tunnel diode and its DP characteristic

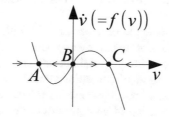

Fig. 10.6 Graphical method applied to Case Study 2

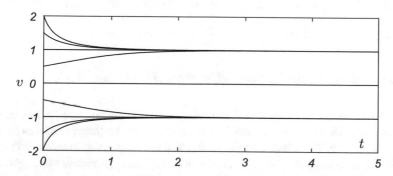

Fig. 10.7 Numerical solutions of Case Study 2 for several initial conditions

An equilibrium is

- **stable and attractive** if the flow is directed toward it in both directions;
- **Lyapunov-stable** if the flow is null in an interval containing (strictly) the equilibrium point, such that trajectories starting close to the equilibrium point are neither attracted nor repelled;
- **unstable** if the flow makes the state variable move away from it in both directions;

> • **metastable** if the flow direction is the same at the left and at the right.
>
> The set of all initial conditions whose corresponding trajectories converge to a given stable and attractive equilibrium point is called the **basin of attraction** of that point.

For instance, in Case Study 2, the outer equilibrium points are stable and attractive, with basins of attraction $(-\infty, 0)$ (A) and $(0, +\infty)$ (C). The central equilibrium point (B) is unstable.

An example of a metastable equilibrium point can be easily obtained by shifting up the vector field of Fig. 10.6, as shown in Fig. 10.8a. From a circuit standpoint, this shifting can be done by acting on the values of E_0 and A_0. In this case, we have two equilibria, one stable and attractive (K), with basin of attraction $(v_1, +\infty)$, and one metastable (H $= (v_1, 0)$).

Figure 10.8b shows an example of a vector field with infinite Lyapunov-stable equilibrium points, since the flow is null everywhere.

In the linear case (see Case Study 1), a stable and attractive equilibrium is always **globally stable**, because its basin of attraction coincides with the state space, due to the shape of f. In the nonlinear case, there are instead two distinct kinds of attractive stability. If the basin of attraction of an equilibrium point coincides with the state space, the equilibrium point is globally stable; otherwise, it is **locally stable**. For instance, in Fig. 10.6, the central equilibrium is unstable, whereas the outer equilibria are locally stable.

10.1.1 First-Order Circuits with More Than One WSV

What if the circuit to be analyzed contains one SSV and one or more elements with memory whose WSVs are not SSVs? Similarly to the linear case described in Sect. 9.14, each WSV of this kind can be algebraically expressed in terms of the SSV and the input voltages/currents generated by independent sources. Moreover, each equation of this kind allows us to replace the corresponding component (capacitor

(a)   **(b)**

Fig. 10.8 Examples of vector fields with a metastable equilibrium point (**a**) and Lyapunov-stable equilibrium points (**b**)

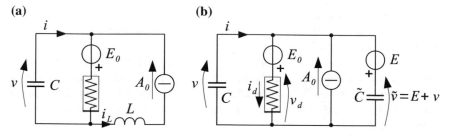

Fig. 10.9 Two examples of first-order circuits with two WSVs (and one SSV)

or inductor) with an equivalent model made up of proper controlled sources, thus generalizing the procedure described in Sect. 9.3.2 for linear circuits.

Then we can always find the state equation in the canonical form Eq. 10.3 and then find an equivalent circuit as in Fig. 10.2.

For instance, if we modify the circuit of Fig. 10.5 as shown in Fig. 10.9a, we trivially obtain again the state equation 10.6:

$$\dot{v} = -\frac{i_d(v + E_0)}{C} + \frac{A_0}{C} = -\frac{\gamma}{C}v^3 + \frac{\alpha}{C}v = f(v). \qquad (10.7)$$

The inductor does not change the complementary component's behavior, because $i_L = A_0$.

If we consider the circuit of Fig. 10.5 modified as shown in Fig. 10.9b, we obtain $\tilde{v} = v + E$ and $i = i_d(v + E_0) - A_0 + \tilde{C}\dfrac{dv}{dt} + \underbrace{\tilde{C}\dfrac{dE}{dt}}_{=0}$. Thus the canonical form of the state equation in this case is

$$\dot{v} = -\frac{i_d(v + E_0)}{C + \tilde{C}} + \frac{A_0}{C + \tilde{C}} = \tilde{f}(v). \qquad (10.8)$$

This corresponds to having an equivalent representation as in Fig. 10.2a, with a nonlinear resistor with descriptive equation $i = \hat{i}_R(v) = -C\tilde{f}(v)$.

10.1.2 Impossibility of Oscillations

Equilibrium points are the most important elements when one is analyzing first-order circuits with constant sources. It is evident that in each interval of the state space contained within two equilibria or between an equilibrium and $\pm\infty$, the vector field maintains a constant sign and that it annihilates at the equilibrium points. Correspondingly, the flow is directed toward the right (positive vector field) or left (negative vector field) or is stagnant (equilibrium points). Therefore, each trajectory is forced either to evolve *monotonically* toward an equilibrium (or to infinity) or to remain constant. This means that **in a first-order circuit with constant sources, there cannot be oscillations**.

This result generalizes what we already noticed for linear circuits with constant sources, where the solution, starting from a given initial condition, can only exponentially converge to the DC regime solution (if the circuit is *asymptotically stable*) or exponentially diverge (if the circuit is *unstable*) or remain constant (if the circuit is *simply stable*).

10.1.3 Equilibrium Stability Analysis Through Linearization

It can be proved [1] that the stability of an equilibrium point x^* (satisfying the condition $f(x^*) = 0$) can be simply checked, both analytically and graphically.

Graphically, it is sufficient to analyze the flow directions around the equilibrium point, as stated above. This is equivalent to analyzing the slope $df/dx = f'(x)$ of the vector field f at the equilibrium point.

> An equilibrium point x^* of a system $\dot{x} = f(x)$ is **hyperbolic** if $f'(x^*) \neq 0$. Otherwise, it is nonhyperbolic.

For hyperbolic equilibrium points:

- If $f'(x^*) > 0$, it is obvious that $f(x) > 0$ at the right of x^* and $f(x) < 0$ at the left of x^* (see, e.g., the central equilibrium point B in Fig. 10.10). Then the flow is directed toward the right to the right of x^* and toward the left to the left of x^*. This means that x^* is *unstable*.
- If $f'(x^*) < 0$, it is obvious that $f(x) < 0$ to the right of x^* and $f(x) > 0$ to the left of x^* (see, e.g., the equilibrium points A and C in Fig. 10.10). Then the flow is directed toward the left to the right of x^* and toward the right to the left of x^*. This means that x^* is (at least locally) *stable* and attractive.

For a nonhyperbolic equilibrium point ($f'(x^*) = 0$), everything is possible: x^* can be globally stable (as in Fig. 10.11a), locally stable (as in Fig. 10.11b), unstable (as in Fig. 10.11c), metastable (like H in Fig. 10.8a), or Lyapunov-stable (as in Fig. 10.8b).

Fig. 10.10 Example of a vector field with hyperbolic equilibrium points

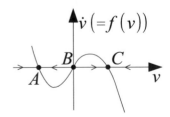

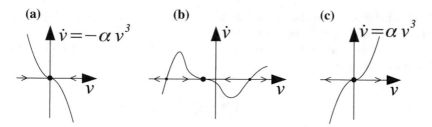

Fig. 10.11 Examples of vector fields with a nonhyperbolic equilibrium point: **a** globally stable equilibrium point; **b** locally stable equilibrium point; **c** unstable equilibrium point

Analytically, we can draw the same conclusions by analyzing the evolution of a small perturbation $w = x - x^*$ away from x^*. Since $\dot{w} = \dot{x} = f(x) = f(w + x^*)$, we can use a Taylor expansion, thus obtaining

$$\dot{w} = f(x^*) + f'(x^*)w + \mathcal{O}(w^2). \tag{10.9}$$

But $f(x^*) = 0$ (by definition of the equilibrium point), and then

$$\dot{w} = f'(x^*)w + \mathcal{O}(w^2). \tag{10.10}$$

The higher-order terms $O(w^2)$ can be neglected if x^* is hyperbolic ($f'(x^*) \neq 0$) and for low perturbation values. Under these assumptions, the perturbation evolution is approximately governed by the linear first-order ODE

$$\dot{w} = f'(x^*)w. \tag{10.11}$$

Then the perturbation either grows (if $f'(x^*) > 0$) or decays (if $f'(x^*) < 0$) in time. In the first case, x^* is unstable, whereas in the second case, x^* is stable (locally, within the limits imposed by approximation Eq. 10.11). Notice that $f'(x^*)$ **is nothing but the eigenvalue of the linearized equation** 10.11.

In contrast, if $f'(x^*) = 0$, we cannot draw any conclusion from Eq. 10.11 about the stability of x^*, because the perturbation evolution is ruled by the higher-order terms $O(w^2)$, neglected by the linearization.

10.2 Equilibrium Points and Potential Functions for First-Order Circuits

In Chap. 4 of Vol. 1 of [2], the descriptive equation of a two-terminal voltage-controlled (current-controlled) resistor was related to the cocontent (content) potential function. For the class of circuits considered in Sect. 10.1 (see Fig. 10.2), the nonlinear resistor's potential function allows an interesting interpretation of the physical behavior of the corresponding circuit, concerning both its transient dynamics and the meaning of its equilibrium points.

The voltage-controlled, time-invariant resistor in Fig. 10.2a admits a cocontent function $\bar{G}(v)$ such that

$$i_R = \frac{d\bar{G}}{dv};$$

then, considering Eq. 10.1, we obtain

$$\frac{d\bar{G}}{dt} = \frac{d\bar{G}}{dv}\frac{dv}{dt} = -C\left(\frac{dv}{dt}\right)^2 \leq 0. \tag{10.12}$$

Therefore, for each value of v out of equilibrium, the time evolution of the circuit makes the value of $\bar{G}$ decrease monotonically until $\bar{G}$ reaches a minimum, when $\frac{dv}{dt} = 0$. This minimum condition corresponds to a stable and attractive equilibrium state for the circuit.

Remark: We exclude that $\bar{G}$ can be "trapped" in a maximum (corresponding to an unstable equilibrium point) or an inflection point (corresponding to a metastable equilibrium point) for physical reasons; in a circuit, there is indeed always some *noise* that provides a perturbation and allows the flow to drive the state away from nonattractive equilibrium points.

Similarly, the current-controlled, time-invariant resistor in Fig. 10.2b admits a content function $G(i_R)$ for which, recalling Eq. 10.2, we have

$$\frac{dG}{dt} = \frac{dG}{di_R}\frac{di_R}{dt} = -L\left(\frac{di_R}{dt}\right)^2 = -L\left(\frac{di}{dt}\right)^2 \leq 0, \tag{10.13}$$

which implies the monotonic decrease of G over time toward a minimum corresponding to a stable and attractive equilibrium point.

To highlight the above concepts, we see how they apply to the two case studies already considered in the previous section.

Case Study 1
Consider the circuit shown in Fig. 10.12a (Case Study 1 of Sect. 10.1 for $a(t) = A_1$). Find the cocontent function of the equivalent two-terminal resistor connected to the capacitor. Then discuss the equilibrium solution for the circuit in terms of the property related to Eq. 10.12.

The parameters of the equivalent resistor (see Fig. 10.12b) are easily found by recasting Eq. 10.5 as follows:

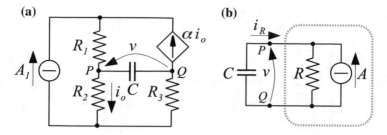

Fig. 10.12 Case Study 1. **a** Linear circuit and **b** its equivalent representation

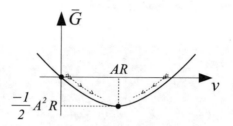

Fig. 10.13 Cocontent function $\bar{G}(v)$ of the resistive two-terminal shown in Fig. 10.12b

$$i_R = \underbrace{\frac{1-\alpha}{R_2+R_3}}_{\frac{1}{R}} v - \underbrace{\frac{R_2+\alpha R_3}{R_2+R_3} A_1}_{A} = \frac{v}{R} - A. \qquad (10.14)$$

The corresponding cocontent function expression is

$$\bar{G}(v) = \int_0^v \left(\frac{x}{R} - A\right) dx = \frac{v^2}{2R} - Av, \qquad (10.15)$$

that is, the parabola with curvature $\frac{1}{R} > 0$ shown in Fig. 10.13. The minimum of this curve is attained for $v = AR = \dfrac{R_2+\alpha R_3}{1-\alpha} A_1$, which is the equilibrium point obtained in the previous section through the equilibrium condition $f(v) = 0$ (see also Fig. 10.4). For every initial capacitor voltage $v \neq AR$, v changes with time until $\bar{G}(v)$ reaches its minimum at $v = AR$.

As a final remark, recall that $\bar{G}$ is defined up to an additive constant. Therefore, the curve $\bar{G}(v)$ may be shifted vertically without any consequences for the equilibrium condition; thus its minimum value $-\frac{1}{2}A^2R$ (see Fig. 10.13) is not important.

(a)   **(b)**

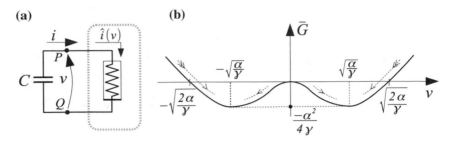

Fig. 10.14 Case Study 2. **a** Equivalent representation of the nonlinear circuit of Fig. 10.5a; **b** cocontent function $\bar{G}(v)$ of the two-terminal resistor

Case Study 2
Given the circuit shown in Fig. 10.14a, find the cocontent function for the nonlinear resistor $i = \hat{i}(v)$ defined in Case Study 2 of Sect. 10.1. Then discuss the equilibrium solutions for the circuit in terms of Eq. 10.12.

From the descriptive equation

$$\hat{i}(v) = \gamma v^3 - \alpha v \quad \text{with } \gamma > 0, \alpha > 0, \tag{10.16}$$

we obtain the cocontent function

$$\bar{G}(v) = \int_0^v \left(\gamma x^3 - \alpha x\right) dx = \frac{\gamma v^4}{4} - \frac{\alpha v^2}{2}. \tag{10.17}$$

This function is plotted in Fig. 10.14b; it displays a characteristic *double-well* shape. The curve has a local maximum at $v = 0$ and two local minima at $v = \pm\sqrt{\frac{\alpha}{\gamma}}$. These three points are equilibria for the circuit; however, for every initial condition different from equilibrium, v changes with time until $\bar{G}(v)$ attains one of its two minima, which are, therefore, stable equilibrium points. For the same reason, $v = 0$ is an unstable equilibrium point. The basins of attraction for $v = \pm\sqrt{\frac{\alpha}{\gamma}}$ are $(0, +\infty)$ for the positive equilibrium point and $(-\infty, 0)$ for the negative one. This circuit is called *bistable*.

Case Study 3
Figure 10.15 shows the composite two-terminal resistor (a) and the resulting PWL characteristic (b) obtained in Vol. 1, Sect. 3.2.5, Case Study 2. Making

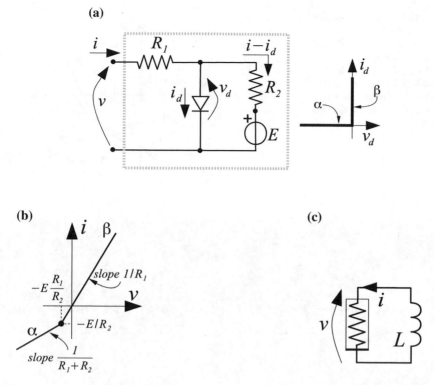

Fig. 10.15 Case Study 3. **a** Composite two-terminal resistor (left) and diode's DP characteristic (right); **b** resulting PWL DP characteristic for $E > 0$ (obtained in Vol. 1, Sect. 3.2.5, Case Study 2); **c** nonlinear circuit under study

use of these results, discuss the equilibrium solution for the nonlinear circuit shown in Fig. 10.15c, where the resistor is connected to an inductor L.

We first obtain the content function G for the two branches α and β of the two-terminal DP characteristic. For both branches, this can be done by taking as reference the intersection point at $i = -\dfrac{E}{R_2}$ (see Fig. 10.15b).

- For branch α, defined for $i < -\dfrac{E}{R_2}$, the resistor equation is $v = E + (R_1 + R_2)i$. Thus the corresponding expression of G is

$$G = \int_{-\frac{E}{R_2}}^{i} [E + (R_1 + R_2)x]\,dx = \frac{1}{2}(R_1 + R_2)i^2 + Ei + \frac{E^2}{2R_2^2}(R_2 - R_1).$$

$$(10.18)$$

Fig. 10.16 Content function $G(i)$ of the resistive two-terminal in Fig. 10.15a. The function is the connection of two arcs of a parabola that are tangent at $i = -\frac{E}{R_2}$. Parabola A was drawn on the assumption that $R_1 > R_2$

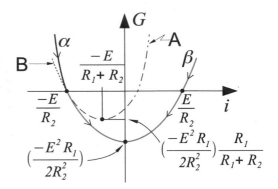

This expression describes the parabola A (with minimum at $i = \dfrac{-E}{R_1 + R_2}$) plotted in Fig. 10.16. The α-branch of G is the black solid-line portion of A for $i < -\dfrac{E}{R_2}$.

- The branch β of G is defined for $i \geq -\dfrac{E}{R_2}$, and the corresponding resistor equation is $v = R_1 i$. The related expression of the content function G is

$$G = \int_{-\frac{E}{R_2}}^{i} R_1 x\, dx = \frac{1}{2} R_1 \left(i^2 - \frac{E^2}{R_2^2} \right). \tag{10.19}$$

This expression describes another parabola B, whose axis of symmetry is the G-axis itself. This parabola is tangent to the previous one at $i = -\dfrac{E}{R_2}$, which is the starting point for the β-branch of G. This branch is represented in Fig. 10.16 as the gray solid-line portion of the parabola B for $i \geq -\dfrac{E}{R_2}$.

Therefore, the content function $G(i)$ has an absolute minimum at $i = 0$.

For the circuit of Fig. 10.15c, we can write

$$\frac{dG}{dt} = -L \left(\frac{di}{dt} \right)^2 \leq 0.$$

For every initial condition $i \neq 0$, i changes with time until $G(i)$ reaches its minimum at $i = 0$. Once more, inasmuch as G is defined up to an additive constant, its minimum value $-\dfrac{E^2 R_1}{2R_2^2}$ is not important.

10.3 Analysis of First-Order Circuits with PWL Memoryless Components

In this section we analyze some circuits containing one memory two-terminal and nonlinear components with PWL DP characteristics. In these cases, the analysis can be carried out by properly composing the analyses of linear circuits that are equivalent to the original one only within specific ranges of some circuit variables.

10.3.1 Clamper

The first circuit is shown in Fig. 10.17a. Its nonlinear element is a diode with the DP characteristic shown in Fig. 10.17b.

We assume that at $t = 0$, the capacitor is uncharged and that $e(t)$ starts from 0. Thus, the initial value of $v - e$ is also zero, and the diode operates on the branch α of its DP characteristic. If we increase $e(t)$ (up to a value E), the voltage $v - e$ across the diode becomes negative (branch α, $i = 0$), and the circuit is equivalent to the one shown in Fig. 10.18a. The capacitor's voltage v holds its null value (the natural frequency of the circuit is $\lambda = 0$), and the output voltage is $v_o = e$.

Now we assume that $e(t)$ decreases monotonically from E to $-E$. The diode keeps working on the horizontal branch α of its characteristic until its voltage reaches the value V_T, which occurs when $e = -V_T$. From this point on, the diode operates on the branch β ($v - e = V_T$, $i \geq 0$), and the original circuit is equivalent to the one shown in Fig. 10.18b. Then v is no longer a state variable ($v = e + V_T$), and the output voltage remains constant ($v_o = -V_T$).

This holds until the diode current $i = -C\dfrac{dv}{dt}$ is positive, that is, until $\dfrac{dv}{dt}$ is negative. Since $v = e + V_T$, this requisite implies that $\dfrac{dv}{dt} = \dfrac{d}{dt}(e + V_T) = \dfrac{de}{dt} < 0$. Thus the equivalent circuit of Fig. 10.18b remains valid until $e(t)$ decreases, that is, on the basis of our assumptions, until $e(t)$ reaches the value $-E$. The corresponding value of the capacitor voltage v is $-E + V_T$.

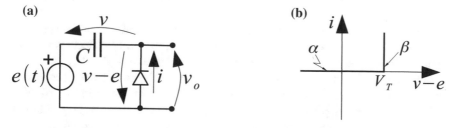

(a)   **(b)**

Fig. 10.17 a Clamper and **b** diode's DP characteristic

(a)

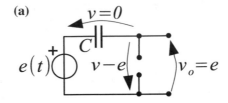

(b)

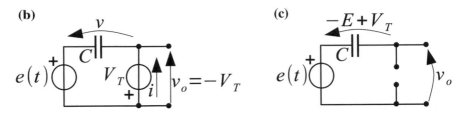

(c)

Fig. 10.18 Equivalent circuits for the Case Study of Fig. 10.17a

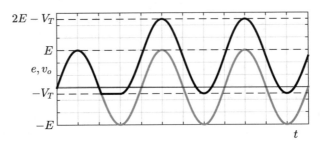

Fig. 10.19 Input $e(t)$ (gray line) and output $v_o(t)$ (black line) waveforms for the circuit of Fig. 10.17a

Now we assume that $e(t)$ increases monotonically from $-E$ to E. The condition for staying on the β branch is no longer satisfied, and the diode begins to work on the α branch. The voltage across the diode becomes less than V_T, and the original circuit is equivalent to the one shown in Fig. 10.18c. Thus v holds its value $-E + V_T$, and the output voltage is $v_o = e + E - V_T$. In particular, at $e(t) = E$, we have $v_o = 2E - V_T$.

Since $v = -E + V_T$, the condition for the diode to work on the α branch is $v - e = -E + V_T - e < V_T$, that is, $e > -E$. Therefore, a new decrease of $e(t)$ from $+E$ to $-E$ keeps the diode on the α branch and $v_o = e + E - V_T$.

In conclusion, if $e(t)$ is a sinusoid of amplitude E, the circuit output is again a sinusoid, but (after one period) with an offset of $E - V_T$, as shown in Fig. 10.19.

This is the working principle of an electronic circuit called a *clamping circuit* or *clamper*, commonly used to shift a sinusoid without altering its amplitude.

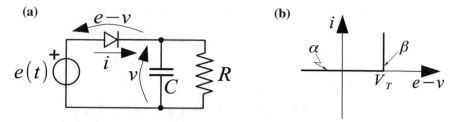

Fig. 10.20 a Half-wave rectifier **b** and diode's DP characteristic

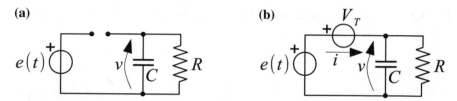

Fig. 10.21 Equivalent circuits for the Case Study of Fig. 10.20a

10.3.2 Half-Wave Rectifier

Consider the circuit shown in Fig. 10.20a, containing a diode with the DP characteristic shown in Fig. 10.20b.

We assume that initially, the capacitor is uncharged and that $e(t)$ starts from 0 (α branch) and increases (up to a value E). The voltage $e - v$ across the diode starts from 0 and increases; the circuit is equivalent to the one shown in Fig. 10.21a, and v holds its null value. When $e = V_T$, the diode begins to work on the β branch, so the original circuit becomes equivalent to the one shown in Fig. 10.21b, and v is no longer a state variable, being algebraically related to $e(t)$ ($v = e - V_T$). This remains true until $i > 0$, that is, until $e(t)$ increases.

When (after reaching the value E at a time instant t_0) $e(t)$ starts decreasing, the diode works again on the α branch of its DP characteristic, and the original circuit is once again equivalent to the one shown in Fig. 10.21a. Then the capacitor discharges through the linear resistor R, starting from the initial condition $v(t_0) = E - V_T$:
$v(t) = (E - V_T)e^{-\frac{t-t_0}{RC}}$.

If $e(t)$ is a sinusoid of proper amplitude E, the diode remains open (α branch) until the condition $e(t) - v(t) < V_T$ holds. Thus as long as $e(t)$ decreases, the diode operating point moves to the left along the α branch. When $e(t)$ starts increasing again, the operating point moves to the right, until, at time t_1, $e(t)$ reaches the value
$e(t_1) = v(t_1) + V_T = (E - V_T)e^{-\frac{t_1-t_0}{RC}} + V_T$. At t_1, the operating point reaches the β branch and moves toward the top, the half-wave rectifier becomes again equivalent to the one shown in Fig. 10.21b, and $v(t) = e(t) - V_T$.

The circuit output $v(t)$ is the waveform shown in Fig. 10.22, black line.

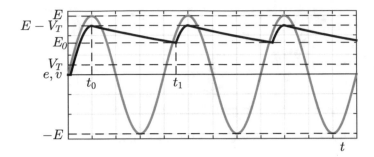

Fig. 10.22 Input $e(t)$ (gray line) and state $v(t)$ (black line) waveforms for the circuit of Fig. 10.20a, with $E_0 = (E - V_T)e^{-\frac{t_1-t_0}{RC}}$

Fig. 10.23 Equivalent circuit of the half-wave rectifier for $R \to \infty$

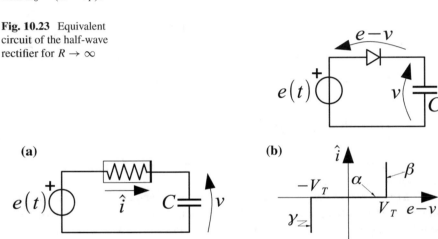

Fig. 10.24 a Hysteretic circuit and **b** PWL DP characteristic of its nonlinear two-terminal resistor

You can easily check that for $R \to \infty$ (see Fig. 10.23), the voltage v will remain constant.

This is the working principle of an electronic circuit called a *half-wave rectifier*, commonly used to convert a sinusoid (AC signal) into an almost constant waveform (DC signal).

10.3.3   Hysteretic Circuit

We analyze the seemingly simple circuit shown in Fig. 10.24a, where the nonlinear resistor has the PWL DP characteristic $\hat{i}(e - v)$ shown in Fig. 10.24b.

(a)   **(b)**

Fig. 10.25 a Composite two-terminal resistor with the DP characteristic shown in Fig. 10.24b and **b** diode's PWL DP characteristic

You can check that this nonlinear resistor can be obtained by connecting diodes and constant-voltage sources as shown in Fig. 10.25a, where each diode has the PWL DP characteristics shown in Fig. 10.25b.

The equation governing the circuit is

$$C\frac{dv}{dt} = \hat{i}\,(e - v).$$ (10.20)

- We assume that $v(0) = e(0) = 0$ and that $e(t)$ increases. As long as $e - v < V_T$, the nonlinear resistor works in the central region α of its DP characteristic, where $\hat{i} = 0$. This implies that $dv/dt = 0$, i.e., that v holds its initial value. In other words, under these conditions, the circuit is equivalent to the one shown in Fig. 10.26a, which has one SSV (voltage v) and natural frequency $\lambda = 0$.
- By further increasing $e(t)$, the working point of the nonlinear resistor moves to the right, along the central region of its DP characteristic, until it reaches the threshold voltage V_T. At this point, the resistor works on the β branch of its characteristic, and the original circuit becomes equivalent to the one shown in Fig. 10.26b. This circuit has no state: v is a WSV, but not an SSV, due to the algebraic constraint imposed by KVL. Then there is no dynamics, and $v = e - V_T$. From this point on, v "follows" $e(t)$ accordingly. Notice that $\hat{i} = C\dfrac{dv}{dt} = C\dfrac{de}{dt} > 0$.
- Now we assume that $e(t)$ reaches a maximum value, say E, and starts decreasing. Before this change, the working point of the PWL resistor is on the β-branch of its DP characteristic. At the turning point (maximum value of $e(t)$) we have $\dfrac{de}{dt} = 0$, that is, $\hat{i} = 0$; as a consequence, the working point of the PWL resistor lies at the intersection point between the α and β branches. When $e(t)$ starts decreasing, the resistor working point moves on the α branch, and we have $e - v < V_T$ and $\hat{i} = 0$, and the original circuit is once more equivalent to the one shown in Fig. 10.26a. Therefore, v holds its initial value $E - V_T$. The working point of the nonlinear resistor moves to the left, along the central region of its DP characteristic, until it reaches (for $e = E - 2V_T$) the threshold voltage $-V_T$. From this point on (keeping on decreasing $e(t)$), the original circuit becomes equivalent to the one shown in Fig. 10.26c (resistor working on the γ branch of its

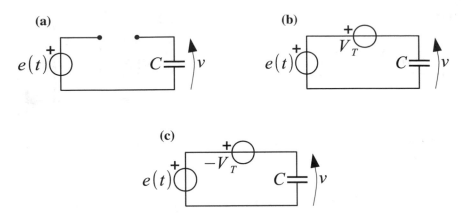

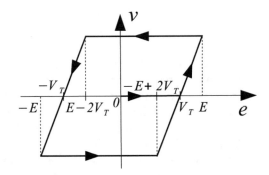

Fig. 10.26 Equivalent circuits when the two-terminal resistor works either **a** in the central region, or **b** on the right or **c** left branch of its PWL DP characteristic

Fig. 10.27 $v(e)$ relationship for the hysteretic circuit, assuming that $V_T < E < 2V_T$

DP characteristic), and v "follows" $e(t)$ according to the KVL $v = e + V_T$. Here we have $\hat{i} = C\dfrac{dv}{dt} = C\dfrac{de}{dt} < 0$.

- Now we assume that $e(t)$ reaches a minimum value, say $-E$, and starts increasing again. The original circuit again becomes equivalent to the one shown in Fig. 10.26a, and v holds its initial value $-E + V_T$. The working point of the nonlinear resistor moves to the right, along the central region α of its DP characteristic, until it reaches again (for $e = -E + 2V_T$) the threshold voltage V_T. From this point on (keeping on increasing $e(t)$), the resistor operates in the β branch. Thus, the original circuit becomes equivalent to the one shown in Fig. 10.26b, and v "follows" $e(t)$ according to the KVL $v = e - V_T$.

Figure 10.27 shows the obtained I/O relationship for the analyzed circuit. It describes a *hysteresis* loop.[1] We remark that this behavior can be obtained from two "ingredients" of the circuit:

---

[1] Hysteresis is a ubiquitous phenomenon in science and technology; it has been observed in many seemingly unrelated scientific areas, such as physics, chemistry, engineering, biology, and eco-

- presence of a nonlinear component;
- DP characteristic of the nonlinear resistor such that v switches from being a state variable in a circuit with natural frequency $\lambda = 0$ (which provides *memory* to the circuit) to being a nonstate variable (which allows changing the capacitor voltage).

10.3.4 Circuit Containing an Operational Amplifier

In the previous examples, we analyzed circuits containing nonlinear two-terminal resistors. Now we want to add a couple of examples containing multi-terminal elements.

The circuit shown in Fig. 10.28a contains an *operational amplifier* (op-amp). We already met this device in Vol. 1 (see Sect. 5.2.2), where we introduced a two-port (called a *nullor*) that can model the op-amp under suitable assumptions.

Actually, the op-amp is a nonlinear device, characterized by a nonlinear relationship between voltages v_o and v_x, which saturates at two values corresponding to the negative and positive power supplies. There are many ways to model this relationship (see also Sect. 10.5), and here we choose the PWL input–output characteristic shown in Fig. 10.28b. The nonlinear behavior of the device is modeled by the two horizontal lines $v_o = \pm V_S$. In the central region, instead, the op-amp behaves like a linear component, and its I/O relationship has slope A.

In many applications, A is assumed to be infinite. Here, we consider a finite value of the slope A, and the behavior of the circuit is studied also outside the linear zone. The input voltage is $e(t) = Eu(t)$, and we assume that the capacitors C_1 and C_2 are initially uncharged: $v_1(0^-) = v_2(0^-) = 0$.

Our aim in this example is to show that due to the presence of a feedback (through the capacitor C_2) from the output to the op-amp inverting input, labeled by $-$, the circuit attempts to do whatever is necessary to bring the voltage v_x to zero, even if it means starting in the positive or negative saturation region.

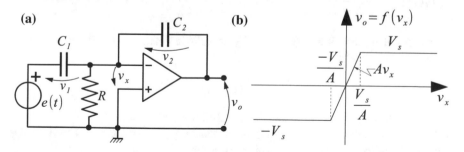

Fig. 10.28 **a** Op-amp circuit to be studied and **b** nonlinear I/O characteristic of the op-amp

nomics. Hysteretic systems are characterized by the fact that their output at a certain time depends on the past history of their input.

We still assume (assumption of infinite input impedance, as for the nullor) that the op-amp input terminals draw no current.

The whole analysis can start from the circuit equations:

$$\begin{cases} e = v_1 - v_x, \\ v_x + v_2 + v_o = 0, \\ v_o = f(v_x), \\ C_1 \dfrac{dv_1}{dt} = \dfrac{v_2 + v_o}{R} + C_2 \dfrac{dv_2}{dt}. \end{cases} \qquad (10.21)$$

It is easy to show that only one of the two capacitor voltages v_1 and v_2 is an SSV. From the first of the previous equations we can write v_x as $v_1 - e$; thus, taking into account the next two equations, we easily obtain

$$v_1 - e + v_2 + f(v_1 - e) = 0,$$

which is an algebraic (nonlinear) constraint among the voltages v_1, v_2, e. Therefore, in the following we choose v_2 as an SSV.

Now the analysis can be carried out in a few steps.

- First of all, we want to establish under what conditions we have $v_o(0^+) = -V_S$. Given that $e(0^+) = E$, the first two of Eq. 10.21 imply

$$E = v_1(0^+) - v_x(0^+); \quad v_x(0^+) + v_2(0^+) - V_S = 0.$$

By integrating the last of Eq. 10.21 between $t = 0^-$ and $t = 0^+$, we have

$$C_1 \left(v_1(0^+) - 0\right) = C_2 \left(v_2(0^+) - 0\right), \quad \text{that is,} \quad C_1 v_1(0^+) = C_2 v_2(0^+).$$

On solving these equations, we obtain

$$v_x(0^+) = \frac{V_S C_2 - E C_1}{C_1 + C_2}; \quad v_2(0^+) = \frac{C_1 (E + V_S)}{C_1 + C_2}; \quad v_1(0^+) = \frac{C_2 (E + V_S)}{C_1 + C_2}.$$

The initial assumption $v_o(0^+) = -V_S$ is valid as long as $v_x(0^+) < -\frac{V_S}{A}$ (see Fig. 10.28b), that is,

$$\frac{E C_1}{C_1 + C_2 (1 + A)} > \frac{V_S}{A}.$$

In the following, we assume that this condition is satisfied.

- The circuit analysis for $t > 0$ begins by setting $v_o = -V_S$ and $e(t) = E$ in the circuit equations, Eq. 10.21. By doing so, we easily obtain

$$E = v_1 + v_2 - V_S; \quad C_1 \frac{dv_1}{dt} = \frac{v_2}{R} - \frac{V_S}{R} + C_2 \frac{dv_2}{dt}.$$

The first of the above equations implies that $\dfrac{dv_1}{dt} + \dfrac{dv_2}{dt} = 0$; thus by eliminating $\dfrac{dv_1}{dt}$ from the second equation, we get the I/O relationship for the state variable v_2:

$$(C_1 + C_2)\frac{dv_2}{dt} + \frac{v_2}{R} = \frac{V_S}{R}.$$

In other words, as long as the op-amp works on the same branch (negative saturation) of its I/O relationship, it is equivalent to a linear circuit with natural frequency $\lambda = -\dfrac{1}{R(C_1 + C_2)}$. With this in mind, and recalling the expression of $v_2(0^+)$, the solution can be written as

$$v_2(t) = \frac{EC_1 - V_S C_2}{C_1 + C_2} e^{\lambda t} + V_S.$$

The voltage v_x is immediately obtained as

$$v_x = -v_2 + V_S = -\frac{EC_1 - V_S C_2}{C_1 + C_2} e^{\lambda t} < 0.$$

As time grows, v_x increases toward zero; therefore, there exists a time T such that $v_x(T) = -\dfrac{V_S}{A}$. At $t = T$, the op-amp begins to work in the linear region of $f(v_x)$. The value of T can be easily found:

$$e^{\lambda T} = \frac{V_S}{A}\frac{C_1 + C_2}{EC_1 - V_S C_2} \quad \Rightarrow \quad T = R(C_1 + C_2)\ln\left(\frac{A(EC_1 - V_S C_2)}{V_S(C_1 + C_2)}\right).$$

In order to continue the circuit analysis in the linear zone $v_o = Av_x$ of the op-amp (for $t > T$), we "photograph" the values of v_2 and v_1 in $t = T$:

$$v_2(T) = -v_x(T) - v_o(T) = V_S\left(1 + \frac{1}{A}\right); \quad v_1(T) = E + v_x(T) = E - \frac{V_S}{A}.$$

In the following, we make reference to the shifted time variable $t' = t - T$. With this choice, the above voltage values are written as $v_2(0)$ and $v_1(0)$, respectively.

• Now the circuit can be studied by considering Eq. 10.21 for $v_o = Av_x$ and $dt' = dt$. After a few manipulations, we get

$$E = v_1 + \frac{v_2}{1 + A}; \quad v_x = -\frac{v_2}{1 + A}; \quad C_1\frac{dv_1}{dt'} = \frac{v_2 + Av_x}{R} + C_2\frac{dv_2}{dt'}.$$

The first of the above equations implies $\dfrac{dv_1}{dt'} = -\dfrac{1}{1 + A}\dfrac{dv_2}{dt'}$, which leads us to recast the last equation as

$$R\left(C_1 + C_2\left(1 + A\right)\right)\frac{dv_2}{dt'} + v_2 = 0.$$

This is the (homogeneous) state equation, which holds in this region and corresponds to another equivalent linear circuit. Owing to the continuity of $f(v_x)$ at the transition point to the linear zone, the initial value of v_2 is not subject to jumps. The corresponding natural frequency is

$$\hat{\lambda} = -\frac{1}{R\left(C_1 + C_2\left(1 + A\right)\right)}$$

Notice that $|\hat{\lambda}| < |\lambda|$. The solution can be written as

$$v_2(t') = V_S\left(1 + \frac{1}{A}\right)e^{\hat{\lambda}t'}.$$

Therefore,

$$v_x = -\frac{V_S}{A}e^{\hat{\lambda}t'}.$$

Thus $v_x < 0$ for all $t' \in (0, +\infty)$: the op-amp remains in its linear zone.

- The qualitative plots of $v_x(t)$ and $v_o(t)$ are shown in Fig. 10.29. The curve $v_x(t)$ consists of two exponentials. The first lies in the time interval $(0, T)$ and is governed by the natural frequency λ; the second lies in the interval $(T, +\infty)$ and is governed by the natural frequency $\hat{\lambda}$. The corresponding output voltage $v_o(t)$ maintains the value $-V_S$ over the first time interval and then goes exponentially to zero.

(a)   **(b)**

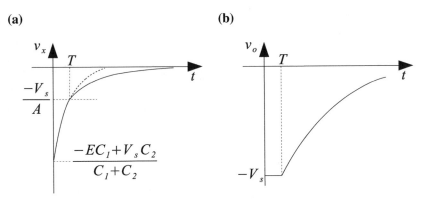

Fig. 10.29 Plots of (a) $v_x(t)$ and (b) $v_o(t) = f(v_x(t))$

10.3.5 Circuit Containing a BJT

Consider the circuit shown in Fig. 10.30. It contains a bipolar junction transistor (BJT) whose descriptive equations $i_b = \hat{i}_b(v_{be})$ and $i_c = \hat{i}_c(i_b, v_{ce})$ are approximated by the PWL characteristics represented in Fig. 10.31 and analytically described as follows.

- Input characteristic (Fig. 10.31a)

  (a): $i_b = 0$ for $v_{be} \leq V_T$

  (b): $i_b = \dfrac{v_{be} - V_T}{h}$ for $v_{be} > V_T$

- Output characteristic (Fig. 10.31b)

  (I): $v_{ce} = 0$; $0 \leq i_c \leq \beta i_b$

  (II): $i_c = \beta i_b$; $v_{ce} > 0$

where h, V_T, β are assigned parameters.

We assume that at $t = 0$, the capacitor is uncharged: $v(0) = 0$. Moreover, we have $e(t) = Eu(t)$ and $E > V_T$. The voltage source E_0 is constant.

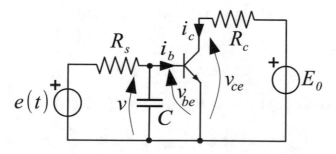

Fig. 10.30 Circuit to be studied

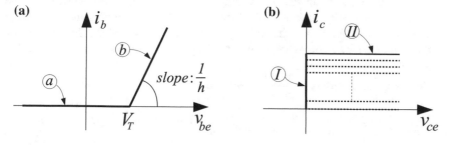

Fig. 10.31 **a** Input and **b** output PWL characteristics of the BJT

The analysis can be carried out in a few steps, based on the PWL equations defined above and on the circuit equations:

$$\begin{cases} v = v_{be} \\ e(t) - R_s \left(C\dfrac{dv}{dt} + i_b \right) - v = 0, \\ E_0 - R_c i_c = v_{ce}. \end{cases} \tag{10.22}$$

1. Analysis at $t = 0$: initial operating point.

   - Since $v(0) = 0$, we trivially have $v_{be} = 0 < V_T$, that is, the BJT initially works on the a branch ($i_b = 0$) of its input characteristic.
   - As far as the output characteristic is concerned, we firstly observe that the BJT cannot work initially in region I, because this would imply $v_{ce} = 0$ and $i_c (\leq \beta i_b) = 0$. But if we had $i_c = 0$, the circuit equations would give $v_{ce} = E_0 - R_c \cdot 0 = E_0$, which contradicts the requirement $v_{ce} = 0$.
     Instead, assuming that the BJT works in region II, we obtain $i_c = \beta i_b = 0$ and $v_{ce} = E_0 > 0$. These results are fully compatible.

   Summing up, the initial working zones of the BJT are a and II. As time increases, the working point remains in region a (with $i_b = 0$) as long as $v_{be} = v < V_T$.

2. Analysis for $t > 0$.
   From Eq. 10.22, with $i_b = 0$, $v_{be} \leq V_T$, $e(t) = E$, and $i_c = 0$ we have

$$E = R_s C \frac{dv}{dt} + v; \quad v_{ce} = E_0.$$

Therefore,

$$v(t) = E \left(1 - e^{\frac{-t}{R_s C}} \right).$$

This result holds up to the time T such that $v(T)(= v_{be}(T)) = V_T$, that is,

$$T = R_s C \ln \left(\frac{E}{E - V_T} \right),$$

as is easy to check. In the following, we make reference to the shifted time variable $t' = t - T$. In terms of t', $v = V_T$ will be thought of as the voltage value at $t' = 0$. For $t' > 0$, the working point (v, i_b) moves along the b branch of the input characteristic, whereas the working point on the output characteristic remains fixed.
From Eq. 10.22 with $i_b = \dfrac{v - V_T}{h}$, we can write

$$E - R_s C \frac{dv}{dt'} - R_s \frac{v - V_T}{h} - v = 0 \quad \text{or} \quad \frac{Eh + V_T R_s}{h R_s} = C \frac{dv}{dt'} + v \frac{h + R_s}{h R_s}.$$

The eigenvalue of this differential equation (i.e., the natural frequency of the corresponding equivalent linear circuit) is $\lambda = -\dfrac{1}{C}\left(\dfrac{h + R_s}{h R_s}\right)$. For $v(0) = V_T$, the solution can be written, after few manipulations, as

$$v = \frac{h\,(V_T - E)}{R_s + h}e^{\lambda t'} + \frac{Eh + V_T R_s}{h + R_s}.$$

The expression for the current i_b follows immediately:

$$i_b = \frac{v - V_T}{h} = \frac{E - V_T}{R_s + h}\left(1 - e^{\lambda t'}\right).$$

Since $i_b > 0$ for all $t' > 0$, the working point on the input characteristic remains on the b branch. The corresponding variables for the output characteristic are

$$i_c = \beta i_b; \quad v_{ce} = E_0 - R_c i_c = E_0 - R_c \beta \frac{E - V_T}{R_s + h}\left(1 - e^{\lambda t'}\right).$$

The BJT works in region II provided that $v_{ce}(t') > 0$ for all $t' > 0$. This means that we must have

$$\lim_{t' \to \infty} v_{ce}(t') = E_0 - R_c \beta \frac{E - V_T}{R_s + h} > 0, \quad \text{that is,} \quad E_0 > R_c \beta \frac{E - V_T}{R_s + h}.$$

Under this assumption, the resulting plots of $i_b(t)$ and $v_{ce}(t)$ are shown in Fig. 10.32.

3. It can be shown that if $E_0 < R_c \beta \dfrac{E - V_T}{R_s + h}$, then $v_{ce}(t')$ reaches the zero value at a time $T_1 < \infty$. After T_1, we have $v_{ce} = 0$ and $i_c (= \dfrac{E_0}{R_c}) < \beta i_b$ (region I). You can easily check this by reasoning as done up to now in analyzing this circuit.

As a concluding remark, consider the equilibrium value of v_{be},

(a)   **(b)**

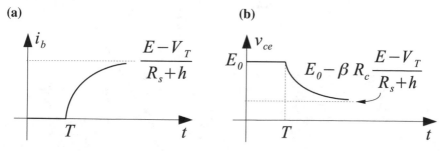

Fig. 10.32 Plots of **(a)** $i_b(t)$ and **(b)** $v_{ce}(t)$ for $t > 0$

$$v_{be}^* = V_T + h \underbrace{\lim_{t' \to \infty} i_b(t')}_{\dfrac{E - V_T}{R_s + h}} = \frac{V_T R_s + Eh}{R_s + h}.$$

This is the equilibrium voltage on the capacitor C when $e(t) = E$. In order to show this, we first observe that the capacitor behaves as if it were connected to an equivalent resistor between the P and Q terminals, as shown in Fig. 10.33a. The voltage-controlled DP characteristic $i = \hat{i}(v)$ of this resistor can be easily found by taking into account the BJT input characteristic and recalling that $v = v_{be}$. The resulting PWL descriptive equation is

$$i = v\left(\frac{1}{h} + \frac{1}{R_s}\right) - \frac{V_T}{h} - \frac{E}{R_s} \quad \text{for} \quad v > V_T$$

$$i = \frac{v - E}{R_s} \quad \text{for} \quad v < V_T,$$

and the corresponding DP characteristic is shown in Fig. 10.33b. The voltage value corresponding to $i = 0$ is $v = v_{be}^*$, as expected. This is a stable equilibrium point for the circuit, inasmuch as $\dot{v} = -\dfrac{\hat{i}(v)}{C}$.

Fig. 10.33 Equivalent resistor between P and Q: descriptive variables (**a**) and PWL DP characteristic (**b**)

(a)

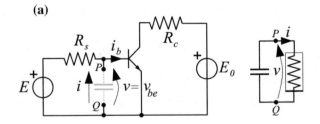

(b)

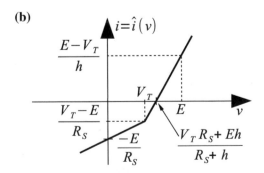

10.4 Bifurcations

In this section, we introduce the concept of *bifurcation*, working with the class of circuits considered in Sect. 10.1 and assuming that one or more circuit parameters (also called *control parameters*) can be varied. According to the definitions of the circuit components, a circuit parameter can be the resistance of a resistor, the voltage generated by a constant voltage source, the gain of a controlled source, the inductance of an inductor, and so forth.

We want to see what happens to a given circuit by changing some (say, M) control parameters. Thus, the state equation in canonical form will be written by pointing out the dependence on these parameters, in addition to the dependence on the state variable:

$$\frac{dx}{dt} = \dot{x} = f(x, p), \tag{10.23}$$

where x is the state variable ($x \in \mathbb{R}$), p is a vector of M parameters ($p \in \mathbb{R}^M$), and f is the vector field ($f : \mathbb{R}^{M+1} \to \mathbb{R}$).

By changing the values of these parameters, the qualitative structure of the system flow can change. For instance, an equilibrium point can appear or disappear, or its stability can change. If this happens, the system flow changes qualitatively, and consequently the system dynamics also has a qualitative change.

A qualitative change in the dynamics of a given circuit (or system, in general) is called a **bifurcation**, and the parameter values at which this change occurs are called **bifurcation** (or **critical**) **points**.

Treating bifurcations deeply is beyond the scope of this book. The interested reader is referred to the books by Strogatz [1] (for beginners) and Kuznetsov [3] (advanced level). Here, we will just propose some examples of circuits where bifurcations occur.

10.4.1 Linear Case

In first-order linear circuits (systems) belonging to the class considered in Sect. 10.1, stability of the circuit and stability of the equilibrium (say x^*) coincide, because $f'(x^*)$ coincides with the circuit's natural frequency λ. Indeed, the generic I/O relationship in this case can be written as

$$a_1 \frac{dx}{dt} + a_0 x = h, \tag{10.24}$$

or in canonical form, as

$$\dot{x} = -\frac{a_0}{a_1}x + \frac{h}{a_1} = f(x). \tag{10.25}$$

From Eq. 10.24 it follows that the natural frequency is $\lambda = -\frac{a_0}{a_1}$, and from Eq. 10.25 it follows that $f'(x) = -\frac{a_0}{a_1} = \lambda$.

In this case, the only possible bifurcation the equilibrium can undergo is a change in its own stability. This can be easily illustrated by returning to Case Study 1 of Sect. 10.1, whose state equation is

$$\dot{v} = \frac{\alpha - 1}{(R_2 + R_3)C}v + \frac{R_2 + \alpha R_3}{(R_2 + R_3)C}A_1 = f(v, \alpha). \tag{10.26}$$

We remove the assumption $0 < \alpha < 1$ and consider α to be a control parameter. It is evident that the natural frequency $\lambda = \dfrac{\alpha - 1}{(R_2 + R_3)C}$ changes its sign when α reaches the critical value $\alpha = 1$:

- for $\alpha < 1$, the natural frequency is negative, and thus the circuit is absolutely stable and the equilibrium point $v = \dfrac{R_2 + \alpha R_3}{1 - \alpha}A_1$ is (globally) stable;
- for $\alpha > 1$, the natural frequency is positive, and thus the circuit is unstable and the equilibrium point is unstable as well.

Thus $\alpha = 1$ is a bifurcation value for this circuit, causing the change in stability of its equilibrium point.

10.4.2   Nonlinear Case

As stated in Sect. 10.1, in the nonlinear case the stability is no longer a property of the whole circuit, but of each equilibrium point.

For first-order circuits, one of the most common bifurcations is the so-called **fold** (or saddle-node) **bifurcation**, by which equilibrium points are created or destroyed. When we analyze a first-order dynamical system through the graphical method, a fold bifurcation corresponds to a collision between two equilibrium points, one stable and one unstable, caused by a change in the vector field, which is caused in turn by a change of value in a control parameter. This can be illustrated by returning to Case Study 2 of Sect. 10.1.

Here we assume that there is a second current source, as shown in Fig. 10.34, whose impressed current A is chosen as bifurcation parameter. The state equation in canonical form becomes

$$\dot{v} = -\frac{\gamma}{C}v^3 + \frac{\alpha}{C}v + \frac{A}{C} = f(v, A). \tag{10.27}$$

Fig. 10.34 Circuit to be
analyzed (nonlinear case)

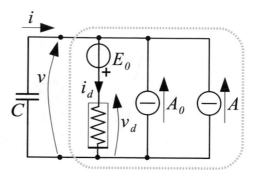

The change in the stability condition $\dfrac{\partial f}{\partial v} = 0$ (eigenvalue of the linearized system crossing the edge of stability; see Sect. 10.1.3) turns out to be

$$-3\frac{\gamma}{C}v^2 + \frac{\alpha}{C} = 0, \tag{10.28}$$

that is,

$$v = \pm\sqrt{\frac{\alpha}{3\gamma}}. \tag{10.29}$$

For values of A ranging in the interval $\left(-\sqrt{\dfrac{\alpha}{3\gamma}}, \sqrt{\dfrac{\alpha}{3\gamma}}\right)$ (recall that α and γ are assumed to be positive), we have three equilibrium points—two stable, P_1 and P_3, and one unstable, P_2—as shown in Fig. 10.35a.

For $A > \sqrt{\dfrac{\alpha}{3\gamma}}$, only P_3 survives (see Fig. 10.35b), whereas for $A < -\sqrt{\dfrac{\alpha}{3\gamma}}$, the only equilibrium point is P_1 (see Fig. 10.35c).

It is easy to be convinced that we have two bifurcation points:

- $A = \sqrt{\dfrac{\alpha}{3\gamma}}$, corresponding to the collision of P_1 and P_2;

- $A = -\sqrt{\dfrac{\alpha}{3\gamma}}$, corresponding to the collision of P_2 and P_3.

These are fold bifurcations, through which two equilibria collide and disappear (or are created, depending on the direction of the variation in the control parameter). We remark that the points $A = \pm\sqrt{\dfrac{\alpha}{3\gamma}}$ have been obtained by imposing the condition $\dfrac{\partial f}{\partial v} = 0$, that is, by imposing that the linearized system have a null eigenvalue.

Remark: Close to a bifurcation value (in the parameter space) and close to an equilibrium point (in the state space), the system dynamics is slower. This is due to the fact that under these assumptions we have both $f(x^*)$ and $f'(x^*)$ close to zero;

(a)

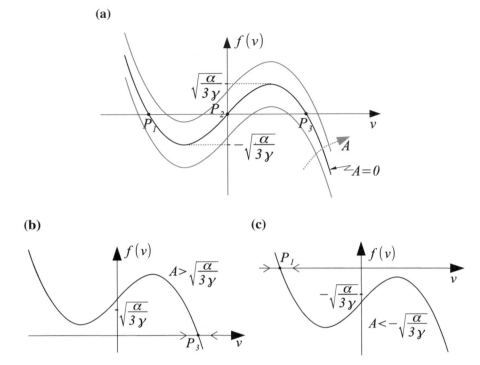

(b)   **(c)**

Fig. 10.35 Graphical method for different values of the control parameter A: **a** $A \in$ $\left(-\sqrt{\dfrac{\alpha}{3\gamma}}, \sqrt{\dfrac{\alpha}{3\gamma}}\right)$; **b** $A > \sqrt{\dfrac{\alpha}{3\gamma}}$; **c** $A < -\sqrt{\dfrac{\alpha}{3\gamma}}$

thus the dynamics of a perturbation with respect to the equilibrium (see Eq. 10.11) is slow.

If we represent the equilibrium points as functions of A in the plane (A, v) (for instance by numerically solving the equilibrium equation $-\gamma v^3 + \alpha v + A = 0$), we obtain the **bifurcation diagram** shown in Fig. 10.36.

The plane (A, v) is the product of the state space (domain of the state variable v) and the parameter space (domain of the control parameter A). This is what is usually called the **control space**.

The bifurcation diagram showing the equilibria in the control space is a smooth curve with two folds, one (point F_1) corresponding to the fold bifurcation (collision) between P_2 and P_3, and the other one (point F_2) corresponding to the fold bifurcation (collision) between P_1 and P_2. In the central (gray) region, two stable equilibria (P_1 and P_3) coexist: the DC steady state reached by the circuit depends on the initial condition, which lies in one of the two corresponding basins of attraction.

The vertical dashed lines show the flows on the state space (v axis) for fixed values of A in the three regions of the parameter space defined by the two critical values $-\sqrt{\dfrac{\alpha}{3\gamma}}$ and $\sqrt{\dfrac{\alpha}{3\gamma}}$.

Fig. 10.36 Bifurcation diagram showing the equilibria in the control space (A, v)

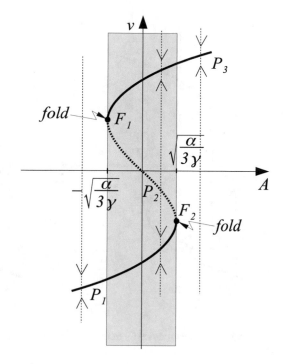

Remark 1 A fold bifurcation could be achieved by tuning a single parameter and is therefore called a **codimension-1 bifurcation.**

Remark 2 It is evident from Fig. 10.36 that in the bistability (gray) region, the unstable equilibrium P_2 separates the basins of attraction of P_1 and P_3.

Remark 3 The presence of bistability in the central region of the bifurcation diagram is due to the presence of the two fold bifurcations F_1 and F_2. Moreover, the *bistability* is related to the presence of *hysteresis* in the control space. Indeed, if you imagine that A is set to a value lower than $-\sqrt{\dfrac{\alpha}{3\gamma}}$, the circuit state variable v, starting from any initial condition, converges to the equilibrium P_1 (path ① in Fig. 10.37a). If we now take the circuit as is and increase A, the equilibrium state is slightly perturbed at each variation of A, but it still converges to P_1, whose position changes with A (path ② in Fig. 10.37a). This remains true up to the fold bifurcation F_2, at which P_1 disappears. After this value, a further increase of A forces the state to jump to the "far" equilibrium P_3. If we keep on increasing A, the equilibrium state remains in P_3, whose position changes with A (path ③ in Fig. 10.37a). Now we decrease A. The DC state of the circuit remains in P_3 (path ④ in Fig. 10.37b) until we reach the fold bifurcation F_1, at which P_3 disappears. After this value, a further decrease of A forces the state to jump again to the equilibrium P_1 (path ⑤ in Fig. 10.37b). Thus the equilibrium state reached by the circuit has described a hysteresis cycle.

(a)   **(b)**

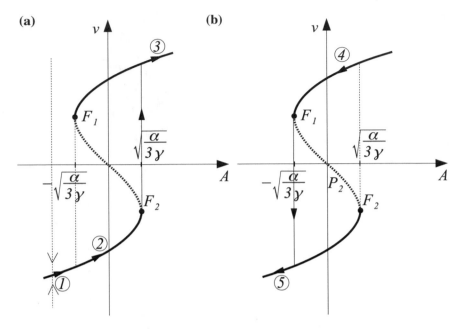

Fig. 10.37 Hysteresis caused by the bistability region in the control space (A, v)

10.5 A Summarizing Example

In this section, we analyze the circuit shown in Fig. 10.38a using the conceptual tools introduced in this chapter.

This circuit contains an op-amp. We already know that the op-amp can be modeled by a *nullor* under suitable assumptions. The first assumption is that there is a signal path of some sort feeding back from the output to the inverting input, labeled by $-$ in the op-amp symbol. This circuit configuration is also known as a *closed loop with negative feedback*, and the output attempts to do whatever is necessary to bring the voltage v_x to zero (see Sect. 10.3.4).

The second assumption is that the input terminals draw no current (assumption of *infinite input impedance*).

Here the first assumption is not satisfied, since the output terminal is connected (through R_1) to the noninverting input, labeled by $+$ in the op-amp symbol. Therefore, in this case, the op-amp cannot be modeled by a nullor. This is usually called a *closed loop with positive feedback* configuration, and in this case, v_x is not constrained to zero.

Voltages v_o and v_x are related by a nonlinear relationship, which saturates at two values corresponding to the negative and positive power supply. Figure 10.38b shows two functions commonly used to characterize the op-amp (as an alternative to the PWL function introduced in Sect. 10.3.4), one continuous (black line) and one discontinuous (thick gray line).

(a)

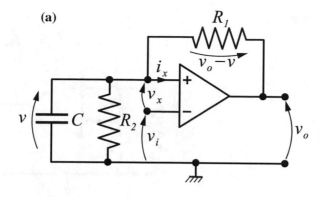

(b)

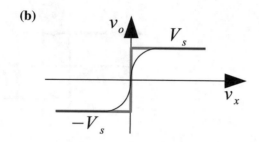

Fig. 10.38 **a** Circuit to be analyzed and **b** two possible $v_o(v_x)$ relationships for the operational amplifier, one continuous (black line) and one discontinuous (thick gray line)

The second assumption, however, still holds, and thus we can assume that $i_x = 0$.

10.5.1 Inverting Schmitt Trigger

First of all, we analyze the circuit shown in Fig. 10.39, which is commonly known as an *inverting Schmitt trigger*.[2]

This is the circuit of Fig. 10.38 without the capacitor. We assume, for the sake of simplicity, that the op-amp I/O relationship is the gray one in Fig. 10.38b.

Since $i_x = 0$, R_1, and R_2 are actually connected in series and act as a voltage divider for v_o, it follows that $v_2 = v_o \dfrac{R_2}{R_1 + R_2}$.

---

[2]Notice that if we swap the $+$ and $-$ input terminals, the behavior of this circuit changes completely, becoming the noninverting amplifier already analyzed in Sect. 5.2.2 in Vol. 1.

Fig. 10.39 Inverting
Schmitt trigger

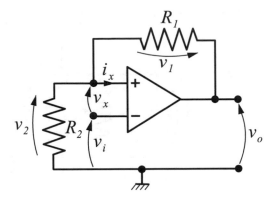

Fig. 10.40 Hysteresis cycle
generated by the inverting
Schmitt trigger

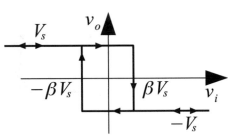

Owing to our assumption about the op-amp I/O relationship, the output voltage
can be either $+V_S$ or $-V_S$ (see Fig. 10.38b). Consequently, v_2 is either $+\beta V_S$ or
$-\beta V_S$, with $\beta = \dfrac{R_2}{R_1 + R_2}$.

Let us assume that $v_o = +V_S$, that is, $v_x > 0$, and that the input voltage v_i is
negative and starts increasing monotonically. Inasmuch as $v_x = v_2 - v_i$, as long as
$v_i < \beta V_S$, v_x remains positive and nothing changes for v_o and v_2. When v_i surpasses
the threshold βV_S, however, v_x becomes negative, thus causing the changes $v_o = -V_S$
and $v_2 = -\beta V_S$.

If we now decrease v_i monotonically, nothing happens to v_o and v_2 as long as
$v_i > -\beta V_S$. When v_i surpasses the threshold $-\beta V_S$, however, v_x becomes positive,
thus causing the changes $v_o = V_S$ and $v_2 = \beta V_S$. Thus the plot of v_o versus v_i is the
hysteresis cycle shown in Fig. 10.40.

Notice that the switching thresholds depend on β, that is, on the balancing of
resistances R_1 and R_2 in the voltage divider.

10.5.2   Dimensionless Formulation

Now we analyze the complete circuit by assuming that the op-amp I/O relationship is the black one in Fig. 10.38b in order to have a smooth vector field:

$v_o = V_S \tanh\left(\dfrac{v_x}{V_T}\right)$, with parameter V_T. According to these assumptions, the circuit equations are

$$v_o = V_S \tanh\left(\frac{v_x}{V_T}\right)$$
$$v_x = v - v_i,$$
$$C\frac{dv}{dt} = \frac{v_o - v}{R_1} - \frac{v}{R_2}. \qquad (10.30)$$

Thus the state equation is

$$R_1 C\frac{dv}{dt} = V_S \tanh\left(\frac{v - v_i}{V_T}\right) - v\frac{R_1 + R_2}{R_2}. \qquad (10.31)$$

Now we define dimensionless time τ and state variable x:

$$\tau = \frac{t}{R_1 C}, \qquad x = \frac{v - v_i}{V_T}. \qquad (10.32)$$

By substituting in Eq. 10.31 and by assuming that v_i is constant, we obtain

$$\frac{R_1 C}{R_1 C}V_T\frac{dx}{d\tau} = V_S \tanh(x) - \frac{R_1 + R_2}{R_2}(V_T x + v_i), \qquad (10.33)$$

that is,

$$\dot{x} = \underbrace{\frac{V_S}{V_T}}_{\doteq \alpha} \tanh(x) - \underbrace{\frac{R_1 + R_2}{R_2}}_{\doteq a} x - \underbrace{\frac{R_1 + R_2}{R_2 V_T}v_i}_{\doteq b}. \qquad (10.34)$$

We remark that nondimensionalizing the equation reduces the number of parameters, by lumping them together into three new dimensionless parameters α, a, and b.

10.5.3   Analysis with Constant Input

Now we assume that $v_i = E$, $V_S = 5\,\text{V}$, and $V_T = 25\,\text{mV}$, and thus $\alpha = 200$. The parameter a controls the voltage divider balancing, whereas b controls the input value.

The state equation is

$$\dot{x} = \alpha \tanh(x) - ax - b = f(x, p), \qquad (10.35)$$

where the parameter vector p contains the bifurcation parameters a and b.
The equilibrium condition ($f(x, p) = 0$) is

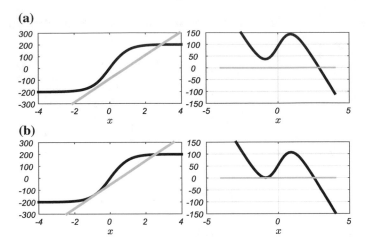

Fig. 10.41 Plots obtained for $a = 100$ and different values of b: **a** -90; **b** -54. Left panels: functions $g(x) = \alpha \tanh(x)$ (black lines) and $h(x) = ax + b$ (gray lines). Right panels: functions $f(x, p) = \alpha \tanh(x) - ax - b$ (black lines, with reference value 0 marked in gray)

$$\alpha \tanh(x) = ax + b, \tag{10.36}$$

which can be solved either numerically or graphically, by looking for intersections between the sigmoid $g(x) = \alpha \tanh(x)$ and the straight line $h(x) = ax + b$.

The fold bifurcation condition ($\dfrac{\partial f}{\partial x} = 0$) is

$$a = \alpha[1 - \tanh^2(x)]. \tag{10.37}$$

Figures 10.41 and 10.42 show some examples of application of the graphical method, for $a = 100$ and different values of b.

The locus of the fold bifurcation points in the plane (a, b) (shown in Fig. 10.43) can be found by solving the system

$$\begin{cases} a = \alpha[1 - \tanh^2(x)], \\ b = \alpha \tanh(x) - \alpha x[1 - \tanh^2(x)]. \end{cases} \tag{10.38}$$

We remark that the two fold bifurcation curves meet tangentially at $(a, b) = (200, 0)$; such a point is called a **cusp point**. The two curves partition the parameter plane into two regions, labeled **A** and **B**. In the *bistability* region **A**, the circuit has two stable equilibrium points (and another one unstable), whereas in the *monostability* region **B**, there is only one stable equilibrium point. Fold bifurcations occurs all along the boundary of the two regions, except at the cusp point.

At the cusp point, the central equilibrium (say P_2, as in Sect. 10.4.2) undergoes both fold bifurcations, colliding with both stable equilibrium points (say P_1 and

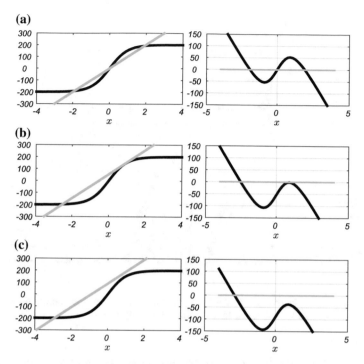

Fig. 10.42 Plots obtained for $a = 100$ and different values of b: **a** 0; **b** 54; **c** 90. Left panels: functions $g(x) = \alpha \tanh(x)$ (black lines) and $h(x) = ax + b$ (gray lines). Right panels: functions $f(x, p) = \alpha \tanh(x) - ax - b$ (black lines, with reference value 0 marked in gray)

Fig. 10.43 Fold bifurcation curves for the considered example

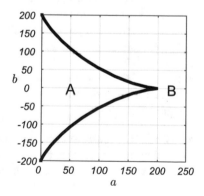

P_3). This is what we call a **codimension-2 bifurcation**, for whose achievement it is necessary to tune two parameters.

Figure 10.44 shows some curves $x(\tau)$ obtained for different initial conditions (15 values uniformly spaced in the range $[-3, 3]$), $a = 100$, and the three nonnegative values of b considered above; the plots for the remaining values of b are completely symmetric.

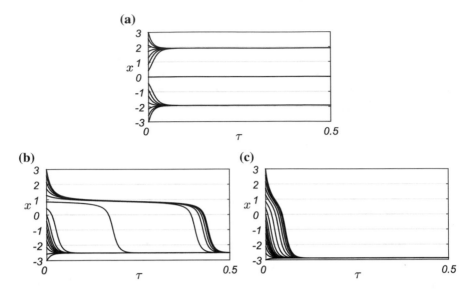

Fig. 10.44 Plots of x versus τ obtained for different initial conditions (15 values uniformly spaced in the range $[-3, 3]$), $a = 100$, and different values of b: **a** 0; **b** 54; **c** 90

- For $b = 0$, the point (a, b) lies within region **A**, and the time evolution of the state variable evidences the presence of two stable equilibria and one unstable equilibrium at $x = 0$.
- For $b = 54$, the point (a, b) lies within region **B**, but close to the fold bifurcation curve; the time evolution of the state variable evidences the presence of a negative stable equilibrium point and a slowing down of the dynamics when x is close to 1, corresponding to the local maximum of $f(x, p)$ very close to 0 (but negative) in Fig. 10.42b.
- For $b = 90$, the point (a, b) lies again within region **B**, but far from the fold bifurcation curve. Thus the state variable converges more quickly to the stable equilibrium point.

Remark 1 The *analysis* carried out can also help in circuit *design*. Indeed, the bifurcation diagram provides a map to set the values of the circuit parameters R_1, R_2, and E in order to have one or two stable equilibria with prescribed value(s). For instance, suppose that we want to obtain two symmetric stable equilibrium points. Since the model, by definition, does not perfectly correspond to the physical circuit, we design the circuit by setting its parameters reasonably far from bifurcation curves. In this way, our design is more robust with respect to model inaccuracies. We choose $a = 21$ and $b = 0$. This means setting $E = 0\,\mathrm{V}$ and $\dfrac{R_1 + R_2}{R_2} = 21$, that is, $R_1 = 20R_2$. As an example, we can choose $R_1 = 2\,\mathrm{k}\Omega$ and $R_2 = 100\,\Omega$. If we want to have just one stable equilibrium point, we can set $a = 51$ and $b = 204$. Thus we obtain $R_1 = 50R_2$ and $E = 0.1\,\mathrm{V}$.

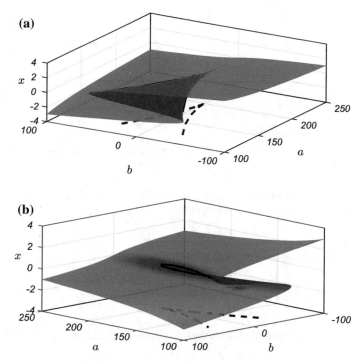

Fig. 10.45 Two different views of the equilibrium manifold (transparent surface) corresponding to Eq. 10.36; stable equilibria are shown in green, unstable equilibria in red. Black solid curve: bifurcation curve in the control space (a, b, x), corresponding to Eq. 10.38. Black dashed curve: bifurcation curve in the parameter space (a, b) (projection of the solid curve onto a plane with constant x, here $x = -4$)

Remark 2 From a geometric standpoint, Eq. 10.38 means working in the control space (a, b, x) and intersecting the manifold (surface described by Eq. 10.36) containing the equilibria (gray transparent manifold in Fig. 10.45) with the manifold (surface described by Eq. 10.37) containing the fold bifurcation points. This intersection is a curve in the control space (black curve in the 3D control space in Fig. 10.45), and its projection on the parameter plane (a, b) provides the fold bifurcation curves (black curves in the parameter plane (a, b) for $x = -4$ in Fig. 10.45).

10.5.4 Potential Functions

The conclusions drawn in the previous section can also be obtained from the perspective of potential functions. We assume that the op-amp has the I/O relationship $v_o = V_S \tanh\left(\dfrac{v_x}{V_T}\right)$ shown in Fig. 10.46a and analyze the nonlinear resistor connected in parallel to the capacitor (see Fig. 10.46b). You can easily check that the

(a)   **(b)**

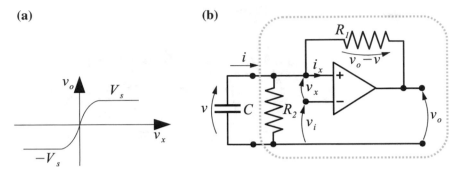

Fig. 10.46 a Op-amp I/O relationship and **b** circuit of Fig. 10.38 redesigned to evidence the non-linear resistor connected in parallel to the capacitor

descriptive equation of the nonlinear resistor is

$$i = \hat{i}(v) = v\left(\frac{1}{R_2} + \frac{1}{R_1}\right) - \frac{1}{R_1}V_S \tanh\left(\frac{v - v_i}{V_T}\right). \tag{10.39}$$

The cocontent function of this voltage-controlled resistor can be defined as

$$\overline{G}(v) = \int_0^v \hat{i}(\xi)d\xi, \tag{10.40}$$

where the lower limit of the integral has been arbitrarily set to zero (see Chap. 4 in Vol. 1). The integral of the first term on the right-hand side of Eq. 10.39 is trivial; as regards the second term, by a change of variable we have

$$\int_0^v \tanh\left(\frac{\xi - v_i}{V_T}\right)d\xi = \int_{-\frac{v_i}{V_T}}^{\frac{v - v_i}{V_T}} \tanh(z)V_T dz = V_T \ln\left(\frac{\cosh\left(\frac{v - v_i}{V_T}\right)}{\cosh\left(\frac{v_i}{V_T}\right)}\right).$$

Therefore,

$$\overline{G}(v) = \frac{1}{2}v^2\left(\frac{1}{R_2} + \frac{1}{R_1}\right) - \frac{V_S V_T}{R_1} \ln\left(\frac{\cosh\left(\frac{v - v_i}{V_T}\right)}{\cosh\left(\frac{v_i}{V_T}\right)}\right). \tag{10.41}$$

From this expression we can easily obtain a dimensionless formulation follow-ing the guidelines given in Sect. 10.5.2. Setting $x = \dfrac{v - v_i}{V_T}$, $a = \dfrac{R_1 + R_2}{R_2}$, $b =$

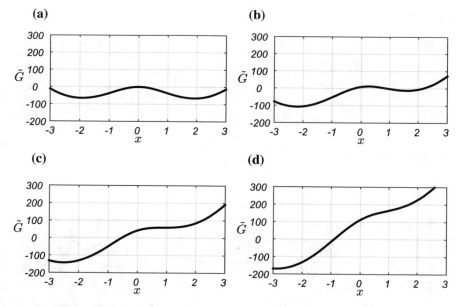

Fig. 10.47 Normalized cocontent function $\tilde{G}(x)$ for $a = 100$ and different values of b: **a** 0, **b** 25, **c** 54, **d** 90

$a\dfrac{v_i}{V_T}$, $V_S = \alpha V_T$, we obtain

$$\overline{G} = \frac{V_T^2}{R_1}\tilde{G}; \quad \tilde{G}(x; \alpha, a, b) = \frac{1}{2}a\left(\frac{b}{a} + x\right)^2 - \alpha \ln\left(\frac{\cosh(x)}{\cosh\left(\dfrac{b}{a}\right)}\right). \quad (10.42)$$

The dimensionless term $\tilde{G}$ can now be studied intrinsically or to analyze the time behavior of the circuit. In terms of the normalized time τ (see Sect. 10.5.3), we have

$$\frac{dx}{d\tau} = -\frac{d\tilde{G}}{dx} = \alpha \tanh(x) - ax - b,$$

that is, Eq. 10.35, as expected.

Figure 10.47 shows the shape of the cocontent $\tilde{G}(x)$ for $a = 100$ and different values of b. Panel a ($b = 0$) shows a double-well symmetric cocontent, corresponding to Figs. 10.42a and 10.44a; the two coexisting stable equilibrium points have symmetric basins of attraction. Panel b ($b = 25$) shows an asymmetric double-well cocontent. Panels c ($b = 54$) and d ($b = 90$) show single-well cocontent functions, corresponding to panels b and c of Figs. 10.42 and 10.44; the only equilibrium point,

(a)   **(b)**

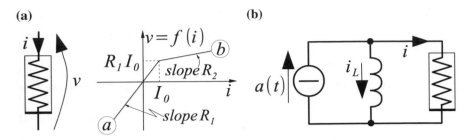

Fig. 10.48 Problem 10.1

(a)   **(b)**

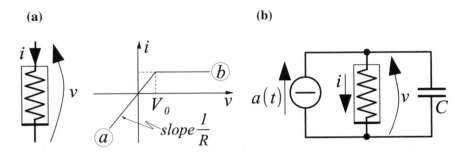

Fig. 10.49 Problem 10.2

corresponding to the minimum of the well, visible on the left edge of each panel, is globally stable.

10.6 Problems

10.1 Assume that the characteristic $v = f(i)$ of the nonlinear resistor shown in Fig. 10.48a has $R_2 < R_1$ and $I_0 > 0$. This element is part of the circuit of Fig. 10.48b, where $a(t) = A \cdot u(t)$ and $i_L(0^-) = 0$. Find:

1. the condition ensuring that in $t = 0^+$ the resistor works on the branch ⓑ of $f(i)$;
2. the currents $i_L(t)$ and $i(t)$ for $t \in (0, T)$, where T denotes the instant when the resistor begins to work on the branch ⓐ (i.e., such that $i(T) = I_0$);
3. an expression for T;
4. the currents $i_L(t)$ and $i(t)$ for $t > T$.

10.2 Assume that the characteristic of the nonlinear resistor shown in Fig. 10.49a has $V_0 > 0$ and that the slope of the branch ⓐ is $\dfrac{1}{R}$, while branch ⓑ is horizontal. This element is part of the circuit of Fig. 10.48b, where $a(t) = Q \cdot \delta(t)$ and $v(0^-) = 0$. Find:

(a)

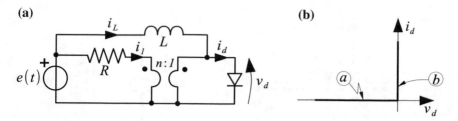

(b)

Fig. 10.50 Problem 10.3

(a)

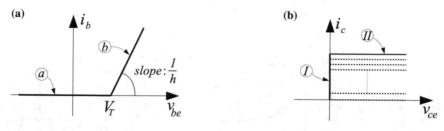

(b)

(c)

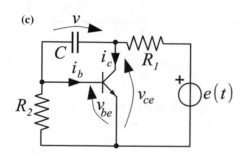

Fig. 10.51 Problem 10.4

1. the condition ensuring that in $t = 0^+$ the resistor works on the branch ⓑ of its characteristic;
2. the voltage $v(t)$ for $t \in (0, T)$, where T denotes the instant when the resistor begins to work on the branch ⓐ (i.e., such that $v(T) = V_0$);
3. an expression for T;
4. the voltage $v(t)$ for $t > T$.

10.3 Given the circuit shown in Fig. 10.50a, assume that $e(t) = \Phi \cdot \delta(t)$, $\Phi > 0$, $n > 1$, and $i_L(0^-) = 0$. The diode's characteristic is shown in Fig. 10.50b.

1. Show that at $t = (0^+)$, the diode works on the branch ⓑ of its characteristic;
2. find $i_L(0^+)$ and $i_d(0^+)$;
3. find $i_L(t)$ and $i_d(t)$ for $t > 0$.

(a)

(b)

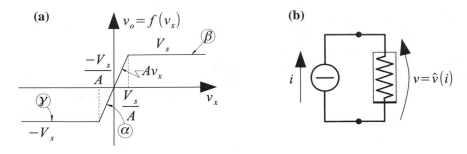

(c)

(d)

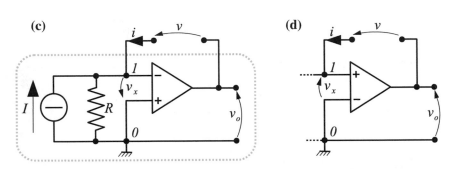

Fig. 10.52 Problem 10.5

10.4 The PWL characteristics represented in Fig. 10.51a, b approximate the behavior of the BJT included in the circuit of Fig. 10.51c. The PWL regions are described as follows:

$$\text{ⓐ} : i_b = 0; \quad v_{be} \leq V_T; \qquad \text{ⓑ} : i_b = \frac{v_{be} - V_T}{h}; \quad v_{be} > V_T$$
$$\text{Ⓘ} : v_{ce} = 0; \quad 0 \leq i_c \leq \beta i_b; \qquad \text{Ⓘ} : i_c = \beta i_b; \quad v_{ce} > 0,$$

where V_T, h, β are positive assigned parameters. The voltage source $e(t) = E \cdot u(t)$ has $E > 0$, and the capacitor is initially uncharged: $v(0) = 0$. Under these assumptions:

1. show that at $t = 0^+$, the BJT can work only in the two pairs of regions Ⓘ – ⓐ and Ⓘ – ⓑ and that the occurrence of one or the other of these depends on mutually exclusive conditions.
2. Assuming Ⓘ – ⓐ as the initial pair in the circuit, find $v_{ce}(t)$ and $v_{be}(t)$.

10.5 The PWL characteristic in Fig. 10.52a represents the behavior of the device included in the circuits of Fig. 10.52c, d. The current source has $I > 0$. The "d" circuit differs from "c" only for the inversion of the input terminals $+$, $-$ of the device. Each of these circuits individuates a nonlinear two-terminal resistor, whose characteristic is assumed to be current-controlled: $v = \hat{v}(i)$, as shown in Fig. 10.52b. The PWL branches ⓐ, ⓑ, ⓨ are described as follows:

$$\text{\textcircled{a}}: v_o = Av_x, \ |v_x| < \frac{V_s}{A}; \quad \text{\textcircled{b}}: v_o = V_s, \ v_x \geq \frac{V_s}{A}; \quad \text{\textcircled{y}}: v_o = -V_s, \ v_x \leq -\frac{V_s}{A}.$$

Find $v = \hat{v}(i)$ for both "c" and "d" and discuss the possibility of its working as the equivalent resistor in the complementary-component circuits of Fig. 10.2a, b.

References

1. Strogatz S (2014) Nonlinear dynamics and chaos with applications to physics, biology, chemistry, and engineering. Westview Press, Boulder
2. Parodi M, Storace M (2018) Linear and nonlinear circuits: basic & advanced concepts, vol 1. Springer, New York
3. Kuznetsov Yuri (2004) Elements of applied bifurcation theory. Springer, New York

Part VI
Second- and Higher-Order Dynamical Circuits

Chapter 11
Basic Concepts: Linear Two-Ports with Memory and Higher-Order Linear Circuits

> *Nothing in life is to be feared, it is only to be understood. Now is the time to understand more, so that we may fear less.*
> Marie Skłodowska-Curie

Abstract In this chapter, we introduce a two-port component with memory, called coupled inductors. The dynamics of higher-order circuits are analyzed by resorting to the complementary component representation and to the state variables method, thus generalizing the concepts introduced for first-order circuits. Methods for the analysis of stability and for finding the circuit response to various inputs, including also discontinuous functions, are provided.

11.1 Coupled Inductors

Here we introduce the descriptive equations for a largely used two-port with memory. The coupled inductors are an electrical device, whose model is shown in Fig. 11.1a. If the voltage reference is the same for both ports, we obtain the three-terminal configuration shown in Fig. 11.1b.

The **coupled inductors descriptive equations** are

$$\begin{cases} v_1(t) = L_1 \dfrac{di_1}{dt} + M \dfrac{di_2}{dt} \\[2mm] v_2(t) = M \dfrac{di_1}{dt} + L_2 \dfrac{di_2}{dt} \end{cases} \tag{11.1}$$

© Springer Nature Switzerland AG 2020

M. Parodi and M. Storace, *Linear and Nonlinear Circuits: Basic and Advanced Concepts*,
Lecture Notes in Electrical Engineering 620,
https://doi.org/10.1007/978-3-030-35044-4_11

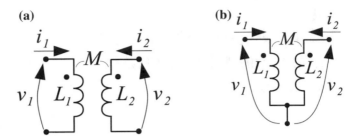

Fig. 11.1 Coupled inductors: **a** two-port configuration; **b** three-terminal configuration

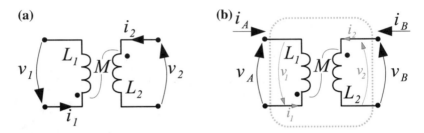

Fig. 11.2 A two-port containing coupled inductors (**a**); auxiliary two-port (**b**) showing the correspondence between the port variables of the coupled inductors (v_1, v_2, i_1, i_2) and of the assigned two-port (v_A, v_B, i_A, i_B)

L_1 is a parameter called *primary inductance*, L_2 is called *secondary inductance* and M *mutual inductance*. All of them are measured in *henry* [H], but L_1 and L_2 are always positive for physical reasons (passivity, see Sect. 11.2.2), whereas M can be either positive or negative. Moreover, again for physical reasons due to passivity, the following constraint holds: $M^2 \leq L_1 L_2$.

Remark: As for the ideal transformer introduced in Vol. 1 (Sect. 5.7.1), the dots determine the orientation of each port; that is, they denote the upper terminal of each port. For instance, Fig. 11.2a shows the relationships between 2-port descriptive variables and position of the dots. In any case, M can be either positive or negative.

Case Study 1
Find the descriptive equations for the two-port shown in Fig. 11.2b in terms of variables v_A, v_B, i_A, i_B.

The port-1 variables of the coupled inductors can be expressed in terms of the port-A variables of the assigned two-port: $v_1 = -v_A$ and $i_1 = -i_A$. Similarly, on port 2 we have $v_2 = v_B$ and $i_2 = i_B$. Thus, from Eq. 11.1, the two-port descriptive equations are

$$\begin{cases} -v_A = -L_1 \dfrac{di_A}{dt} + M \dfrac{di_B}{dt} \\[2mm] v_B = -M \dfrac{di_A}{dt} + L_2 \dfrac{di_B}{dt} \end{cases}$$

that is,

$$\begin{cases} v_A = L_1 \dfrac{di_A}{dt} - M \dfrac{di_B}{dt} \\[2mm] v_B = -M \dfrac{di_A}{dt} + L_2 \dfrac{di_B}{dt} \end{cases}$$

This is obviously equivalent to having coupled inductors in the standard configuration of Fig. 11.1a, but with mutual inductance $-M$.

The above case study suggests a method (illustrated in Fig. 11.3) to assign the dots in such a way that $M > 0$. The dot on the primary inductor can be assigned arbitrarily. The port variables v_1 and i_1 are set accordingly, with v_1 arrow on the dot. Now we can assign $i_1 = Ktu(t)$ (ramp current), with $K > 0$. A voltmeter is connected to the secondary inductor in both possible ways (see cases α and β in Fig. 11.3), thus measuring $\pm|M|K$. By placing the dot in correspondence of the '+' voltmeter's terminal when the measured voltage is $+|M|K$, as shown in the figure for both α and β cases, we are sure that the mutual inductance is $|M|$ and thus positive.

It is apparent from the descriptive equations that also this component *keeps memory* of the past. Thus, it is linear, time-invariant (assuming that the inductances L_1, L_2, M are constant), with memory.

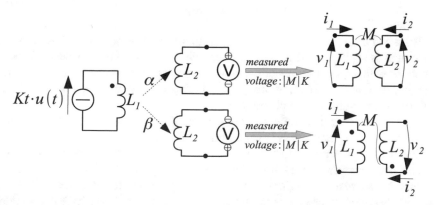

Fig. 11.3 Method to assign the dots in such a way that $M > 0$

The shapes and sizes of real devices vary widely, but in their most basic version they are made up of two close coils of conductor wire, often wound around a magnetic core; both models in Fig. 11.1 sketch this physical situation.

The physical principle at the basis of coupled inductors behavior is as follows. When a current i_1 flows through the primary coil, it induces not only a flux ϕ_{11} in the same coil, as for a simple inductor (see Sect. 9.1.2), but also a flux ϕ_{21} in the nearby secondary coil. Similarly, when a current i_2 flows through the secondary coil, it induces both a flux ϕ_{12} in the primary coil and a flux ϕ_{22} in the secondary coil itself.

By assuming that the device is linear, the total magnetic flux induced by both currents on each coil is

$$\begin{cases} \phi_1 = \phi_{11} + \phi_{12} = L_1 i_1 + M i_2 \\ \phi_2 = \phi_{21} + \phi_{22} = M i_1 + L_2 i_2 \end{cases} \tag{11.2}$$

The ratio of the magnetic flux $\phi_1(t)$ ($\phi_2(t)$) to the current $i_1(t)$ ($i_2(t)$) in the absence of $i_2(t)$ ($i_1(t)$) is the primary (secondary) inductance L_1 (L_2). Moreover, the ratio of both $\phi_1(t)$ and $\phi_2(t)$ to the current $i_2(t)$ and $i_1(t)$, respectively, in the absence of the other current, is the mutual inductance M. In other terms

$$\begin{aligned} L_1 &= \left. \frac{\phi_1}{i_1} \right|_{i_2=0} \\ L_2 &= \left. \frac{\phi_2}{i_2} \right|_{i_1=0} \\ M &= \left. \frac{\phi_1}{i_2} \right|_{i_1=0} = \left. \frac{\phi_2}{i_1} \right|_{i_2=0} \end{aligned} \tag{11.3}$$

In matrix terms, Eq. 11.2 can be recast as

$$\begin{pmatrix} \phi_1 \\ \phi_2 \end{pmatrix} = \begin{pmatrix} L_1 & M \\ M & L_2 \end{pmatrix} \begin{pmatrix} i_1 \\ i_2 \end{pmatrix} \tag{11.4}$$

These could be chosen as the coupled inductors descriptive equations, but, as usual, we want to use voltages and currents as descriptive variables. Then, by deriving these equations with respect to t, under the assumption that inductances L_1, L_2, M do not change with time, we obtain the coupled inductors descriptive equations (Eq. 11.1) in terms of port voltages and currents.

Once more, we assume (unless otherwise stated) that the inductances are constant and then that the component is linear and time invariant.

One of the main reasons for the large use of coupled inductors is that, when primary and secondary coils are linked only by magnetic fields—namely, in the absence of external connections—there is an important safety feature, called the *galvanic isolation*, which helps to prevent high voltages from reaching places they should not go.

Remark: Coupling may be either intentional or unintentional. In the latter case, it is usually an undesired phenomenon, which, for instance, can cause signals from one circuit (part) to affect other variables in a nearby circuit (part): this is often called *cross-talk*. By contrast, intentional coupling is widely used in electric motors and generators, inductive charging devices, induction heating/cooking systems, metal detectors, Radio-Frequency Identification Devices (RFID), wireless power transfer, etc.

Particular case: when the port currents are constant, the descriptive equations reduce to $v_1(t) = v_2(t) = 0$. In other words, the coupled inductors become equivalent to a pair of short circuits, apart from the aspects related to the stored energy (see the next section). This is the typical situation of the *DC steady state*.

11.2 Properties of Coupled Inductors

The coupled inductors are a reciprocal and, in general (for $L_1 \neq L_2$), non-symmetrical two-port. Other properties are discussed in the following.

11.2.1 Series Connection

We want to find the equivalent inductance L resulting from the series connection of the coupled inductors shown in Fig. 11.4a.

The equation of the resulting two-terminal inductor can be found by referring to the set of variables shown in Fig. 11.4b. By substituting $i_1 = i$ and $i_2 = -i$ into Eq. 11.1, we obtain

$$\begin{cases} v_1 = L_1 \dfrac{di}{dt} - M \dfrac{di}{dt}, \\ v_2 = M \dfrac{di}{dt} - L_2 \dfrac{di}{dt}. \end{cases} \tag{11.5}$$

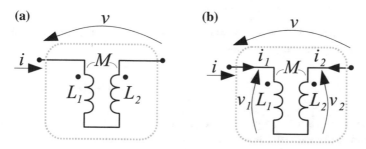

Fig. 11.4 Coupled inductors connected in series (**a**); convention adopted for the descriptive variables (**b**)

Moreover, $v = v_1 - v_2$, and thus

$$v = [(L_1 - M) - (M - L_2)] \frac{di}{dt} = \underbrace{(L_1 + L_2 - 2M)}_{L} \frac{di}{dt}.$$

Therefore, $L = L_1 + L_2 - 2M$. Given that L results from the series connection between parts of a passive element, we can expect that also L is passive, that is, $(L_1 + L_2 - 2M) > 0$. We can now prove that this property follows from the inequality $L_1 L_2 - M^2 \geq 0$ and that it holds regardless of the sign of M. We limit the proof to the $M > 0$ case (for $M < 0$ it is obviously satisfied). Because $M \leq \sqrt{L_1 L_2}$, we have $-M \geq -\sqrt{L_1 L_2}$, which implies that

$$L_1 + L_2 - 2M \geq L_1 + L_2 - 2\sqrt{L_1 L_2} = (\sqrt{L_1} - \sqrt{L_2})^2 > 0,$$

that is, $L_1 + L_2 - 2M > 0$, as expected.

11.2.2 Passivity

The energy absorbed by coupled inductors is

$$w(t) = \int_{-\infty}^{t} p(\tau) d\tau = \int_{-\infty}^{t} \left[\left(L_1 \frac{di_1}{d\tau} + M \frac{di_2}{d\tau} \right) i_1(\tau) + \left(M \frac{di_1}{d\tau} + L_2 \frac{di_2}{d\tau} \right) i_2(\tau) \right] d\tau,$$

(11.6)

that is,

$$w(t) = L_1 \int_{-\infty}^{t} i_1 \frac{di_1}{d\tau} d\tau + L_2 \int_{-\infty}^{t} i_2 \frac{di_2}{d\tau} d\tau + M \int_{-\infty}^{t} \underbrace{i_1(\tau) \frac{di_2}{d\tau} + i_2(\tau) \frac{di_1}{d\tau}}_{= \frac{d(i_1 i_2)}{d\tau}} d\tau.$$

(11.7)

Thus, by assuming as usual that in the past, both inductors were uncharged $(i_1(-\infty) = i_2(-\infty) = 0)$, we obtain

$$w(t) = \frac{1}{2} L_1 i_1^2(t) + \frac{1}{2} L_2 i_2^2(t) + M i_1(t) i_2(t).$$

(11.8)

Inasmuch as M can be positive or negative, from this expression it is not easy to determine the sign of the absorbed energy. But the component is passive due to its physical nature, and thus we can find under what conditions such passivity is guaranteed.

Equation 11.8 shows that $w(t)$ is a quadratic form with respect to the values assumed by i_1 and i_2 and can be recast in the matrix form

$$w(t) = \frac{1}{2} (i_1 \ i_2) \begin{pmatrix} L_1 & M \\ M & L_2 \end{pmatrix} \begin{pmatrix} i_1 \\ i_2 \end{pmatrix}. \tag{11.9}$$

Because the component is passive, we must have $w(t) \geq 0$ for all t. This occurs when the quadratic form w, viewed as a function of (i_1, i_2), falls into one or the other of the following categories:

- *positive definite*: w is positive for every vector $(i_1, i_2) \neq (0, 0)$ and vanishes only when $i_1 = i_2 = 0$. According to Sylvester's criterion [1], this occurs if and only if the matrix elements are such that $L_1 > 0$ and $L_1 L_2 - M^2 > 0$ (which implies, by the way, that also $L_2 > 0$).
- *positive semidefinite*: $w \geq 0$ for every vector (i_1, i_2). In this case, the value $w = 0$ can also occur for some values of the vector (i_1, i_2) other than the trivial value $(0, 0)$. In this case, the matrix must be singular, that is, $L_1 L_2 - M^2 = 0$.

11.2.3 Coupling Coefficient and Closely Coupled Inductors

The mutual inductance is a measure of the coupling between primary and secondary inductors. As an alternative, we can use the so-called *coupling coefficient*
$$k = \frac{|M|}{\sqrt{L_1 L_2}}.$$
Owing to the physical constraint $M^2 \leq L_1 L_2$, k ranges in the interval $[0, 1]$.

When $k = 0$, there is no coupling between the two inductors: $M = 0$ and the two-port degenerates to a pair of simple (uncoupled) inductors. This can be due to the (large) distance between L_1 and L_2 or to the presence of shielding or to an orthogonal position of the two inductors, which prevents/reduces the induction of flow in the other inductor. Indeed, the latter is one of the most common techniques to prevent undesired inductive coupling in small circuits.

When $k = 1$, there is maximum coupling (*closely coupled inductors*) between the two inductors. This is usually obtained by winding both primary and secondary coils on the same core.

For closely coupled inductors, the matrix $\begin{pmatrix} L_1 & M \\ M & L_2 \end{pmatrix}$ is singular, and thus the two port voltages become proportional. For instance, by setting $M = +\sqrt{L_1 L_2}$, we obtain

$$v_1 = \sqrt{\frac{L_1}{L_2}} v_2 = n v_2. \tag{11.10}$$

Thus, this two-port can be used to scale the voltage level.[1] Moreover, in this condition we have (by substitution of Eq. 11.10 into one of the descriptive equations of the coupled inductors, Eq. 11.1)

$$v_1(t) = \sqrt{L_1} \frac{d}{dt}(\sqrt{L_1} i_1 + \sqrt{L_2} i_2), \qquad (11.11)$$

or equivalently,

$$v_2(t) = \sqrt{L_2} \frac{d}{dt}(\sqrt{L_1} i_1 + \sqrt{L_2} i_2). \qquad (11.12)$$

Therefore, the descriptive equations of closely coupled inductors with $M = +\sqrt{L_1 L_2}$ are

$$\begin{cases} v_1 = \sqrt{\frac{L_1}{L_2}} v_2, \\ v_2 = \sqrt{L_2} \frac{d}{dt}(\sqrt{L_1} i_1 + \sqrt{L_2} i_2), \end{cases} \qquad (11.13)$$

and *this degenerate component has only one WSV*, which is a proper linear combination of i_1 and i_2 (that is, $\sqrt{L_1} i_1 + \sqrt{L_2} i_2$), whereas neither i_1 nor i_2 can be a WSV by itself. This reflects the fact that in this condition, the absorbed energy can be expressed in terms of this WSV:

$$w(t) = \frac{1}{2} L_1 i_1^2(t) + \frac{1}{2} L_2 i_2^2(t) + \sqrt{L_1 L_2} i_1(t) i_2(t) = \frac{1}{2}(\sqrt{L_1} i_1 + \sqrt{L_2} i_2)^2. \quad (11.14)$$

The above system (Eq. 11.13) also describes the equivalent model shown in Fig. 11.5a. You can easily check that for $M = -\sqrt{L_1 L_2}$, the descriptive equations are

$$\begin{cases} v_1 = -\sqrt{\frac{L_1}{L_2}} v_2, \\ v_2 = -\sqrt{L_2} \frac{d}{dt}(\sqrt{L_1} i_1 - \sqrt{L_2} i_2). \end{cases} \qquad (11.15)$$

In this case, the WSV is $\sqrt{L_1} i_1 - \sqrt{L_2} i_2$ and the absorbed energy is $w(t) = \frac{1}{2}(\sqrt{L_1} i_1 - \sqrt{L_2} i_2)^2$. The equivalent model is shown in Fig. 11.5b.

11.2.4 Energy Conservation

Coupled inductors, as well as capacitors and inductors, are a *conservative* element. In other words, the energy $w(t)$ stored by the component is uniquely identified by

---

[1] For instance, real closely coupled inductors are devices largely used in electric grids to step up the voltage exiting a power plant (AC voltage with amplitude on the order of 10^4 V) to a high voltage (with amplitude on the order of 10^5 V) so that it can be transmitted with great efficiency, and to step it down at the distribution end to levels (on the order of 10^2 V) appropriate to homes and factories.

(a)　　　　　　　　　　　　　　　**(b)**

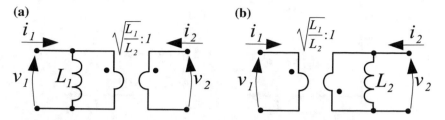

Fig. 11.5 Equivalent models of closely coupled inductors

the values of the descriptive currents i_1, i_2; in particular, the values $i_1 = 0$, $i_2 = 0$ indicate the zero-level of stored energy.

This requirement finds its mathematical correspondent in the symmetry of the matrix $\begin{pmatrix} L_1 & M \\ M & L_2 \end{pmatrix}$. This can be easily evidenced by considering the expression $w(t)$ would assume if we had an asymmetric matrix $\begin{pmatrix} L_1 & M_{12} \\ M_{21} & L_2 \end{pmatrix}$, that is, if we had $M_{12} = M$, $M_{21} = M + M_0$. In this case, the expression of the power entering the two-port is

$$
\begin{aligned}
p(t) = v_1 i_1 + v_2 i_2 &= \left(L_1 \frac{di_1}{dt} + \underbrace{M_{12}}_{M} \frac{di_2}{dt} \right) i_1 + \left(\underbrace{M_{21}}_{M+M_0} \frac{di_1}{dt} + L_2 \frac{di_2}{dt} \right) i_2 = \\
&= \frac{d}{dt} \left[\frac{1}{2} L_1 i_1^2 + M i_1 i_2 + \frac{1}{2} L_2 i_2^2 \right] + M_0 i_2 \frac{di_1}{dt}.
\end{aligned}
$$

Therefore,

$$
p(t)dt = d \left[\frac{1}{2} L_1 i_1^2 + M i_1 i_2 + \frac{1}{2} L_2 i_2^2 \right] + M_0 i_2 di_1. \tag{11.16}
$$

The first term on the right-hand side of Eq. 11.16 is well known and represents the differential of energy in the case of a symmetric matrix. As for the second term $M_0 i_2 di_1$, it can be shown that $M_0 \equiv 0$ due to the conservative nature of this component. To this end, consider the plane (i_1, i_2) and choose a point $P = (I_1, I_2)$. Starting from the origin O of the plane (corresponding to the zero energy level), in a conservative component the energy value at P must be independent of the path followed by the currents $i_1(t)$ and $i_2(t)$ to reach P. Whatever the path, denoting by t_0 and T the time instants corresponding to the points O and P, we can integrate both sides of Eq. 11.16 and write

$$
\int_{t_0}^{T} p(\tau)d\tau = \frac{1}{2} L_1 I_1^2 + M I_1 I_2 + \frac{1}{2} L_2 I_2^2 + M_0 \int_{O}^{P} i_2 di_1.
$$

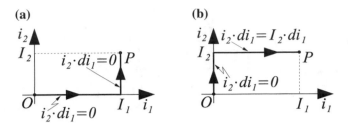

Fig. 11.6 Two different paths connecting the zero-level energy point O with a point P. In a conservative component, the energy value at P must be path-independent

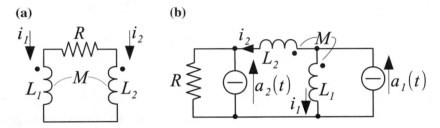

Fig. 11.7 Case Studies **a** 1 and **b** 2

This result is path-dependent unless $M_0 = 0$. This can be seen by considering two different paths connecting O and P. For instance, along the path of Fig. 11.6a, $i_2 di_1 = 0$ everywhere; in contrast, following the path of Fig. 11.6b, we have $i_2 di_1 = I_2 di_1$ along the horizontal segment ending in P. Therefore, we conclude that $M_0 = 0$ and the consequent symmetry of the matrix ensures that the two-port of coupled inductors is conservative.

It can be shown that the matrix symmetry is also a necessary condition for the coupled inductors to be a reciprocal component. This reciprocity property, as well as those of capacitors and inductors, will be established in a later chapter.

Case Study 1
Find the energy dissipated for $t > t_0$ by the resistor in the circuit shown in Fig. 11.7a, assuming that $i_1(t_0) = I_0$ and that the inductors are not closely coupled.

A first way to solve this problem would be to find an expression for $i_1(t)$ for $t > t_0$ (notice that in this circuit we have two WSVs, but only one SSV, due to the KCL constraint $i_2 = -i_1$) and computing the required energy as the integral of the power absorbed by the resistor ($p(t) = Ri_1^2(t)$) over the time interval $(t_0, +\infty)$. This is a correct solution, of course, but requiring some work.

An alternative way to answer the question more easily is based on the knowledge of the energetic behaviors of the components. The resistor dissipates all the absorbed energy, whereas the coupled inductors conserve all the absorbed energy. Moreover, due to the assumption of not closely coupled inductors, $L_1 L_2 - M^2 > 0$, which implies $L_1 + L_2 - 2M > 0$, as shown in Sect. 11.2.1. Inasmuch as the circuit is asymptotically stable (you can easily check that its natural frequency is $\lambda = -\dfrac{R}{L_1 + L_2 - 2M} < 0$), the current i_1 will tend to 0 as time goes by. In other words, the energy initially stored in the coupled inductors will be transferred to the resistor and thus dissipated. This means that the energy dissipated for $t > t_0$ by the resistor is nothing but the energy stored in $t = t_0$ in the coupled inductors, that is (see Eq. 11.8),

$$w(t_0) = \frac{1}{2} L_1 I_0^2 + \frac{1}{2} L_2 I_0^2 - M I_0^2 = \frac{1}{2}(L_1 + L_2 - 2M) I_0^2. \qquad (11.17)$$

Case Study 2
Find the state variable $i_2(t)$ in the circuit shown in Fig. 11.7b, which works at steady state for $t < 0$, with $a_1(t) = A_0 u(-t)$ and $a_2(t) = Q_0 \delta(t) + A u(-t)$. Determine the circuit stability condition, assuming that $L_1 L_2 - M^2 > 0$ and $L_1, L_2, R > 0$. Finally, find the energy dissipated by the circuit for $t > 0$, by assuming that the circuit is absolutely stable.

You can check that there is only one SSV, due to the algebraic constraint $i_1 + i_2 = a_1$, and the state equation for i_2 is

$$L \frac{di_2}{dt} + R i_2 = -R a_2 + (L_1 - M) \frac{da_1}{dt},$$

where $L = L_1 + L_2 - 2M$.

The natural frequency is $\lambda = -\dfrac{R}{L} = -\dfrac{R}{L_1 + L_2 - 2M}$, and thus the absolute stability condition is $L_1 + L_2 > 2M$. This condition, however, is automatically fulfilled for $L_1 L_2 - M^2 > 0$, as shown in Sect. 11.2.1.

For $t < 0$, $a_1(t) = A_0$, and $a_2(t) = A$, you can therefore easily check that $i_2(t) = -A$, which is also the value of $i_2(0^-)$.

From the discontinuity balance, we get that $i_2(0^+) \neq i_2(0^-)$. By integrating the state equation between 0^- and 0^+, we obtain

$$i_2(0^+) = -\frac{R Q_0}{L} - \frac{L_1 - M}{L} A_0 + i_2(0^-) = -\frac{R Q_0}{L} - \frac{L_1 - M}{L} A_0 - A.$$

For $t > 0$, we have $a_1(t) = a_2(t) = 0$; therefore $i_2(t) = i_2(0^+) e^{\lambda t}$.

The energy dissipated by the circuit for $t > 0$ is the energy initially stored in the coupled inductors, namely

$$w(0^+) = \frac{1}{2}L_1 i_1^2(0^+) + \frac{1}{2}L_2 i_2^2(0^+) + M i_1(0^+) i_2(0^+).$$

But $i_1(0^+) = a_1(0^+) - i_2(0^+) = -i_2(0^+)$, and therefore

$$w(0^+) = \frac{1}{2}(L_1 + L_2 - 2M) i_2^2(0^+) = \frac{1}{2}(L_1 + L_2 - 2M) \left(\frac{RQ_0}{L} + \frac{L_1 - M}{L} A_0 + A \right)^2.$$

11.2.5 Equivalent Models

Here we introduce two equivalent models of coupled inductors and an equivalent model of *real* coupled inductors.

11.2.5.1 Model 1

The first equivalent model is shown in Fig. 11.8a. The dashed short circuit can be either present (three-terminal configuration) or not. The two-port descriptive equations can be easily obtained from Fig. 11.8b:

$$\begin{cases} v_1(t) = (L_s + L_p)\dfrac{di_1}{dt} + \dfrac{L_p}{n}\dfrac{di_2}{dt}, \\[2mm] v_2(t) = \dfrac{L_p}{n}\dfrac{di_1}{dt} + \dfrac{L_p}{n^2}\dfrac{di_2}{dt}. \end{cases} \tag{11.18}$$

By comparing Eqs. 11.1 and 11.18, we easily identify

$$L_1 = L_s + L_p,$$

$$M = \frac{L_p}{n}, \tag{11.19}$$

$$L_2 = \frac{L_p}{n^2}.$$

Fig. 11.8 a Equivalent
model 1 of coupled inductors
and **b** hints for deriving its
descriptive equations

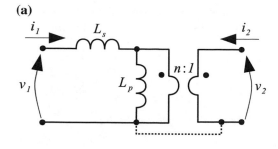

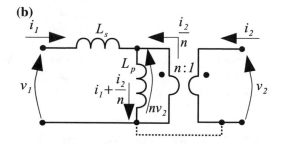

The inverse relationships are

$$L_s = (1 - k^2)L_1,$$
$$L_p = k^2L_1, \tag{11.20}$$
$$n = k\sqrt{\frac{L_1}{L_2}}.$$

It is evident that for closely coupled inductors (where $k = 1$), $L_s = 0$, which is
perfectly consistent with the model shown in Fig. 11.5a.

11.2.5.2 Model 2 (T-Model)

A second equivalent model (valid only for coupled inductors in the three-terminal
configuration), also called a T-model, is shown in Fig. 11.9.

Fig. 11.9 Equivalent model
2 (T-model) of coupled
inductors

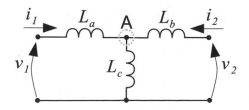

Finding the descriptive equations is left to the reader as an exercise. Once you have found the descriptive equations, you can easily identify

$$
\begin{aligned}
L_1 &= L_a + L_c, \\
L_2 &= L_b + L_c, \\
M &= L_c.
\end{aligned}
\tag{11.21}
$$

The inverse relationships are

$$
\begin{aligned}
L_a &= L_1 - M, \\
L_b &= L_2 - M, \\
L_c &= M.
\end{aligned}
\tag{11.22}
$$

Notice that in this equivalent model we have three inductors, but only two of their currents are WSVs, due to the presence of a cut-set (node A in Fig. 11.9) involving only inductors.

11.2.5.3 Model of Real Coupled Inductors

The concrete realizations of coupled inductors are subject to some undesirable behaviors. The most important, particularly in industrial applications, is the power dissipation associated with the resistivity of winding conductors and parasitic currents inside the core. The structure of a model that takes into account all of this is shown in Fig. 11.10: the resistor R_0 models power dissipation in the core, whereas R_1 and R_2 model power dissipations in the windings. This model neglects the effects of nonlinearity of the core and the influence of other parasitic parameters, such as capacitances.

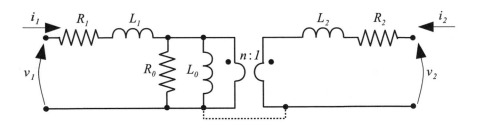

Fig. 11.10 Model of real coupled inductors

Case Study 1
In full analogy with coupled inductors, the descriptive equations for coupled capacitors can be defined as

$$\begin{cases} i_1 = C_1 \dfrac{dv_1}{dt} + C_m \dfrac{dv_2}{dt}, \\[2mm] i_2 = C_m \dfrac{dv_1}{dt} + C_2 \dfrac{dv_2}{dt}. \end{cases} \qquad (11.23)$$

As for the physical parameters, we must have $C_1 > 0$, $C_2 > 0$, while C_m can be either positive or negative and such that $\Delta \doteq C_1 C_2 - C_m^2 \geq 0$. Assuming $\Delta > 0$, show that the circuits of Fig. 11.11a and b can be equivalent models for the three-terminal and for the two-port component respectively, and find the corresponding values for the model parameters in terms of C_1, C_2, C_m.

The currents i_1, i_2 entering the three-terminal model follow immediately on writing the KCL at the upper terminals and simplifying:

$$\begin{cases} i_1 = C_a \dfrac{d}{dt} v_1 + C \dfrac{d}{dt} (v_1 - v_2) = (C_a + C) \dfrac{d}{dt} v_1 - C \dfrac{d}{dt} v_2, \\[2mm] i_2 = C \dfrac{d}{dt} (v_2 - v_1) + C_b \dfrac{d}{dt} v_2 = -C \dfrac{d}{dt} v_1 + (C_b + C) \dfrac{d}{dt} v_2. \end{cases}$$

These equations coincide with Eq. 11.23, provided that the values of the circuit parameters are $C = -C_m$, $C_a = C_1 + C_m$, $C_b = C_2 + C_m$.

Taking into account the equations for the ideal transformer, the currents i_1, i_2 for the two-ports are

$$\begin{cases} i_1 = C_p \dfrac{d}{dt} v_1 + C_q \dfrac{d}{dt} (v_1 - nv_2) = (C_p + C_q) \dfrac{d}{dt} v_1 - nC_q \dfrac{d}{dt} v_2, \\[2mm] i_2 = -nC_q \dfrac{d}{dt} (v_1 - nv_2) = -nC_q \dfrac{d}{dt} v_1 + n^2 C_q \dfrac{d}{dt} v_2. \end{cases}$$

By comparison with Eq. 11.23, the values of the circuit parameters must be $C_p = C_1 + \dfrac{C_m}{n} = \dfrac{\Delta}{C_2}$, $C_q = -\dfrac{C_m}{n} = \dfrac{C_m^2}{C_2}$, $n = -\dfrac{C_2}{C_m}$. The possible negative sign for the value of n is fully compatible with the descriptive equations of the transformer.

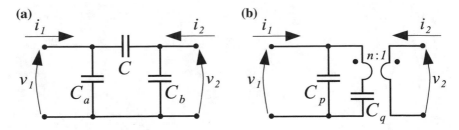

Fig. 11.11 Equivalent coupled capacitors models for **a** three-terminals and **b** two-ports

11.2.6 Thévenin and Norton Equivalent Representations of Charged Coupled Inductors

Let us consider coupled inductors whose currents, at time t_0^-, assume values $i_1(t_0^-)$ and $i_2(t_0^-)$. For this component, we can obtain two equivalent representations by following the same line of reasoning as for a single inductor; see Sect. 9.13.2. For brevity, we denote by Δ the term $L_1 L_2 - M^2$, and we assume $\Delta > 0$. Under this assumption, Eq. 11.1 can be recast as

$$
\begin{cases}
\dfrac{di_1}{dt} = \dfrac{L_2}{\Delta} v_1 - \dfrac{M}{\Delta} v_2, \\[2mm]
\dfrac{di_2}{dt} = -\dfrac{M}{\Delta} v_1 + \dfrac{L_1}{\Delta} v_2.
\end{cases}
\tag{11.24}
$$

We can now integrate both equations from t_0^- to a generic time instant $t > t_0$. Similarly to the case of single inductor, we can write

$$
\begin{cases}
i_1(t) = i_1(t_0^-)u(t-t_0) + \underbrace{\left[\dfrac{L_2}{\Delta} \int_{t_0^-}^{t} v_1(\tau)d\tau - \dfrac{M}{\Delta} \int_{t_0^-}^{t} v_2(\tau)d\tau \right]}_{\hat{\imath}_1(t)}, \\[6mm]
i_2(t) = i_2(t_0^-)u(t-t_0) + \underbrace{\left[-\dfrac{M}{\Delta} \int_{t_0^-}^{t} v_1(\tau)d\tau + \dfrac{L_1}{\Delta} \int_{t_0^-}^{t} v_2(\tau)d\tau \right]}_{\hat{\imath}_2(t)}.
\end{cases}
\tag{11.25}
$$

These equations can be viewed as KCLs. In each of them, the first current on the right-hand side represents the contribution of the initial condition, while the bracketed term is the current flowing into the initially uncharged coupled inductors. The corresponding equivalent circuit is shown in Fig. 11.12. Its Norton-like structure needs no comment.

An equivalent formulation in terms of KVL can now be obtained by simply observing that in the circuit of Fig. 11.12, v_1 and v_2 can be expressed as

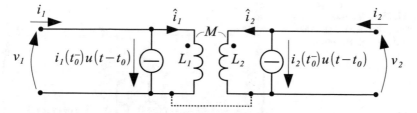

Fig. 11.12 Norton equivalent representation for charged coupled inductors

$$\begin{cases} v_1(t) = L_1\dfrac{d\hat{i}_1}{dt} + M\dfrac{d\hat{i}_2}{dt}, \\[2mm] v_2(t) = M\dfrac{d\hat{i}_1}{dt} + L_2\dfrac{d\hat{i}_2}{dt}. \end{cases} \tag{11.26}$$

The terms $\hat{i}_1$ and $\hat{i}_2$ can now be replaced by their corresponding expressions, obtained from Eqs. 11.25:

$$\begin{cases} \hat{i}_1(t) = i_1(t) - i_1(t_0^-)u(t - t_0), \\[2mm] \hat{i}_2(t) = i_2(t) - i_2(t_0^-)u(t - t_0). \end{cases} \tag{11.27}$$

In this way, we immediately obtain the equations

$$\begin{cases} v_1(t) = L_1\dfrac{di_1}{dt} + M\dfrac{di_2}{dt} - \phi_1(t_0^-)\delta(t - t_0), \\[2mm] v_2(t) = M\dfrac{di_1}{dt} + L_2\dfrac{di_2}{dt} - \phi_2(t_0^-)\delta(t - t_0), \end{cases} \tag{11.28}$$

where

$$\phi_1(t_0^-) \doteq L_1 i_1(t_0^-) + M i_2(t_0^-), \qquad \phi_2(t_0^-) \doteq M i_1(t_0^-) + L_2 i_2(t_0^-),$$

are the initial flux values for the component and constitute the "areas" of two voltage impulse terms. The corresponding (Thévenin) equivalent representation is shown in Fig. 11.13.

11.3 Higher-Order Linear Circuits

In this section we consider LTI circuits of generic order and analyze them by defining a common framework that generalizes what was already stated for first-order LTI circuits. As shown in Sect. 9.3.3, a linear circuit with N wide-sense state variables (WSVs) and $M\,(< N)$ independent algebraic constraints has $N - M$ strict-sense state variables (SSVs) and is said to be of order $N - M$.

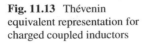

Fig. 11.13 Thévenin
equivalent representation for
charged coupled inductors

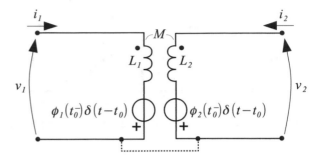

11.3.1 General Method

An LTI circuit having $2L$ descriptive variables is described by a system of $2L$ equations. Some of them are topological, whereas the rest are the descriptive equations of the circuit components (see Vol. 1, Sect. 3.1). In the presence of one or more memory components, the circuit has order $N - M$. If $N - M > 0$, the system of $2L$ equations is of algebraic-differential nature. For a given set of initial conditions, the circuit variables as functions of time are obtained by solving the initial value problem (briefly, IVP) associated with the differential equations of the circuit.

A first way of getting a solution for an IVP is the numerical approach, which is based on the discretization of the circuit variables. Discrete variables are processed by numerical algorithms that generate, at each time instant, a system of algebraic equations to be solved. In the presence of nonlinear elements, this approach is almost always the only one possible. In contrast, in the case of LTI circuits, it is also possible to use analytical methods, which offer great advantages in the study of the general properties of the circuit.

An efficient way of setting the analytic treatment of the IVP is based on the partition of the space of the descriptive variables into a subspace containing only the $N - M$ state variables and the complementary subspace formed by the remaining variables.

By extending the complementary component method considered in Chap. 9 to the case of multiple state variables (organized in a vector), it is concluded that the state vector satisfies a canonical equation whose structure is the same as for the case of a single variable. Additionally, the nonstate variables (complementary subspace) depend algebraically on the state vector components and on the inputs in a way similar to that encountered in Chap. 9.

The canonical form of the state equations enables us to formulate the "paper and pencil" solution of the IVP for the circuit in the simplest way. This is done by determining the natural frequencies of the circuit. Every other variable is obtained algebraically, namely, without solving further differential problems. Therefore, all the $2L$ descriptive variables evolve over time based on the inputs and on the values of the natural frequencies. This generalizes the result found for first-order circuits.

As a final remark, the canonical form of the state equations also has advantages from the standpoint of numerical treatment of the solution, which can rely on algorithms specifically designed for this form and as such, are more efficient for speed and accuracy.

11.3.2 Complementary Component Method and State Equations

Every dynamical circuit can be represented as the connection between a memoryless part, containing all the circuit's memoryless components (including P independent sources), and a part with memory, containing capacitors, inductors, and coupled inductors, as shown in Fig. 11.14.

We know that the coupled inductors can be replaced by one of their equivalent representations. We choose Model 1, described in Sect. 11.2.5.1, given that it keeps the information about the wide-sense state variables, also for closely coupled inductors. Similarly, coupled capacitors can be replaced by the model shown in Fig. 11.11b. With these substitutions, the memory part contains only capacitors and inductors, N altogether.

Moreover, each memory component whose WSV is not an SSV due to a linear algebraic constraint can be replaced by an equivalent memoryless two-terminal, as discussed in Sect. 9.3.2. This two-terminal, in general, contains sources (both independent and controlled) whose current/voltage involve the time derivatives of one or more of the original P sources, as illustrated in the case studies of Sect. 9.14.

By assuming that there are M independent algebraic constraints, after the substitutions the circuit includes N_C capacitors and N_L inductors, $n = N - M$ altogether, whose WSVs are also SSVs. Therefore, the original circuit can be represented as the connection of $n = N_C + N_L$ components with memory and a memoryless *complementary component*, which is an n-port $\aleph$ containing the rest of the circuit, as shown in Fig. 11.14.

The memoryless n-port is described with the standard choice of descriptive variables. The port variables can be grouped in two sets, that is, the SSVs x (voltages v_C across the capacitors and currents i_L in the inductors) and the other variables y (currents i_C in the capacitors and voltages v_L across the inductors):

$$x = \begin{pmatrix} v_C \\ i_L \end{pmatrix} \qquad y = \begin{pmatrix} i_C \\ v_L \end{pmatrix}. \tag{11.29}$$

Given that there are no longer algebraic constraints relating the variables x, we are guaranteed that $\aleph$ admits the basis x. Therefore, we can write the descriptive equations of $\aleph$ in explicit form. In the simplest case ($M = 0$), only the original P independent sources $\hat{u}(t)$ appear in the descriptive equations:

Fig. 11.14 Dynamical
circuit representation as the
connection between a
memoryless n-port $\aleph$ and
and n memory components

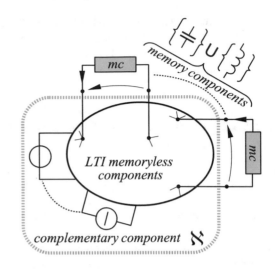

$$y = H^x x + H^u \hat{u}(t), \tag{11.30}$$

where $\hat{u}(t)$ is a P-size vector whose (physically hybrid) entries are either zeros or impressed currents/voltages, whereas H^x and H^u are hybrid matrices of size $n \times n$ and $n \times P$, respectively.

Otherwise, if $M \neq 0$, the time derivatives of one or more of the original P sources can also appear:

$$y = H^x x + H^u \hat{u}(t) + H_1^u \frac{d\hat{u}}{dt} + H_2^u \frac{d^2\hat{u}}{dt^2} + \cdots , \tag{11.31}$$

where H_k^u ($k = 1, 2, \ldots$) is a hybrid matrix of size $n \times P$.

The rest of the circuit can be compactly described using the descriptive equations of capacitors and inductors, paying attention to the orientations of the port currents:

$$\begin{cases} i_C = -[C] \dfrac{dv_C}{dt}, \\[2mm] v_L = -[L] \dfrac{di_L}{dt}, \end{cases} \tag{11.32}$$

where $[C]$ ($[L]$) is a diagonal matrix of size $N_C \times N_C$ ($N_L \times N_L$) whose diagonal entries are the capacitances (inductances) of the capacitors (inductors).

Equation 11.32 can be recast in the following more compact matrix form:

$$\underbrace{\begin{pmatrix} i_C \\ v_L \end{pmatrix}}_{y} = -\underbrace{\begin{pmatrix} [C] & 0_{N_C \times N_L} \\ 0_{N_L \times N_C} & [L] \end{pmatrix}}_{D} \frac{d}{dt} \underbrace{\begin{pmatrix} v_C \\ i_L \end{pmatrix}}_{\dot{x}}. \tag{11.33}$$

Notice that matrix D (of size $n \times n$), being diagonal, is invertible.

In the simplest case ($M = 0$), by comparing Eqs. 11.30 and 11.33, we obtain a system of purely differential equations (*state equations*) in terms of the SSVs (coinciding in this case with the WSVs):

$$-D\dot{x} = H^x x + H^u \hat{u}(t), \tag{11.34}$$

that is,

$$\dot{x} = -D^{-1} H^x x - D^{-1} H^u \hat{u}(t). \tag{11.35}$$

Thus, the dynamics of the *state vector* x is described by the above equation. If $M \neq 0$, the state equations become

$$\dot{x} = -D^{-1} H^x x - D^{-1} \left[H^u \hat{u}(t) + H_1^u \frac{d\hat{u}}{dt} + H_2^u \frac{d^2 \hat{u}}{dt^2} + \cdots \right]. \tag{11.36}$$

11.3.3 State Equations in Canonical Form and I/O Relationships

According to the results obtained in the above section:

In an LTI circuit with P independent sources and with n strict-sense state variables collected in the state vector x, the **state equations** can always be written in one of the following **canonical forms**:
- for $M = 0$ (and $n = N$):
$$\dot{x} = Ax + B\hat{u}(t), \tag{11.37}$$

- for $M \neq 0$ (and $n = N - M$):

$$\dot{x} = Ax + B_0 \hat{u}(t) + B_1 \frac{d\hat{u}}{dt} + B_2 \frac{d^2 \hat{u}}{dt^2} + \cdots . \tag{11.38}$$

In all cases, A is the *state matrix*, of size $n \times n$, whereas B and B_k ($k = 0, 1, 2, \ldots$) are *input matrices*, of size $n \times P$.

A circuit without any input ($P = 0$, or equivalently, $\hat{u}(t) = 0$) is said to be **autonomous** (see also Sect. 10.1).

The above canonical forms can be handled to obtain the I/O relationship for a specific state variable.

For instance, in a circuit with $N = 2$, $M = 0$, and $P = 2$, the state equations in canonical form (Eq. 11.37) can be recast as

$$\begin{pmatrix} \dot{x}_1 \\ \dot{x}_2 \end{pmatrix} = \begin{pmatrix} A_{11} & A_{12} \\ A_{21} & A_{22} \end{pmatrix} \begin{pmatrix} x_1 \\ x_2 \end{pmatrix} + \begin{pmatrix} B_{11} & B_{12} \\ B_{21} & B_{22} \end{pmatrix} \begin{pmatrix} \hat{u}_1 \\ \hat{u}_2 \end{pmatrix}, \tag{11.39}$$

that is,

$$\begin{cases} \dot{x}_1 = A_{11}x_1 + A_{12}x_2 + B_{11}\hat{u}_1 + B_{12}\hat{u}_2, \\ \dot{x}_2 = A_{21}x_1 + A_{22}x_2 + B_{21}\hat{u}_1 + B_{22}\hat{u}_2. \end{cases} \tag{11.40}$$

By differentiating the first equation and substituting $\dot{x}_2$ from the second, we obtain

$$\ddot{x}_1 = A_{11}\dot{x}_1 + A_{12}(A_{21}x_1 + A_{22}x_2 + B_{21}\hat{u}_1 + B_{22}\hat{u}_2) + B_{11}\frac{d\hat{u}_1}{dt} + B_{12}\frac{d\hat{u}_2}{dt}. \tag{11.41}$$

From the first equation, we have

$$A_{12}x_2 = \dot{x}_1 - A_{11}x_1 - B_{11}\hat{u}_1 - B_{12}\hat{u}_2,$$

which can be substituted in Eq. 11.41. After tedious but straightforward manipulations, we finally obtain

$$\ddot{x}_1 - \tau\dot{x}_1 + \Delta x_1 = (A_{12}B_{21} - A_{22}B_{11})\hat{u}_1 + (A_{12}B_{22} - A_{22}B_{12})\hat{u}_2 + B_{11}\frac{d\hat{u}_1}{dt} + B_{12}\frac{d\hat{u}_2}{dt},$$

where $\tau = Tr(A) = A_{11} + A_{22}$ is the trace of the state matrix A and $\Delta = A_{11}A_{22} - A_{12}A_{21}$ is its determinant.

This example points out that an I/O relationship can contain the time derivatives of the inputs, even in the absence of algebraic constraints.

In general, every differential system of n scalar equations of order 1 whose unknowns are the n state variables can be recast as a single differential equation of order n whose only unknown is one of the state variables.

The **I/O relationship** for an LTI circuit with a state variable x (one among n) and a single input $\hat{u}(t)$ can always be written as follows:

$$a_n \frac{d^n x}{dt^n} + \cdots + a_1 \frac{dx}{dt} + a_0 x = b_m \frac{d^m \hat{u}}{dt^m} + \cdots + b_1 \frac{d\hat{u}}{dt} + b_0 \hat{u}. \qquad (11.42)$$

In general, in the presence of P inputs, we have $\hat{u} = \sum_{k=1}^{P} \beta_k \hat{u}_k$ and

$$\sum_{j=0}^{n} a_j \frac{d^j x}{dt^j} = \sum_{i=0}^{m} \sum_{k=1}^{P} b_{ik} \frac{d^i \hat{u}_k}{dt^i}, \qquad (11.43)$$

where $b_{ik} = b_i \beta_k$.

For instance, in the above example we can easily identify some coefficients in addition to $a_2 = 1$:

$$\ddot{x}_1 \underbrace{-\tau}_{a_1} \dot{x}_1 + \underbrace{\Delta}_{a_0} x_1 = \underbrace{(A_{12}B_{21} - A_{22}B_{11})}_{b_{01}} \hat{u}_1 + \underbrace{(A_{12}B_{22} - A_{22}B_{12})}_{b_{02}} \hat{u}_2$$

$$+ \underbrace{B_{11}}_{b_{11}} \frac{d\hat{u}_1}{dt} + \underbrace{B_{12}}_{b_{12}} \frac{d\hat{u}_2}{dt}.$$

Remark 1: For circuits with $M = 0$, we have $n = N$ and $m = n - 1 = N - 1$. For circuits with $M > 0$, we have $n = N - M$.

Remark 2: Some of the coefficients a_j and b_{ik} can be null, but $a_n \neq 0$.

Case Study 1

Find the state equation in canonical form for the circuit shown in Fig. 11.15a.

There are no inputs, and thus the circuit is autonomous. The voltages v_1 and v_2 are WSVs.

The two-port $\aleph$ is shown in Fig. 11.15b, within the dotted line. You can easily check that its descriptive equations are

$$\begin{cases} i_1 = \dfrac{v_1}{R} - \dfrac{v_2}{R}, \\[3mm] i_2 = -\dfrac{v_1}{R} + \dfrac{v_2}{R}(2 - \alpha). \end{cases}$$

Thus the voltage basis is admitted, and v_1 and v_2 are SSVs, because $M = 0$.

Fig. 11.15 Case Study 1

(a)

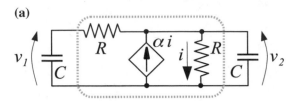

(b)

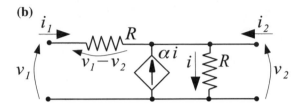

The circuit state equations are

$$-\begin{pmatrix} C & 0 \\ 0 & C \end{pmatrix}\begin{pmatrix} \dot{v}_1 \\ \dot{v}_2 \end{pmatrix} = \frac{1}{R}\begin{pmatrix} 1 & -1 \\ -1 & 2-\alpha \end{pmatrix}\begin{pmatrix} v_1 \\ v_2 \end{pmatrix},$$

that is, in canonical form,

$$\begin{pmatrix} \dot{v}_1 \\ \dot{v}_2 \end{pmatrix} = -\frac{1}{RC}\begin{pmatrix} 1 & 0 \\ 0 & 1 \end{pmatrix}\begin{pmatrix} 1 & -1 \\ -1 & 2-\alpha \end{pmatrix}\begin{pmatrix} v_1 \\ v_2 \end{pmatrix} =$$

$$= \begin{pmatrix} -\dfrac{1}{RC} & \dfrac{1}{RC} \\ \dfrac{1}{RC} & \dfrac{\alpha-2}{RC} \end{pmatrix}\begin{pmatrix} v_1 \\ v_2 \end{pmatrix}.$$

You can check that the I/O relationship for v_1 is

$$(RC)^2\frac{d^2v_1}{dt^2} + (3-\alpha)RC\frac{dv_1}{dt} + (1-\alpha)v_1 = 0.$$

Remark: Similarly to the case of first-order circuits, the left-hand side of Eq. 11.43 can be also written as $\mathcal{L}(x)$, to point out that we are applying a linear operator $\mathcal{L} = \sum\limits_{j=0}^{n} a_j\dfrac{d^j}{dt^j}$ to the variable $x(t)$. The homogeneous differential equation associated with Eq. 11.43 is obtained by setting $\hat{u}_k(t) = 0$ for all k, thus yielding $\mathcal{L}(x) = 0$.

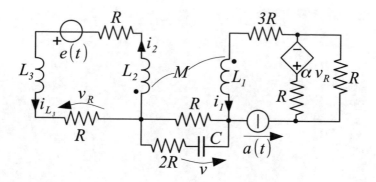

Fig. 11.16 Case Study 2

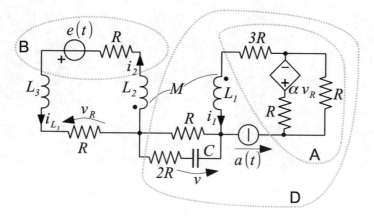

Fig. 11.17 Solution of Case Study 2

Case Study 2

Find the state equation in canonical form for the circuit shown in Fig. 11.16.

Determining the algebraic constraints $i_1 = a(t)$ (KCL for cut-set A shown in Fig. 11.17) and $i_2 = i_{L_3}$ (KCL for cut-set B) is immediate. Therefore, we have $N = 4$ and $M = 2$. We choose i_2 as the second SSV in addition to v.

Using the KCL for cut-set D, we directly obtain the I/O relationship (state equation) for the state variable v, which is uncoupled from the rest of the circuit:

$$3RC\frac{dv}{dt} + v = 0.$$

The I/O relationship for i_2 can be obtained by considering the KVL for the left circuit mesh and using the aforesaid algebraic constraints:

$$M \frac{da}{dt} + L_2 \frac{di_2}{dt} + 2Ri_2 + L_3 \frac{di_2}{dt} = e(t),$$

that is,

$$(L_2 + L_3) \frac{di_2}{dt} + 2Ri_2 = e(t) - M \frac{da}{dt}.$$

The state equations in canonical form can be directly derived from the I/O relationships:

$$\frac{d}{dt} \begin{pmatrix} v \\ i_2 \end{pmatrix} = \underbrace{\begin{pmatrix} -\dfrac{1}{3RC} & 0 \\ 0 & -\dfrac{2R}{L_2 + L_3} \end{pmatrix}}_{A} \begin{pmatrix} v \\ i_2 \end{pmatrix} + \underbrace{\begin{pmatrix} 0 & 0 \\ \dfrac{1}{L_2 + L_3} & 0 \end{pmatrix}}_{B_0} \begin{pmatrix} e(t) \\ a(t) \end{pmatrix} +$$

$$+ \underbrace{\begin{pmatrix} 0 & 0 \\ 0 & -\dfrac{M}{L_2 + L_3} \end{pmatrix}}_{B_1} \frac{d}{dt} \begin{pmatrix} e(t) \\ a(t) \end{pmatrix}.$$

The state matrix A is diagonal due to the fact that the two state variables are uncoupled.

11.4 Discontinuity Balance

The discontinuity balance introduced in Sect. 9.7 can also be applied to n-order I/O relationships, thus generalizing what was already shown in Sect. 9.14 for first-order circuits.

Inasmuch as solving an n-order ordinary differential equation requires n initial conditions, we can analyze the circuit behavior across input discontinuities by applying this balance technique.

Making reference to Eq. 11.42, the discontinuity balance implies that

- if $n = m$, x has the same order of discontinuity as $\hat{u}$;
- if $n > m$, x is "more continuous" than $\hat{u}$;
- if $n < m$, x is "more discontinuous" than $\hat{u}$.

For instance, assume that $n = 2$. Equation 11.43 takes the form

$$a_2 \frac{d^2 x}{dt^2} + a_1 \frac{dx}{dt} + a_0 x = \sum_{i=0}^{m} \sum_{k=1}^{P} b_{ik} \frac{d^i \hat{u}_k}{dt^i}.$$

If the highest-order discontinuity on $\dfrac{d^m \hat{u}_k}{dt^m}$ at a given time t_0 is a step, this means that at t_0,

- $\dfrac{d^2 x}{dt^2}$ must exhibit a step discontinuity as well;
- $\dfrac{dx}{dt}$ and $x(t)$ are continuous across t_0.

In this case, the two initial conditions needed for the analysis for $t > 0$ are $x(t_0^+) = x(t_0^-)$ and $\left. \dfrac{dx}{dt} \right|_{t_0^+} = \left. \dfrac{dx}{dt} \right|_{t_0^-}$.

If the highest-order discontinuity on $\dfrac{d^m \hat{u}_k}{dt^m}$ at t_0 is a Dirac impulse $\delta(t - t_0)$ with coefficient b, then

- $\dfrac{d^2 x}{dt^2}$ must exhibit a δ-discontinuity at t_0 as well;
- $\dfrac{dx}{dt}$ has a corresponding step discontinuity at t_0;
- $x(t)$ is continuous across t_0.

Therefore, the first initial condition needed for the analysis for $t > 0$ is $x(t_0^+) = x(t_0^-)$. We can find the second one, $\left(\left. \dfrac{dx}{dt} \right|_{t_0^+} \right)$, by integrating the I/O relationship between t_0^- and t_0^+:

$$
\underbrace{\int_{t_0^-}^{t_0^+} a_2 \frac{d^2 x}{dt^2}\, dt}_{= a_2 \left(\left. \frac{dx}{dt} \right|_{t_0^+} - \left. \frac{dx}{dt} \right|_{t_0^-} \right)} + \underbrace{\int_{t_0^-}^{t_0^+} a_1 \frac{dx}{dt}\, dt}_{a_1 [x(t_0^+) - x(t_0^-)] = 0} + \underbrace{\int_{t_0^-}^{t_0^+} a_0 x\, dt}_{= 0} = \underbrace{\sum_{i=0}^{m} \sum_{k=1}^{P} \int_{t_0^-}^{t_0^+} b_{ik} \frac{d^i u_k}{dt^i}\, dt}_{= b \int_{t_0^-}^{t_0^+} \delta(t - t_0)\, dt = b},
$$

that is,

$$
\left. \frac{dx}{dt} \right|_{t_0^+} = \frac{b}{a_2} + \left. \frac{dx}{dt} \right|_{t_0^-}.
$$

In other words, the height of the step jump exhibited by $\dfrac{dx}{dt}$ across t_0 is $\dfrac{b}{a_2}$.

Case Study 1

In the circuit shown in Fig. 11.18, the input sources are $e_1(t) = E_0 + E_2 \cos(2\omega t)u(-t)$, $e_2(t) = E_1 \sin(\omega t)u(-t) + 2E_0 u(t)$ and $a(t) = Q_0 \delta(t)$. By assuming that the conditions in 0^- are known, find the conditions in 0^+ needed to determine $v_1(t)$ and $v_2(t)$ for $t > 0$.

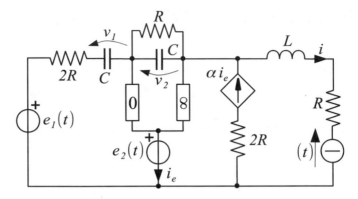

Fig. 11.18 Case Study 1

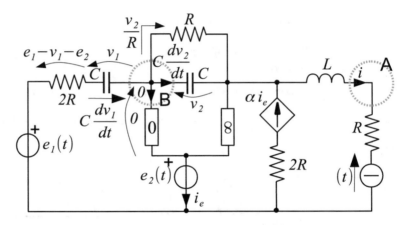

Fig. 11.19 Solution of Case Study 1

By inspecting the circuit, we see that we have three WSVs, and thus $N = 3$. But there is also an algebraic relationship $i = -a(t)$ (due to the cut-set A shown in Fig. 11.19), and thus $M = 1$ and the circuit has $N - M = 2$ SSVs.

Following Fig. 11.19, we immediately obtain the I/O relationship for v_1:

$$2RC\frac{dv_1}{dt} + v_1 = e_1 - e_2. \tag{11.44}$$

Given that the highest-order discontinuity on the input side at $t = 0$ is a step, we conclude that $v_1(0^+) = v_1(0^-)$, which is the only initial condition needed to determine v_1 for $t > 0$.

Now we look for the I/O relationship for v_2. We start from the KCL for cut-set B in Fig. 11.19:

$$C\frac{dv_1}{dt} = C\frac{dv_2}{dt} + \frac{v_2}{R},$$

that is,

$$\frac{dv_1}{dt} = \frac{dv_2}{dt} + \frac{v_2}{RC}. \tag{11.45}$$

By substituting this expression into Eq. 11.44, we obtain

$$2RC\frac{dv_2}{dt} + 2v_2 = e_1 - v_1 - e_2. \tag{11.46}$$

Now we derive Eq. 11.46 with respect to time:

$$2RC\frac{d^2v_2}{dt^2} + 2\frac{dv_2}{dt} = \frac{de_1}{dt} - \frac{dv_1}{dt} - \frac{de_2}{dt}. \tag{11.47}$$

Finally, we substitute Eq. 11.45 into Eq. 11.47, thus obtaining the I/O relationship for v_2,

$$2RC\frac{d^2v_2}{dt^2} + 3\frac{dv_2}{dt} + \frac{v_2}{RC} = \frac{de_1}{dt} - \frac{de_2}{dt}. \tag{11.48}$$

Given that $n(=2) > m(=1)$, v_2 is "more continuous" than the inputs, and therefore it is continuous across $t = 0$, and we conclude that $v_2(0^+) = v_2(0^-)$. Moreover, we obtain $\left.\dfrac{dv_2}{dt}\right|_{0^+}$ by integrating Eq. 11.48 between 0^- and 0^+:

$$2RC\left(\left.\frac{dv_2}{dt}\right|_{0^+} - \left.\frac{dv_2}{dt}\right|_{0^-}\right) + 0 + 0 =$$

$$= [\underbrace{e_1(0^+)}_{E_0} - \underbrace{e_1(0^-)}_{E_0 + E_2}] - [\underbrace{e_2(0^+)}_{2E_0} - \underbrace{e_2(0^-)}_{0}] = -E_2 - 2E_0.$$

Therefore,

$$\left.\frac{dv_2}{dt}\right|_{0^+} = -\frac{E_2 + 2E_0}{2RC} + \left.\frac{dv_2}{dt}\right|_{0^-}. \tag{11.49}$$

Case Study 2
Suppose that a circuit containing coupled inductors has I/O relationship

$$\frac{L_1 L_2 - M^2}{L_2} \frac{di_1}{dt} + R i_1 = e_1 + \frac{M}{L_2} e_2, \tag{11.50}$$

with $e_1(t) = \Phi_0 \delta(t) + E_1 \sin(\omega t) u(t)$ and $e_2(t) = E_0 u(-t) + E_2 \cos(2\omega t) u(t)$.
Find the order of discontinuity of the state variable i_1 in the general case and in the case of closely coupled inductors.

Given that the highest-order discontinuity on the input side at $t = 0$ is an impulse, we conclude that in the general case, i_1 exhibits a step discontinuity (of amplitude $\dfrac{L_2 \Phi_0}{L_1 L_2 - M^2}$) across $t = 0$.

In the case of closely coupled inductors, the coefficient of the term $\dfrac{di_1}{dt}$ becomes null and i_1 is no longer an SSV, because it turns out to be algebraically related to the inputs. Therefore, it has the same order of discontinuity (δ-discontinuity) as the most discontinuous input. In this case, the area of the current impulse is $\dfrac{\Phi_0}{R}$.

11.5   Solution of the State Equations: Free Response and Forced Response

It would be possible to express the general solution of higher-order circuits in terms of matrix functions [3].

However, here we consider a second way to find the (unique) solution $x(t)$ to Eq. 11.43 with n initial conditions, based on the superposition principle. In this case, as for the first-order circuits, the solution is found by summing different contributions as follows:

$$x(t) = x_{fr}(t) + x_{fo}(t) = x_{fr}(t) + \sum_{k=1}^{P} \hat{x}_k(t), \tag{11.51}$$

where $x_{fr}(t)$ is the **free response**, $x_{fo}(t)$ is the **forced response**, and $\hat{x}_k(t)$ is a **particular integral** due to the kth input.

11.5.1   *Free Response*

The **free response**, or **transient response**, $x_{fr}(t)$ is the solution of the homogeneous differential equation $\mathcal{L}(x) = 0$ associated with Eq. 11.43 such that the initial conditions hold.

If the homogeneous differential equation $\mathcal{L}(x) = 0$ (of order n) has q distinct eigenvalues λ_j, the free response has the following structure:

$$x_{fr}(t) = \sum_{j=1}^{q} \left[e^{\lambda_j t} \sum_{k=1}^{\mu(\lambda_j)} c_{jk} \, t^{k-1} \right], \tag{11.52}$$

where $\mu(\lambda_j)$ is the multiplicity of the eigenvalue λ_j, with $\sum_{j=1}^{q} \mu(\lambda_j) = n$, and the coefficients c_{jk} are complex numbers in general.

The terms λ_j are **natural frequencies** of the circuit and contain information about the way in which a state variable naturally behaves starting from a given initial condition.

The jth addend between square brackets in Eq. 11.52 is called the *natural mode* of the circuit.

The natural frequencies are solutions of the *characteristic equation* (or *auxiliary equation*) associated with Eq. 11.43, which can be directly obtained from the homogeneous differential equation by replacing each term $\dfrac{d^j x}{dt^j}$ with λ^j and $x(= \dfrac{d^0 x}{dt^0})$ with $\lambda^0 = 1$.

For instance, the I/O relationship found in Case Study 1 in Sect. 11.3.3 (where there are no inputs) is

$$(RC)^2 \frac{d^2 v_1}{dt^2} + (3 - \alpha)RC \frac{dv_1}{dt} + (1 - \alpha)v_1 = 0.$$

Its characteristic equation is

$$(RC)^2 \lambda^2 + (3 - \alpha)RC\lambda + (1 - \alpha) = 0.$$

By defining $T^2 = (RC)^2$, $\tau = (3 - \alpha)RC$, and $\beta = (1 - \alpha)$, we obtain

$$T^2 \lambda^2 + \tau \lambda + \beta = 0,$$

whose solutions are

$$
\begin{aligned}
\lambda_{1,2} &= \frac{-\tau \pm \sqrt{\tau^2 - 4\beta T^2}}{2T^2} = \\
&= \frac{-(3-\alpha)RC \pm \sqrt{(3-\alpha)^2(RC)^2 - 4(1-\alpha)(RC)^2}}{2(RC)^2} = \\
&= \frac{\alpha - 3 \pm \sqrt{(3-\alpha)^2 - 4(1-\alpha)}}{2RC} = \frac{\alpha - 3 \pm \sqrt{(\alpha-1)^2 + 4}}{2RC},
\end{aligned}
$$

which are real for all parameter values.

> For a given circuit, the complete set of natural frequencies is the set of eigen-
> values of the state matrix A.

We remark once more that the natural frequencies are an intrinsic property of the circuit, independent of its inputs, and as such, they can be determined by turning off all of the P independent sources. This is reflected by the fact that they are the eigenvalues of A (see Eq. 11.37). Moreover, they determine the circuit's stability properties, as shown in the next section.

The eigenvalues of the homogeneous equation $\mathcal{L}(x) = 0$ for a specific state variable are in general a subset of the circuit's natural frequencies. A state variable can be insensitive to one or more natural frequencies. This corresponds to the fact that as shown in the case studies examined up to now, in a circuit of order n, the I/O relationship of every state variable has *at most* order n, but in general can have a lower order.

For instance, in Case Study 1 analyzed in Sect. 11.4, matrix A can be directly obtained from Eqs. 11.44 and 11.46:

$$
A = -\frac{1}{2RC} \begin{pmatrix} 1 & 0 \\ 1 & 2 \end{pmatrix}.
$$

This is a lower triangular matrix, whose eigenvalues are the diagonal elements:
$\lambda_1 = -\dfrac{1}{2RC}$ and $\lambda_2 = -\dfrac{1}{RC}$.

If we analyze the I/O relationships for the two state variables, we find that the characteristic equation for v_1 is

$$
2RC\lambda + 1 = 0.
$$

This means that v_1 is sensitive to λ_1 *only*. In contrast, the characteristic equation for v_2 is

$$
2RC\lambda^2 + 3\lambda + \frac{1}{RC} = 0,
$$

that is,

$$2(RC)^2\lambda^2 + 3RC\lambda + 1 = 0,$$

whose solutions are

$$\lambda_{1,2} = \frac{-3RC \pm \sqrt{(3RC)^2 - 8(RC)^2}}{4(RC)^2}.$$

Therefore, v_2 is sensitive to *both* circuit natural frequencies $\lambda_1 = -\dfrac{1}{2RC}$ and $\lambda_2 = -\dfrac{1}{RC}$.

We remark that each natural mode in Eq. 11.52 is basically governed by the exponential term $e^{\lambda_j t} = e^{\Re\{\lambda_j\}t}[\cos(\Im\{\lambda_j\}t) + j\sin(\Im\{\lambda_j\}t)]$. Therefore, the jth mode converges to zero if and only if $\Re\{\lambda_j\} < 0$.

The reciprocals of the absolute values of the real parts of the natural frequencies have the physical dimension of time and can be considered the **time constants** τ_j of the jth natural mode: $\tau_j = \dfrac{1}{|\Re\{\lambda_j\}|}$.

For instance, in the above example, $\tau_1 = 2RC$ and $\tau_2 = RC$. Moreover, the two natural frequencies are simple, since their multiplicity is one: $\mu(\lambda_1) = \mu(\lambda_2) = 1$. This means, according to Eq. 11.52, that the free response can be written as

$$x_{fr}(t) = K_1 e^{\lambda_1 t} + K_2 e^{\lambda_2 t}, \tag{11.53}$$

that is, the sum of two natural modes, one ($K_2 e^{\lambda_2 t}$, corresponding to the smaller time constant, τ_2) going to zero faster, and one ($K_1 e^{\lambda_1 t}$, corresponding to the larger time constant, τ_1) going to zero more slowly.

In general, we can also have complex conjugate natural frequencies, which correspond to oscillating natural modes, as shown in the following Case Study 1. The same example also points out that even if a solution involves complex numbers, we must finally obtain a *real function of time*, as expected for any circuit variable.

Case Study 1

Consider the circuit shown in Fig. 11.20, where the two-port is described by the conductance matrix $G = \dfrac{1}{R}\begin{pmatrix} 1 & \alpha \\ -\alpha & 0 \end{pmatrix}$, with $R, \alpha > 0$, and find:

Fig. 11.20 Case Study 1

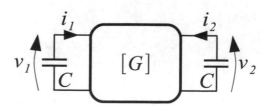

1. *the I/O relationship for $v_1(t)$;*
2. *the condition on circuit parameters that ensures two complex conjugate natural frequencies;*
3. *the state variable v_1 for $t > 0$, by assuming that the above condition is satisfied and that the initial conditions at $t_0 = 0^+$ are known.*

1. By combining the two-port equations and the capacitor descriptive equation, we obtain

$$\begin{cases} -C\dfrac{dv_1}{dt} = \dfrac{v_1}{R} + \dfrac{\alpha v_2}{R} \\ -C\dfrac{dv_2}{dt} = -\dfrac{\alpha v_1}{R}, \end{cases}$$

which corresponds to the state equation $\dot{x} = Ax$ with $x = \begin{pmatrix} v_1 \\ v_2 \end{pmatrix}$ and $A = -\dfrac{1}{RC}\begin{pmatrix} 1 & \alpha \\ -\alpha & 0 \end{pmatrix}$.

Now we derive the first equation with respect to time and substitute the second equation, thus obtaining the I/O relationship for v_1:

$$(RC)^2\frac{d^2v_1}{dt^2} + RC\frac{dv_1}{dt} + \alpha^2 v_1 = 0.$$

The corresponding natural frequencies are

$$\lambda_{1,2} = \frac{-RC \pm \sqrt{(RC)^2 - 4(\alpha RC)^2}}{2(RC)^2} = \frac{-1 \pm \sqrt{1 - 4\alpha^2}}{2RC}.$$

We remark that they can be found also as the eigenvalues of A.
2. The condition such that the natural frequencies are complex conjugates is $1 - 4\alpha^2 < 0$, that is, taking α positive according to the problem assumptions, $\alpha > 0.5$.
3. The circuit is autonomous; therefore, the complete solution coincides with the free response, and we have

$$v_1(t) = v_{1fr}(t) = K_1 e^{\lambda_1 t} + K_2 e^{\lambda_2 t} = K_1 e^{\lambda t} + K_2 e^{\lambda^* t},$$

subject to the initial conditions

$$\begin{cases} v_1(0^+) = K_1 + K_2, \\ \dfrac{dv_1}{dt}\bigg|_{0^+} = K_1\lambda + K_2\lambda^*. \end{cases}$$

Inasmuch as the values of the initial conditions $v_1(0^+)$ and $\left.\dfrac{dv_1}{dt}\right|_{0^+}$ are real, we must have

$$\begin{cases} \Im\{K_1 + K_2\} = 0, \\ \Im\{K_1\lambda + K_2\lambda^*\} = 0, \end{cases}$$

which is equivalent to

$$\begin{cases} (K_1 + K_2) - (K_1 + K_2)^* = 0, \\ (K_1\lambda + K_2\lambda^*) - (K_1\lambda + K_2\lambda^*)^* = 0, \end{cases}$$

that is,

$$\begin{cases} (K_1 - K_2^*) - (K_1^* - K_2) = 0, \\ \lambda(K_1 - K_2^*) - \lambda^*(K_1^* - K_2) = 0. \end{cases}$$

By defining $\beta = K_1 - K_2^*$ for the sake of compactness, these conditions can be recast as

$$\begin{cases} \beta - \beta^* = 0, \\ \lambda\beta - \lambda^*\beta^* = 0, \end{cases}$$

which imply

$$\begin{cases} \beta = \beta^*, \\ (\lambda - \lambda^*)\beta = 0. \end{cases}$$

Therefore, $\beta = 0$, that is, $K_1 = K_2^*$.

Now, taking $\lambda = \gamma + j\Omega$, $K_1 = a + jb$, and $e^{\lambda t} = e^{\gamma t}[\cos(\Omega t) + j\sin(\Omega t)] = c + jd$, we have

$$v_1(t) = 2\Re\{K_1 e^{\lambda t}\} = 2\Re\{(a + jb)(c + jd)\} = 2(ac - bd) = \\ = 2e^{\gamma t}[a\cos(\Omega t) - b\sin(\Omega t)].$$

Given that $\gamma < 0$, this is a damped sinusoid.

We remark that as expected, we finally obtain a real function of time.

We can provide a geometric interpretation of this result. We know that $v_1(t)$ is a linear combination of two exponential functions $e^{\lambda t}$ and $e^{\lambda^* t}$. This means that $e^{\lambda t}$ and $e^{\lambda^* t}$ form a basis. A linear combination of the basis elements is still a basis; in this case, we have taken as new basis elements the *real* functions

$$\frac{e^{\lambda t} + e^{\lambda^* t}}{2} = e^{\gamma t}\cos(\Omega t),$$

$$\frac{e^{\lambda t} - e^{\lambda^* t}}{2j} = e^{\gamma t}\sin(\Omega t).$$

Of course, it is possible that a natural frequency is zero. The number of null natural frequencies in a given circuit coincides with the number of algebraic constraints

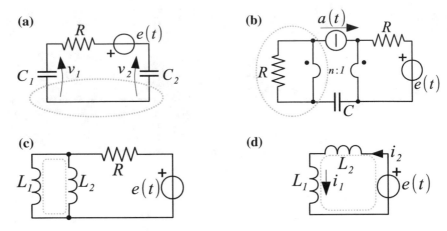

Fig. 11.21 Circuits with null natural frequencies

involving only time derivatives of one or more SSVs and (possibly) inputs, either derived or not. For instance, a cut-set involving only capacitors (as in Fig. 11.21a, where the second natural frequency is $\lambda_2 = -\dfrac{C_1 + C_2}{RC_1C_2}$) or capacitors and current sources (as in Fig. 11.21b) implies the presence of a null natural frequency.

Similarly, a loop involving only inductors (as in Fig. 11.21c, where the second natural frequency is $\lambda_2 = -\dfrac{R(L_1 + L_2)}{L_1L_2}$) or inductors and voltage sources (as in Fig. 11.21d[2]) implies the presence of a null natural frequency.

These circuit structures can be easily identified within a circuit, which allows one to compute the natural frequencies only by inspection, without computing I/O relationships or state equations. For instance, the circuit shown in Fig. 11.22a has three WSVs not related by algebraic constraints and therefore three SSVs and three natural frequencies, two of which (see loop **A** and cut-set **B**) are null. The third one can be easily computed by turning off all independent sources; it is $\lambda_3 = -\dfrac{1}{RC_1}$. Similarly, the circuit shown in Fig. 11.22b has three WSVs, but two algebraic constraints relating them: $e = v_1 + v_2$ (loop **D**) and $i = -a_1$ (cut-set **B**). The only natural frequency is null, due to cut-set **A**.

Remark: We know that the natural frequencies are obtained by turning off every independent source of the circuit. In this condition, the circuit's state equation is $\dot{x} = Ax$. The presence of linearly dependent rows in matrix A implies that $det(A) = 0$ and therefore the presence of a null eigenvalue of A, that is, a null circuit natural frequency.

---

[2]In this circuit there is only one SSV and therefore one natural frequency, due to the inductor cut-set, which determines the algebraic constraint $i_1 = i_2$.

(a) **(b)**

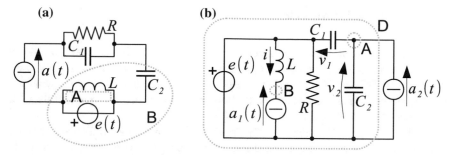

Fig. 11.22 Circuits with easy-to-compute natural frequencies

Fig. 11.23 Case Study 2

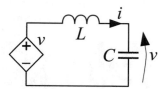

Case Study 2

Consider the circuit shown in Fig. 11.23. By assuming that the initial conditions at $t_0 = 0^+$ are known, find the state variable v for $t > 0$.

Owing to the unitary-gain VCVS, the I/O relationship for v is

$$LC\frac{d^2v}{dt^2} = 0.$$

Therefore, we have a natural frequency $\lambda = 0$ with multiplicity $\mu(\lambda) = 2$. The circuit is autonomous, and so the solution for $t > 0$ coincides with the free response:

$$v(t) = v_{fr}(t) = K_1 + K_2 t,$$

where the coefficients are real (with proper physical dimensions) and can be determined by exploiting the (known) initial conditions:

$$v(0^+) = K_1,$$
$$\frac{dv}{dt}\bigg|_{0^+} = K_2.$$

11.5.2 Forced Response

The forced response can be obtained in the main cases (DC and AC inputs) by applying the superposition principle and the similarity principle, as for first-order circuits. The DC and AC cases do not require any further analysis beyond what was already stated for first-order circuits. Particular attention should be paid, as usual, in the presence of null frequencies.

In the AC case, provided that the frequency of the sinusoidal input does not coincide with a natural frequency, the solution strategy is once more based on the *similarity criterion* already described in Sects. 9.4.2 and 9.5. However, for I/O relationships of order greater than one, it is usually simpler to find the forced response using the phasor-based method described in Sect. 9.5.2.

The following case studies provide some examples.

Case Study 1
Find the forced response $v_{fo}(t)$ for the state variable v in the circuit shown in Fig. 11.24, which works at steady state with $a(t) = A_0 + A_1 \sin(\omega t) + A_2 \cos(2\omega t)$. The 2-port is described by its hybrid matrix $H = \begin{pmatrix} R & 0 \\ \alpha & g \end{pmatrix}$, with $\alpha \neq 0, R > 0$ and $g > 0$.

You can check that the I/O relationship for v is

$$LC\frac{d^2v}{dt^2} + (Lg + RC)\frac{dv}{dt} + Rgv = -\alpha L\frac{da}{dt}.$$

The natural frequencies are $\lambda_1 = -\dfrac{R}{L}$ and $\lambda_2 = -\dfrac{g}{C}$, which are real and negative, owing to the problem assumptions. Therefore, the circuit is absolutely stable, which is consistent with the steady-state hypothesis.

By defining $\tau = Lg + RC$ for the sake of compactness, we obtain

$$LC\frac{d^2v}{dt^2} + \tau\frac{dv}{dt} + Rgv = -\alpha L\frac{da}{dt}. \tag{11.54}$$

Fig. 11.24 Case Study 1

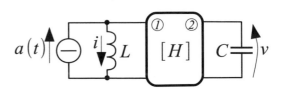

We have three different functions in the input: a constant term corresponding to a DC term v_{DC} in the forced response, a sinusoidal term with angular frequency ω corresponding to an AC term $v_{AC1}(t)$ in the forced response, and a second sinusoidal term with angular frequency 2ω corresponding to another AC term $v_{AC2}(t)$ in the forced response. Therefore, we have to apply the superposition principle, consider one term at time, and sum the three corresponding solutions: $v_{fo}(t) = v_{DC} + v_{AC1}(t) + v_{AC2}(t)$.

DC Term

Given that Eq. 11.54 contains only $\dfrac{da}{dt}$ on the input side, the DC part of the input does not provide any contribution to $v_{fo}(t)$: $v_{DC} = 0$.

You can check that the I/O relationship for the second state variable $i(t)$ is

$$L\frac{di}{dt} + Ri = Ra$$

and that the DC part of the forced response $i_{fo}(t)$ is $i_{DC} = A_0$.

AC1 Term

For this term of the input, the angular frequency is ω; therefore, Eq. 11.54 in the phasor domain becomes

$$LC(j\omega)^2\dot{V} + \tau(j\omega)\dot{V} + Rg\dot{V} = -\alpha L(j\omega)(-jA_1),$$

that is,

$$(Rg - \omega^2 LC + j\omega\tau)\dot{V} = -\alpha L\omega A_1.$$

For the sake of compactness, we define the dimensionless coefficient $\beta = Rg - \omega^2 LC$; thus the above equation can be recast as

$$(\beta + j\omega\tau)\dot{V} = -\alpha\omega L A_1.$$

We algebraically obtain $\dot{V}$:

$$\dot{V} = \frac{-\alpha\omega L A_1}{\beta + j\omega\tau} = \frac{-\alpha\omega L A_1}{\beta^2 + (\omega\tau)^2}(\beta - j\omega\tau).$$

We apply Eq. 9.38 to obtain $v_{AC1}(t)$:

$$v_{AC1}(t) = \Re\{\dot{V}e^{j\omega t}\} = \frac{-\alpha\omega L A_1}{\beta^2 + (\omega\tau)^2}\Re\{(\beta - j\omega\tau)[\cos(\omega t) + j\sin(\omega t)]\} =$$

$$= \frac{-\alpha\omega L A_1}{\beta^2 + (\omega\tau)^2}[\beta\cos(\omega t) + \omega\tau\sin(\omega t)].$$

AC2 Term

For this term of the input, the angular frequency is 2ω; therefore, Eq. 11.54 in the phasor domain becomes

$$LC(j2\omega)^2\dot{V} + \tau(j2\omega)\dot{V} + Rg\dot{V} = -\alpha L(j2\omega)(-jA_2),$$

that is,

$$(Rg - 4\omega^2 LC + j2\omega\tau)\dot{V} = -\alpha L2\omega A_2.$$

For the sake of compactness, we define the dimensionless coefficient $\gamma = Rg - 4\omega^2 LC$; thus the above equation can be recast as

$$(\gamma + j2\omega\tau)\dot{V} = -2\alpha\omega LA_2.$$

We algebraically obtain $\dot{V}$:

$$\dot{V} = \frac{-2\alpha\omega LA_2}{\gamma + j2\omega\tau} = \frac{-2\alpha\omega LA_2}{\gamma^2 + (2\omega\tau)^2}(\gamma - j2\omega\tau).$$

We apply Eq. 9.38 to obtain $v_{AC2}(t)$:

$$v_{AC2}(t) = \Re\{\dot{V}e^{j2\omega t}\} = \frac{-2\alpha\omega LA_2}{\gamma^2 + (2\omega\tau)^2}\Re\{(\gamma - j2\omega\tau)[\cos(2\omega t) + j\sin(2\omega t)]\} =$$
$$= \frac{-2\alpha\omega LA_2}{\gamma^2 + (2\omega\tau)^2}[\gamma\cos(2\omega t) + 2\omega\tau\sin(2\omega t)].$$

Case Study 2

Assume that the I/O relationship in the phasor domain for a circuit working in AC steady state with a sinusoidal input $e(t)$ with angular frequency ω is

$$(1 - \omega^2 LC + j\omega RC)\dot{V} = (K + j\omega\tau)\dot{E}$$

with $R, L, C > 0$ and find:
1. *the corresponding I/O relationship in the time domain;*
2. *the natural frequencies;*
3. *the absolute stability condition;*
4. *the condition for having real natural frequencies.*

1. We have to apply the inverse transform, that is, replace $j\omega$ with $\dfrac{d}{dt}$ and $-\omega^2$ with $\dfrac{d^2}{dt^2}$. By doing so, we obtain the I/O relationship in the time domain:

$$LC\frac{d^2v}{dt^2} + RC\frac{dv}{dt} + v = Ke(t) + \tau\frac{de}{dt}.$$

2. The natural frequencies are found by solving the characteristic equation

$$LC\lambda^2 + RC\lambda + 1 = 0,$$

whose solutions are

$$\lambda_\pm = \frac{-RC \pm \sqrt{(RC)^2 - 4LC}}{2LC}.$$

3. The absolute stability condition is $\Re\{\lambda_\pm\} < 0$. Given that $LC > 0$, we have $(RC)^2 - 4LC < (RC)^2$. This implies that if $(RC)^2 \geq 4LC$ (i.e., for real natural frequencies), both λ_+ and λ_- are negative. Otherwise, if $(RC)^2 < 4LC$ (i.e., for complex conjugate natural frequencies), $\Re\{\lambda_\pm\} = \dfrac{-RC}{2LC}$. In all cases, owing to the problem assumptions, the absolute stability condition is fulfilled.

4. The condition for having real natural frequencies is $(RC)^2 \geq 4LC$.

Case Study 3
Find the forced response $v_{fo}(t)$ for the state variable v in the circuit shown in Fig. 11.25, which works at steady state, with $e(t) = E$, $a(t) = A_0 + A\cos(\omega t)$, and $\alpha \neq -1$. Find also the stability condition and assume that it is fulfilled.

Fig. 11.25 Case Study 3

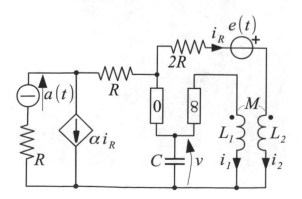

You can check that there is an algebraic constraint $i_2 = \dfrac{a}{\alpha + 1}$ (thus the circuit has only two SSVs, i_1 and v), and the I/O relationship for v is

$$(\alpha + 1)MC\frac{d^2v}{dt^2} + (\alpha + 1)v = -(\alpha + 1)e + 2Ra + L_2\frac{da}{dt}.$$

The natural frequencies are $\lambda_\pm = \pm\sqrt{\dfrac{-1}{MC}}$. In order to ensure simple stability, they must be complex conjugates: if they were real, they would be one positive and one negative, and therefore we would have strong instability. To this end, we impose $M > 0$, by assuming that $C > 0$, as usual.

By defining $\gamma = (\alpha + 1)MC$ and $\beta = \alpha + 1$ for the sake of compactness, we obtain

$$\gamma\frac{d^2v}{dt^2} + \beta v = -\beta e + 2Ra + L_2\frac{da}{dt}. \tag{11.55}$$

To find the forced response, we apply the superposition principle.

DC Terms

We consider only the constant inputs in Eq. 11.55. Therefore, the differential equation to be considered is

$$\gamma\frac{d^2v_{DC}}{dt^2} + \beta v_{DC} = -\beta E + 2RA_0,$$

and the DC part of the forced response $v_{fo}(t)$ is $v_{DC} = -E + \dfrac{2RA_0}{\beta}$.

AC Term

We consider only the sinusoidal inputs in Eq. 11.55. Therefore, the differential equation to be considered is

$$\gamma\frac{d^2v_{AC}}{dt^2} + \beta v_{AC} = 2RA\cos(\omega t) + L_2\frac{d}{dt}[A\cos(\omega t)].$$

In the phasor domain, the above equation becomes

$$\gamma(j\omega)^2\dot{V} + \beta\dot{V} = 2RA + L_2(j\omega)A,$$

that is,

$$(\beta - \omega^2\gamma)\dot{V} = (2R + j\omega L_2)A.$$

We algebraically obtain $\dot{V}$:

$$\dot{V} = \frac{A}{\beta - \omega^2\gamma}(2R + j\omega L_2).$$

We apply Eq. 9.38 to find $v_{AC}(t)$:

$$v_{AC}(t) = \Re\{\dot{V}e^{j\omega t}\} = \frac{A}{\beta - \omega^2\gamma}\Re\{(2R + j\omega L_2)[\cos(\omega t) + j\sin(\omega t)]\} =$$

$$= \frac{A}{\beta - \omega^2\gamma}[2R\cos(\omega t) - \omega L_2\sin(\omega t)].$$

Therefore, the complete forced response is $v_{fo}(t) = v_{DC} + v_{AC}(t)$.

Case Study 4
Given the I/O relationship

$$T^2\frac{d^2v}{dt^2} + \tau\frac{dv}{dt} + \alpha v = L\frac{da}{dt} + Ra(t),$$

find the phasor $\dot{V}$. *Then, assuming* $L = 0$, *find the input current* $a(t)$ *such that the forced response is* $v(t) = 10\cos(200t + \frac{\pi}{2})$.

Transforming the I/O relationship into the phasor domain, we obtain

$$\dot{V} = \frac{R + j\omega L}{\alpha - \omega^2 T^2 + j\omega\tau}\dot{A}.$$

Given that $v(t) = 10\cos(200t + \frac{\pi}{2})$, its phasor is $\dot{V} = 10e^{j\frac{\pi}{2}} = 10j$ and its angular frequency is $\omega = 200\,\text{rad/s}$.
From the expression for $\dot{V}$ with $L = 0$, we obtain

$$\dot{A} = \frac{\alpha - \omega^2 T^2 + j\omega\tau}{R}\dot{V} = \frac{10}{R}[-\omega\tau + j(\alpha - \omega^2 T^2)].$$

Therefore,

$$a(t) = \Re\{\dot{A}e^{j\omega t}\} = -\frac{10}{R}[\omega\tau\cos(\omega t) + (\alpha - \omega^2 T^2)\sin(\omega t)]$$

with $\omega = 200\,\text{rad/s}$.

As an alternative, one could reason from the beginning of the analysis in the phasor domain using the transformed descriptive equations (Table 11.1):

Table 11.1 Descriptive equations in phasor domain

Component	Time domain	Phasor domain
Resistor	$v = Ri$	$\dot{V} = R\dot{I}$
Capacitor	$i = C\dfrac{dv}{dt}$	$\dot{I} = j\omega C \dot{V}$
Inductor	$v = L\dfrac{di}{dt}$	$\dot{V} = j\omega L \dot{I}$
Coupled inductors	$\begin{cases} v_1 = L_1\dfrac{di_1}{dt} + M\dfrac{di_2}{dt} \\[2ex] v_2 = M\dfrac{di_1}{dt} + L_2\dfrac{di_2}{dt} \end{cases}$	$\begin{cases} \dot{V}_1 = j\omega L_1\dot{I}_1 + j\omega M\dot{I}_2 \\[2ex] \dot{V}_2 = j\omega M\dot{I}_1 + j\omega L_2\dot{I}_2 \end{cases}$

Fig. 11.26 Case Study 5

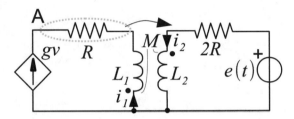

Kirchhoff's equations can be easily transformed in the phasor domain; then we can analyze a circuit in AC steady state working directly in this domain.

Case Study 5

Find the forced response $i_{2fo}(t)$ for the state variable i_2 in the circuit shown in Fig. 11.26, which works at steady state with $e(t) = E\cos(\omega t)$. Find also the stability condition under the assumption of closely coupled inductors.

We first follow the standard sequence of steps.

From KCL for cut-set A, we obtain $i_1 = -gv$. From KVL for the rightmost mesh, we have

$$e = 2Ri_2 + M\frac{di_1}{dt} + L_2\frac{di_2}{dt}, \tag{11.56}$$

and from KVL for the central mesh,

$$L_1\frac{di_1}{dt} + M\frac{di_2}{dt} + M\frac{di_1}{dt} + L_2\frac{di_2}{dt} = v = -\frac{i_1}{g}. \tag{11.57}$$

Equation 11.56 can be recast as

$$\frac{di_1}{dt} = \frac{e}{M} - \frac{2R}{M}i_2 - \frac{L_2}{M}\frac{di_2}{dt}. \tag{11.58}$$

By substituting in Eq. 11.57, we obtain

$$\frac{L_1 + M}{M}e - \frac{L_1 + M}{M}2Ri_2 - \frac{L_1L_2 - M^2}{M}\frac{di_2}{dt} + \frac{i_1}{g} = 0.$$

Now we derive this equation with respect to time and multiply it by M:

$$(L_1 + M)\frac{de}{dt} - (L_1 + M)2R\frac{di_2}{dt} - (L_1L_2 - M^2)\frac{d^2i_2}{dt^2} + \frac{M}{g}\frac{di_1}{dt} = 0. \tag{11.59}$$

By substituting Eq. 11.58 into Eq. 11.59, we obtain, after a few algebraic steps, the I/O relationship

$$g(L_1L_2 - M^2)\frac{d^2i_2}{dt^2} + [2Rg(L_1 + M) + L_2]\frac{di_2}{dt} + 2Ri_2 = e + g(L_1 + M)\frac{de}{dt}.$$

By defining $D = g(L_1L_2 - M^2)\left[\frac{Vs^2}{A}\right]$, $L = 2Rg(L_1 + M) + L_2\left[\frac{Vs}{A}\right]$, and $\tau = g(L_1 + M)$ [s] for the sake of compactness, we obtain

$$D\frac{d^2i_2}{dt^2} + L\frac{di_2}{dt} + 2Ri_2 = e + \tau\frac{de}{dt}. \tag{11.60}$$

For closely coupled inductors, we have $D = 0$, which reflects the fact that one of the two currents i_1 and i_2 is no longer an SSV. The natural frequency in this condition is $\lambda = -\frac{2R}{L}$, and the absolute stability condition is $L > 0$, that is, $2Rg(L_1 + M) + L_2 > 0$.

Equation 11.60 in the phasor domain becomes

$$(-\omega^2 D + j\omega L + 2R)\dot{I}_2 = (1 + j\omega\tau)E.$$

We algebraically obtain $\dot{I}_2$:

$$\dot{I}_2 = \frac{E}{(2R - \omega^2 D)^2 + (\omega L)^2}(2R - \omega^2 D - j\omega L)(1 + j\omega\tau). \tag{11.61}$$

We apply Eq. 9.38 to obtain $i_2(t)$:

$$i_2(t) = \frac{E}{(2R - \omega^2 D)^2 + (\omega L)^2}[a\cos(\omega t) + b\sin(\omega t)]$$

with $a = 2R - \omega^2 D + \omega^2 \tau L$ and $b = \omega[L - \tau(2R - \omega^2 D)]$.

An alternative way of solving the problem consists in directly finding $\dot{I}_2$ from the circuit. The phasor form of Eqs. 11.56 and 11.57 is

$$E = 2R\dot{I}_2 + j\omega M \dot{I}_1 + j\omega L_2 \dot{I}_2$$

$$j\omega L_1 \dot{I}_1 + j\omega M \dot{I}_2 + j\omega M \dot{I}_1 + j\omega L_2 \dot{I}_2 = -\frac{\dot{I}_1}{g}.$$

By finding $\dot{I}_2$ algebraically, you can check that we directly obtain Eq. 11.61.

11.5.2.1 Beyond the Similarity Criterion

In the case studies considered above, the particular integral for the I/O relationship was obtained through the similarity criterion. To go beyond this criterion, we refer to an input

$$\hat{u}(t) = U e^{\bar{\lambda} t} u(t) \quad \text{with} \ \ \bar{\lambda} \in \mathbb{R} \tag{11.62}$$

or to an input

$$\hat{u}(t) = U \cos(\bar{\omega} t + \varphi) u(t) = \Re\left\{\dot{U} e^{\bar{\lambda} t}\right\} u(t) \quad \text{with} \ \ \bar{\lambda} = j\bar{\omega}; \ \ \dot{U} = U e^{j\varphi}. \tag{11.63}$$

In both cases, the I/O relationship is written as $\mathcal{L}(x) = \hat{u}(t)$, where (see Eq. 11.43)

$$\mathcal{L} = a_n \frac{d^n}{dt^n} + \cdots + a_1 \frac{d}{dt} + a_0.$$

Let $\Lambda = \{\lambda_j\}$ be the set of natural frequencies, which are the solutions of the characteristic equation

$$\underbrace{a_n \lambda^n + \cdots + a_1 \lambda + a_0}_{\doteq\, p(\lambda)} = 0.$$

As stated in Sect. 11.5.1, each root (natural frequency) λ_j has its own multiplicity $\mu(\lambda_j)$.

- When $\bar{\lambda} \notin \Lambda$, the similarity criterion can be applied.

  For the input term defined by Eq. 11.62, the simplest particular integral can be obtained as $x_{fo}(t) = D e^{\bar{\lambda} t}$. The constant D follows by imposing $\mathcal{L}(D e^{\bar{\lambda} t}) = U e^{\bar{\lambda} t}$, which implies

$$Dp(\bar{\lambda}) = U \quad \text{or} \quad D = \frac{U}{p(\bar{\lambda})}$$

  because $p(\bar{\lambda}) \neq 0$ by assumption.

Similarly, for the input term of Eq. 11.63 we set $x_{fo}(t) = \Re\left\{\dot{D}e^{j\bar{\omega}t}\right\}$. Here we have

$$\mathcal{L}\left(\Re\left\{\dot{D}e^{j\bar{\omega}t}\right\}\right) = \Re\left\{\mathcal{L}\left(\dot{D}e^{j\bar{\omega}t}\right)\right\} = \Re\left\{\dot{D}p(j\bar{\omega})e^{j\bar{\omega}t}\right\}.$$

Therefore, the condition $\mathcal{L}\left(\Re\left\{\dot{D}e^{j\bar{\omega}t}\right\}\right) = \Re\left\{\dot{U}e^{j\bar{\omega}t}\right\}$ is satisfied on taking

$$\dot{D}p(j\bar{\omega}) = \dot{U} \quad \text{or} \quad \dot{D} = \frac{\dot{U}}{p(j\bar{\omega})} = \frac{\dot{U}}{p(\bar{\lambda})}.$$

- When $\bar{\lambda} = \lambda_k \in \Lambda$, we have $p(\bar{\lambda}) = p(\lambda_k) = 0$, and the similarity criterion can no longer be applied.

  Denoting by $\mu_k = \mu(\lambda_k)$ the multiplicity of λ_k, the general structure of the particular integral when $\lambda_k \in \mathbb{R}$ is

  $$q(t)e^{\lambda_k t},$$

where $q(t)$ is the simplest polynomial of degree μ_k such that

$$\mathcal{L}\left(q(t)e^{\lambda_k t}\right) = Ue^{\lambda_k t};$$

it can be found by substitution.

For $\lambda_k = j\omega_k$ and $\hat{u}(t) = U\cos(\omega_k t + \varphi)u(t)$, the particular integral is

$$q(t)\cos(\omega_k t + \psi).$$

Similar to the previous case, $q(t)$ is the simplest μ_kth-degree polynomial such that

$$\mathcal{L}\left(q(t)\cos(\omega_k t + \psi)\right) = U\cos(\omega_k t + \varphi).$$

Both $q(t)$ and ψ are found by substitution.

As a first example, consider the circuit of Fig. 11.27a. The capacitor is initially uncharged, and the input source is $a(t) = Ae^{-\frac{t}{RC}}u(t)$. The I/O relationship for $v(t)$ is

$$\underbrace{RC\frac{dv}{dt} + v}_{\mathcal{L}(v)} = RAe^{-\frac{t}{RC}}u(t).$$

In this case, $\bar{\lambda} = -\dfrac{1}{RC}$, which is equal to the natural frequency λ of the circuit. This is the only root of the characteristic equation, and it has multiplicity $\mu(\lambda) = 1$. The similarity criterion cannot be applied, and we therefore assume as a particular integral

$$v_p(t) = (Ht + F)e^{-\frac{t}{RC}}.$$

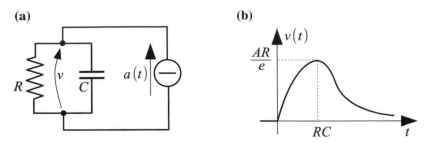

Fig. 11.27 a Circuit with $\lambda = -\dfrac{1}{RC}$; **b** plot of the zero-state response $v(t)$ when $a(t) = Ae^{\lambda t}u(t)$

By imposing $\mathcal{L}(v_p) = RAe^{-\frac{t}{RC}}$, we obtain

$$RC\left(H - \frac{Ht}{RC} - \frac{F}{RC}\right)e^{-\frac{t}{RC}} + (Ht + F)\,e^{-\frac{t}{RC}} = RAe^{-\frac{t}{RC}}.$$

Notice that the terms containing F cancel out. After obvious simplifications, we obtain $H = \dfrac{A}{C}$. Therefore, $v_p(t) = \left(\dfrac{A}{C}t + F\right)e^{-\frac{t}{RC}}$ for any value of F. The most obvious choice for F corresponds to the *simplest* particular integral, that is, $F = 0$. Therefore, we choose

$$v_p(t) = \frac{A}{C}te^{-\frac{t}{RC}}.$$

The solution of the associate homogeneous equation $\mathcal{L}(v_h) = 0$ can be written as

$$v_h(t) = Ke^{-\frac{t}{RC}};$$

then the general expression for $v(t)$ with $t \geq 0$ is

$$v(t) = \left(K + \frac{A}{C}t\right)e^{-\frac{t}{RC}}.$$

The initial condition $v(0) = 0$ implies $K = 0$, and thus the final result for $t \geq 0$ is

$$v(t) = \frac{A}{C}te^{-\frac{t}{RC}},$$

which corresponds to the curve plotted in Fig. 11.27b.

It is instructive to obtain the above expression for $v(t)$ as the result of a limit process applied to a case for which the similarity criterion holds. To this end, we first consider the circuit of Fig. 11.27a, changing its resistance to $R + \delta$ (with $\delta \neq 0$, $|\delta| < R$) and study the corresponding voltage $\hat{v}(t)$; then we consider the limit of this expression when $\delta \to 0$. The I/O relationship is

$$(R + \delta)C\frac{d\hat{v}}{dt} + \hat{v} = (R + \delta)Ae^{-\frac{t}{RC}}.$$

In this case, the natural frequency $-\dfrac{1}{(R + \delta)C}$ does not coincide with $\bar{\lambda} = -\dfrac{1}{RC}$. Therefore, we have a standard forced response term

$$\hat{v}_{fo}(t) = -\frac{AR}{\delta}(R + \delta)e^{-\frac{t}{RC}},$$

and the resulting expression for $\hat{v}(t)$ with $\hat{v}(0) = 0$ is

$$\hat{v}(t) = \frac{AR}{\delta}(R + \delta)\left(e^{-\frac{t}{(R + \delta)C}} - e^{-\frac{t}{RC}}\right).$$

The evaluation of the limit of $\hat{v}(t)$ for $\delta \to 0$ requires applying L'Hospital's rule to the indeterminate form $\dfrac{0}{0}$ contained in the above expression. We have

$$\lim_{\delta \to 0} \hat{v}(t) = AR^2 \lim_{\delta \to 0} \frac{\left(e^{-\frac{t}{(R + \delta)C}} - e^{-\frac{t}{RC}}\right)}{\delta} =$$

$$= AR^2 \lim_{\delta \to 0}\left(\frac{t}{C}\frac{1}{(R + \delta)^2}e^{-\frac{t}{(R + \delta)C}}\right) = \frac{At}{C}e^{-\frac{t}{RC}},$$

as expected. You can easily check that the same result could be achieved using the Taylor expansion of $e^{-\frac{t}{(R + \delta)C}}$ for $\delta \to 0$.

As a second example, consider the LC circuit shown in Fig. 11.28a. Both L and C are initially uncharged, and the input source is $e(t) = E\sin(\omega_0 t)u(t)$ with $\omega_0 = \dfrac{1}{\sqrt{LC}}$. The I/O relationship for $v(t)$ is

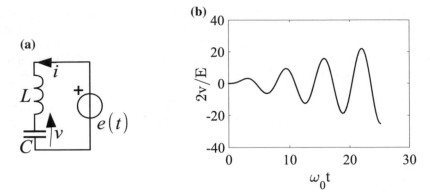

Fig. 11.28 **a** LC circuit; **b** zero-state response $v(t)$ when the input is sinusoidal, with angular frequency $\omega_0 = \dfrac{1}{\sqrt{LC}}$

$$\underbrace{\frac{1}{\omega_0^2}\frac{d^2v}{dt^2} + v}_{\mathcal{L}(v)} = E\sin(\omega_0 t)u(t).$$

Here ω_0 is the angular frequency of the sinusoidal input, but $\pm j\omega_0$ are also roots of the characteristic equation, each with multiplicity 1. Therefore, we assume as particular integral

$$v_p(t) = (At + B)\cos(\omega_0 t + \psi).$$

It is easy to verify that the condition $\mathcal{L}\left(q(t)\cos(\omega_0 t + \psi)\right) = E\sin(\omega_0 t)$, where $q(t) = At + B$, takes the form

$$-\frac{2A}{\omega_0}\sin(\omega_0 t + \psi) = E\sin(\omega_0 t)$$

for every value of B. Then we have $\psi = 0$ and $A = -\dfrac{E}{2}\omega_0$. Setting $B = 0$, we obtain for the simplest particular integral the expression

$$v_p(t) = -\frac{E}{2}\omega_0 t\cos(\omega_0 t).$$

Writing now the free response term as $H\cos(\omega_0 t) + K\sin(\omega_0 t)$, the coefficients H and K can be identified by imposing that the complete solution $v(t)$ satisfy the initial conditions

$$v(0) = 0; \quad \left.\frac{dv}{dt}\right|_0 = 0.$$

These conditions imply that $H = 0$ and $K = \dfrac{E}{2}$. Therefore, the resulting expression for $v(t)$ for $t \geq 0$ is

$$v(t) = \frac{E}{2} \left(\sin(\omega_0 t) - \omega_0 t \cos(\omega_0 t) \right).$$

The amplitude of the second term grows linearly with t. The resulting function $v(t)$ is plotted in Fig. 11.28b. Under these conditions, the circuit is said to be in *resonance*: the increasing oscillations are induced by the voltage source at the resonant frequency ω_0.

This infinite growth in the amplitude of the oscillations must be seen as the result of a limiting physical situation. Any realistic model of a physical circuit containing a voltage source and the LC series shown Fig. 11.28a must include a dissipative term. This term represents the power dissipation present in the real components, and it can be modeled by an equivalent resistor connected in series with L and C. Owing to this resistor, the two natural frequencies of the circuit move from the imaginary axis into the left half of the complex plane, and we have $\lambda_{1,2} = \sigma_0 \pm j\omega_0$ with $\sigma_0 < 0$. In this case, of course, the determination of the particular integral in the $v(t)$ expression satisfies the similarity criterion.

11.6 Circuit Stability

As shown in the previous section, for nth-order circuits with $n > 1$, the natural frequencies can be real or complex conjugates. Moreover, we showed in Case Study 1 of Sect. 11.5.1 that a pair of complex conjugate natural frequencies with unitary multiplicity corresponds, in the solution, to a sinusoid multiplied by an exponential: the imaginary part of the natural frequencies determines the oscillation frequency of the sinusoid, whereas the real part determines the behavior of the exponential.

As already stated for first-order circuits, a real natural frequency with unitary multiplicity corresponds, in the solution, to a single exponential term in the free-response part of the solution.

In any case, the behavior of each natural mode in the general expression of Eq. 11.52 depends primarily on the sign of the real part of λ_j: if $Re\{\lambda_j\} < 0$, the corresponding mode is forced to converge to zero with time due to the decreasing exponential; if $Re\{\lambda_j\} > 0$, the corresponding mode diverges in time due to the increasing exponential.

If $Re\{\lambda_j\} = 0$, we can have either a single natural frequency $\lambda = 0$ or a pair of purely imaginary complex conjugate natural frequencies $\lambda_\pm = \pm j\Omega$. In these cases, the corresponding exponential terms generate either a constant (for $\lambda_j = 0$) or a sinusoidal term (for purely imaginary natural frequencies), and the behavior of the jth natural mode is determined by the multiplicity of λ_j: if $\mu(\lambda_j) > 1$, then the polynomial terms t^{k-1} in Eq. 11.52 indeed determine a divergence of the mode.

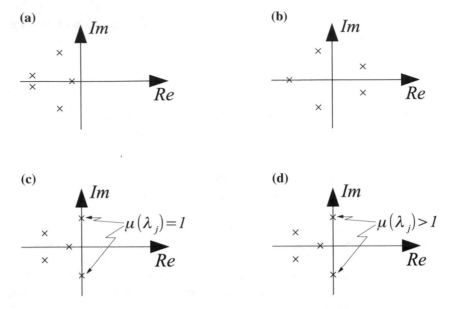

Fig. 11.29 Arrangements of natural frequencies corresponding to different qualitative behaviors:
a absolute stability; **b** strong instability; **c** simple stability; **d** weak instability

Therefore, the circuit's **stability** depends on the position of its natural frequencies
in the complex plane:

- If $Re\{\lambda_j\} < 0$ for any j (as shown in Fig. 11.29a), all natural modes of the free
  response contain a decreasing exponential. This means that after some time (about
  5–10 τ_{max}, where $\tau_{max} = \max_j \{\tau_j\}$), the free response almost vanishes, and the
  circuit is said to be **absolutely stable** or **exponentially stable**.
- If $Re\{\lambda_j\} > 0$ for at least one j (as shown in Fig. 11.29b), there is at least one natural
  mode of the free response containing an increasing exponential. This means that
  the free response diverges, due to the presence of this term. In this case, the circuit
  is **strongly unstable** or **exponentially unstable**.
- If $Re\{\lambda_j\} \leq 0$ for any j and the multiplicity of all the natural frequencies with
  $Re\{\lambda_j\} = 0$ is equal to 1 (as shown in Fig. 11.29c), none of the natural modes
  diverges, even if those of them corresponding to the natural frequencies with
  $Re\{\lambda_j\} = 0$ do not converge to 0. This means that the free response does not
  diverge but does not vanish, and the circuit is said to be **simply stable**.
- If $Re\{\lambda_j\} \leq 0$ for any j and the multiplicity of at least one natural frequency with
  $Re\{\lambda_j\} = 0$ is greater than 1 (as shown in Fig. 11.29d), the corresponding natural
  mode diverges polynomially. This means that the free response diverges, and the
  circuit is said to be **weakly unstable** or **polynomially unstable**.

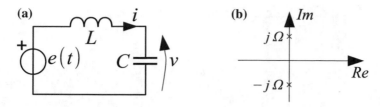

Fig. 11.30 Case Study 1. **a** Circuit to be analyzed and **b** its natural frequencies

Remark 1: For absolutely stable circuits, the natural frequency (or frequencies) closest to the imaginary axis corresponds to the natural mode that lasts the longest, given that it corresponds to the maximum time constant.

Remark 2: Absolute stability requires *dissipation* inside the circuit, that is, the presence, in a proper position, of at least one memoryless passive component. In this way, the energy initially stored in the circuit (by means of the initial values of its state variables) is progressively transformed into heat or other forms of energy. Therefore, as time increases, the energy inside the circuit converges to zero, and the same happens to the values of its state variables.

Case Study 1
Consider the circuit shown in Fig. 11.30a. By assuming that the initial conditions at $t_0 = 0^+$ are known, find:
1. *the I/O relationships for both state variables;*
2. *the state equations in canonical form;*
3. *the circuit's natural frequencies;*
4. *the free response v_{fr} for the state variable v for $t > 0$, by assuming that $e(t) = 0$ for $t > 0$.*

1. The KVL for the unique circuit mesh is

$$L\frac{di}{dt} + v = e(t). \tag{11.64}$$

First, we substitute in Eq. 11.64 the descriptive equation for the capacitor, thus obtaining the I/O relationship for v:

$$LC\frac{d^2v}{dt^2} + v = e(t).$$

Second, we derive Eq. 11.64 with respect to time and substitute in the resulting equation the descriptive equation for the capacitor. Therefore, the I/O relationship for i is

$$LC\frac{d^2i}{dt^2} + i = C\frac{de}{dt}.$$

2. By writing Eq. 11.64 and the descriptive equation for the capacitor in normal form, we obtain the required state equations:

$$\begin{cases} \dfrac{di}{dt} = -\dfrac{v}{L} + \dfrac{e}{L}, \\[2mm] \dfrac{dv}{dt} = \dfrac{i}{C}. \end{cases}$$

In matrix form,

$$\begin{pmatrix} \dot{v} \\ \dot{i} \end{pmatrix} = \underbrace{\begin{pmatrix} 0 & \dfrac{1}{C} \\ -\dfrac{1}{L} & 0 \end{pmatrix}}_{A} \begin{pmatrix} v \\ i \end{pmatrix} + \underbrace{\begin{pmatrix} 0 \\ \dfrac{1}{L} \end{pmatrix}}_{B} e(t).$$

3. The characteristic equation (from any of the I/O relationships found at point 1 or from the state matrix, by imposing $\det(\lambda I - A) = 0$) is

$$LC\lambda^2 + 1 = 0,$$

from which we obtain

$$\lambda_{1,2} = \pm j\frac{1}{\sqrt{LC}} = \pm j\Omega.$$

The two purely imaginary natural frequencies are shown in Fig. 11.30b.
4. The free response is

$$v_{fr}(t) = K_1 e^{j\Omega t} + K_1^* e^{-j\Omega t}$$

with K_1 a complex number whose real and imaginary parts can be determined from the initial conditions.

In the absence of input ($e(t) = 0$), this (autonomous) circuit is also known as a *simple harmonic oscillator*. When $e(t) \neq 0$, the oscillator is *driven* or nonautonomous.

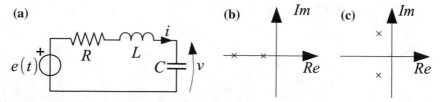

Fig. 11.31 Case Study 2. **a** Circuit to be analyzed; **b** its natural frequencies for $R^2C > 4L$; **c** its natural frequencies for $R^2C < 4L$

Case Study 2

Consider the circuit shown in Fig. 11.31a, which is commonly known as an RLC series circuit. By assuming that the initial conditions at $t_0 = 0^+$ are known, find:

1. *the I/O relationships for both state variables;*
2. *the state equations in canonical form;*
3. *the circuit's natural frequencies.*

   Finally, discuss the circuit's stability with respect to its parameters.

1. The KVL for the unique circuit mesh is

$$L\frac{di}{dt} + Ri + v = e(t). \tag{11.65}$$

First, we substitute in Eq. 11.65 the descriptive equation for the capacitor, thereby obtaining the I/O relationship for v:

$$LC\frac{d^2v}{dt^2} + RC\frac{dv}{dt} + v = e(t).$$

Second, we derive Eq. 11.65 with respect to time and substitute in the resulting equation the descriptive equation for the capacitor. The I/O relationship for i is then seen to be

$$LC\frac{d^2i}{dt^2} + RC\frac{di}{dt} + i = C\frac{de}{dt}.$$

2. By writing Eq. 11.65 and the descriptive equation for the capacitor in normal form, we obtain the required state equations:

$$\begin{cases} \dfrac{di}{dt} = -\dfrac{v}{L} - \dfrac{R}{L}i + \dfrac{e}{L}, \\ \dfrac{dv}{dt} = \dfrac{i}{C}. \end{cases}$$

In matrix form,

$$\begin{pmatrix} \dot{v} \\ \dot{i} \end{pmatrix} = \underbrace{\begin{pmatrix} 0 & \dfrac{1}{C} \\ -\dfrac{1}{L} & -\dfrac{R}{L} \end{pmatrix}}_{A} \begin{pmatrix} v \\ i \end{pmatrix} + \underbrace{\begin{pmatrix} 0 \\ \dfrac{1}{L} \end{pmatrix}}_{B} e(t).$$

3. The characteristic equation (from any of the I/O relationships found at point 1 or by imposing $\det(\lambda I - A) = 0$) is

$$LC\lambda^2 + RC\lambda + 1 = 0,$$

from which we obtain

$$\lambda_{1,2} = \frac{-RC \pm \sqrt{(RC)^2 - 4LC}}{2LC}. \tag{11.66}$$

4. The circuit is always stable, given that it contains only passive components in addition to the independent source. But the natural frequencies can be either real or complex conjugates. The limit condition between these two cases can be easily found by imposing that the argument of the square root in Eq. 11.66 be null:

$$(RC)^2 = 4LC,$$

that is,

$$\zeta = \frac{R}{2}\sqrt{\frac{C}{L}} = 1,$$

where ζ is a dimensionless parameter.

In this case, we have one real and negative natural frequency with multiplicity 2.

If $\zeta > 1$, the natural frequencies are real and negative, as shown in Fig. 11.31b.

If $\zeta < 1$, the natural frequencies are complex conjugates with negative real part, as shown in Fig. 11.31c.

You can check your comprehension by analyzing the RLC parallel circuit shown in Fig. 11.32.

This circuit is also known as a *(driven) damped harmonic oscillator*.

Fig. 11.32 RLC parallel circuit

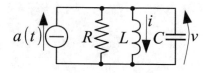

Fig. 11.33 Case Study 3

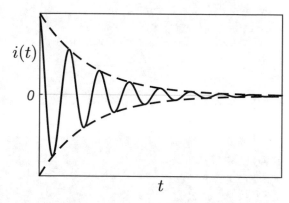

Case Study 3

Consider a circuit that contains only LTI memoryless components, an independent source, two inductors, and a capacitor. From this circuit, we measure the response (for the current $i(t)$ in one of the inductors) shown in Fig. 11.33. Assuming that a, b, and c are real positive coefficients with suitable physical dimensions, which of the following statements are compatible with the measurement?

1. *the circuit has four natural frequencies;*
2. *the circuit has three natural frequencies $\lambda_1 = a$, $\lambda_2 = b$, $\lambda_3 = c$;*
3. *the circuit has two natural frequencies $\lambda_{1,2} = -a \pm jb$;*
4. *the circuit has two natural frequencies $\lambda_{1,2} = \pm a + jb$;*
5. *the circuit has three natural frequencies $\lambda_{1,2} = a \pm jb$, $\lambda_3 = -c$.*

Finally, is it possible that the memoryless and time-invariant part of the circuit contains only an ideal transformer?

The current $i(t)$ is a damped sinusoid. Therefore, the circuit must contain at least two complex conjugate natural frequencies with negative real part. Moreover, the circuit has three WSVs, thus the maximum number of natural frequencies is three. These considerations imply that statements 1, 2, 4, and 5 are false.

Statement 3 can be true and requires that only two out of the three WSVs be SSVs.

Regarding the last question, the correct answer is no. According to the above discussion, the circuit is absolutely stable; this requires that it contain at least one component that dissipates energy. But

- the ideal transformer is a nonenergic component;
- the independent source is turned off, given that there is no forced response in the measured variable;
- the components with memory are conservative elements.

Therefore, in the absence of dissipative components, we cannot have a damped response.

Case Study 4

Find the natural frequencies for the circuit shown in Fig. 11.34 and discuss its stability.

You can check (see Fig. 11.35) that the I/O relationship for the voltage v_2 is

$$n^2 LC_2 \frac{d^2 v_2}{dt^2} + RC_2 \frac{dv_2}{dt} + n(\alpha + n)v_2 = (1 - n)Ra(t). \qquad (11.67)$$

Moreover, the state equation for v_1 (uncoupled from the other state variables) is

$$C_1 \frac{dv_1}{dt} = -a(t),$$

which implies that the third natural frequency is zero, due to the algebraic constraint between the derivative of the SSV v_1 and the input.

Fig. 11.34 Case Study 4

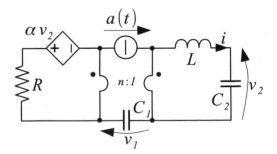

Fig. 11.35 Solution of Case
Study 4

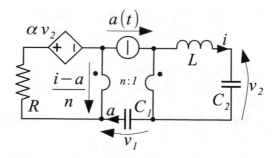

The characteristic equation associated with Eq. 11.67 is

$$n^2 LC_2 \lambda^2 + RC_2 \lambda + n(\alpha + n) = 0,$$

and thus the corresponding natural frequencies are

$$\lambda_{1,2} = \frac{-RC_2 \pm \sqrt{(RC_2)^2 - 4n^3(\alpha + n)LC_2}}{2n^2 LC_2}.$$

The condition for having complex conjugate natural frequencies is

$$(RC_2)^2 < 4n^3(\alpha + n)LC_2.$$

Due to the presence of $\lambda_3 = 0$, the circuit can be at most simply stable. To state something more, we have to discuss the sign of $\Re\{\lambda_{1,2}\}$:

- If $R^2 C_2 < 4n^3(\alpha + n)L$ (complex conjugate natural frequencies), then $\Re\{\lambda_{1,2}\} = \dfrac{-RC_2}{2n^2 LC_2} = \dfrac{-R}{2n^2 L} < 0$, under the usual assumption that $R, L > 0$. Therefore in this case, the circuit is simply stable.
- If $R^2 C_2 > 4n^3(\alpha + n)L$ (real natural frequencies), we have to check the sign of the rightmost natural frequency, that is, the sign of $-RC_2 + \sqrt{(RC_2)^2 - 4n^3(\alpha + n)LC_2}$:

  - if $n^3(\alpha + n) > 0$, the rightmost natural frequency is negative ($\sqrt{(RC_2)^2 - 4n^3(\alpha + n)LC_2} < RC_2$) and the circuit is simply stable;
  - if $n^3(\alpha + n) < 0$, the rightmost natural frequency is positive and the circuit is strongly unstable;
  - finally, if $n^3(\alpha + n) = 0$, the rightmost natural frequency is zero and the circuit is weakly unstable.

11.7 Normalizations and Comparisons with Mechanical Systems

We return to the normalization concept, already introduced in Sect. 9.11, by applying normalizations to the RLC series oscillator (Case Study 2 of Sect. 11.6) and to a corresponding mechanical oscillator.

We start with the circuit, where the I/O relationship for v is

$$LC\frac{d^2v}{dt^2} + RC\frac{dv}{dt} + v = e(t).$$

We consider the autonomous case (by setting $e(t) = 0$), define a reference voltage V_0 [V], and use as reference time the term $\sqrt{LC}$, thus defining two new dimensionless variables

$$\tilde{t} = \frac{t}{\sqrt{LC}} \qquad \text{and} \qquad \tilde{x} = \frac{v}{V_0}.$$

Conversely, the physical circuit variables can be expressed as

$$t = \sqrt{LC}\tilde{t} \qquad \text{and} \qquad v = V_0\tilde{x}. \tag{11.68}$$

By substituting these expressions in the I/O relationship (with $e(t) = 0$) and dividing each term by V_0, we obtain the normalized I/O relationship

$$\frac{d^2\tilde{x}}{d\tilde{t}^2} + R\sqrt{\frac{C}{L}}\frac{d\tilde{x}}{d\tilde{t}} + \tilde{x} = 0.$$

Using the normalized coefficient (see step 4 of the above-mentioned case study) $\zeta = \frac{R}{2}\sqrt{\frac{C}{L}}$ (called the *damping coefficient*), the above equation can be recast as follows:

$$\frac{d^2\tilde{x}}{d\tilde{t}^2} + 2\zeta\frac{d\tilde{x}}{d\tilde{t}} + \tilde{x} = 0.$$

Now we consider the mechanical system shown in Fig. 11.36. It represents a mass suspended through a spring and moving in a medium with viscous friction.

We assume that the physical system is characterized through its mass M [Kg], acceleration of gravity g [m/s^2], spring constant k [Kg/s^2], and viscous friction coefficient γ [Kg/s], which are the system parameters. The main variables are mass position x [m] and time t [s]. The resting position of the mass (Fig. 11.36a) is denoted by 0 [m] and corresponds to an extension ℓ of the spring, i.e., to a distance ℓ from the suspension point of the system. In this position, the force exerted by the spring exactly counterbalances the weight force: $k\ell = M g$. When we perturb the mass from its resting position, Newton's law describes the action of the frictional force and of the restoring force exerted by the spring, which oppose to the force of gravity:

Fig. 11.36 Mechanical system: **a** resting configuration; **b** out of resting configuration and acting forces

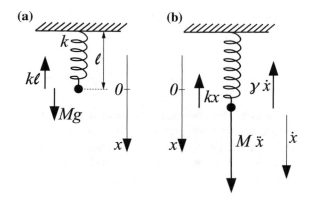

$$M\frac{d^2x}{dt^2} = -\gamma\frac{dx}{dt} - k(x+\ell) + Mg,$$

where $(x + \ell)$ is the overall elongation of the spring, as shown in Fig. 11.36b. Given that $Mg - k\ell = 0$, the differential equation describing the dynamics of this system can be written as

$$\frac{d^2x}{dt^2} + \frac{\gamma}{M}\frac{dx}{dt} + \frac{k}{M}x = 0. \tag{11.69}$$

By defining

$$\omega_0 = \sqrt{\frac{k}{M}} \qquad \text{and} \qquad \zeta = \frac{\gamma}{2\sqrt{kM}},$$

we obtain

$$\frac{d^2x}{dt^2} + 2\zeta\omega_0\frac{dx}{dt} + \omega_0^2 x = 0.$$

In this case, we use as a reference time the term $1/\omega_0$ [s]. Therefore, also in this case we can define two dimensionless variables

$$\tilde{t} = \omega_0 t \qquad \text{and} \qquad \tilde{x} = \frac{x}{\ell}.$$

Conversely, the physical system variables can be expressed as

$$t = \frac{\tilde{t}}{\omega_0} \qquad \text{and} \qquad x = \ell\tilde{x}. \tag{11.70}$$

By substituting these expressions in Eq. 11.69 and dividing each term by $\omega_0^2\ell$, we obtain once more the normalized differential equation

$$\frac{d^2\tilde{x}}{d\tilde{t}^2} + 2\zeta\frac{d\tilde{x}}{d\tilde{t}} + \tilde{x} = 0.$$

You can easily check that the corresponding state equations in matrix form are

$$\begin{pmatrix} \dot{\tilde{x}} \\ \dot{\tilde{y}} \end{pmatrix} = \underbrace{\begin{pmatrix} 0 & 1 \\ -1 & -2\zeta \end{pmatrix}}_{A} \begin{pmatrix} \tilde{x} \\ \tilde{y} \end{pmatrix} \qquad (11.71)$$

Therefore, the same equation can be used to describe the dynamics of two physical systems that at a first glance, seem to be completely different! The actual variables can be obtained by applying the conversion formulas defined in Eqs. 11.68 and 11.70.

Since the time is normalized in both examples, for the normalized system the dimensionless natural frequencies are

$$\lambda_{1,2} = -\zeta \pm \sqrt{\zeta^2 - 1}. \qquad (11.72)$$

Therefore, assuming that ζ (commonly known as the *damping ratio*) is a positive parameter in both systems:

- If $\zeta > 1$ (*overdamping*), the two natural frequencies are real and negative, and thus each system, starting from any initial condition, tends to its equilibrium solution without oscillations. This is due to the fact that the energy dissipation exerted by the resistor in the circuit and by the friction in the mechanical system is very strong with respect to the energy storing processes provided by the memory elements in the circuit and by the spring in the mechanical system.
- If $\zeta < 1$ (*underdamping*), the two natural frequencies are complex conjugates with negative real part, and thus each system, starting from any initial condition, tends to its equilibrium solution with oscillation. This is due to the fact that the energy dissipation exerted by the resistor in the circuit and by the friction in the mechanical system is quite weak with respect to the energy storing processes provided by the memory elements in the circuit and by the spring in the mechanical system.
- If $\zeta = 1$ (*critical damping*), there is only one real and negative natural frequency, with multiplicity 2.

Excluding the case of critical damping, the solution for $t > 0$, given the initial conditions $\tilde{x}(0^+)$ and $\tilde{y}(0^+) = \left. \dfrac{d\tilde{x}}{d\tilde{t}} \right|_{0^+}$, can always be expressed as follows:

$$\tilde{x}(\tilde{t}) = K_1 e^{\lambda_1 \tilde{t}} + K_2 e^{\lambda_2 \tilde{t}}. \qquad (11.73)$$

In the overdamped case, the coefficients K_1 and K_2 are real, as well as λ_1 and λ_2, and the solution is the sum of two exponentials converging to 0, as shown in Fig. 11.37a. In the underdamped case, K_1 and K_2 are complex conjugates, as well as λ_1 and λ_2, and the solution is a damped sinusoid converging to 0, as shown in Fig. 11.37b.

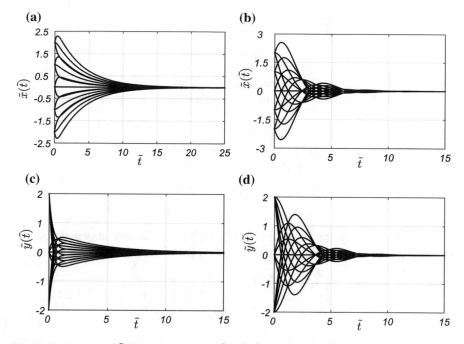

Fig. 11.37 Solutions $\tilde{x}(\tilde{t})$ (upper panels) and $\tilde{y}(\tilde{t})$ (lower panels) obtained by numerically integrating Eq. 11.71 for **a, c** $\zeta = 2$ (overdamped system) and **b, d** $\zeta = 0.5$ (underdamped system), starting from a grid of initial conditions ($\tilde{x}(0^+)$ from -2 to 2 with step 1, $\tilde{y}(0^+)$ from -2 to 2 with step 2)

11.7.1 A Double Mass–Spring Chain and Its Circuit Model

The mechanical system shown in Fig. 11.38a consists of two masses m_1 and m_2 suspended in series through springs and subject to the force of gravity. Each mass can oscillate vertically, and the viscous friction of the medium is ignored. The spring constants are k_1 and k_2 respectively. Denoting by x_1 and x_2 the positions of the two masses, m_2 is subject to the gravitational force $m_2 g$ and to the upward-directed spring force $k_2(x_2 - x_1)$; this same force acts, directed downward, even on m_1, together with the gravitational force $m_1 g$ and with the upward directed spring force $k_1 x_1$. With this in mind, we can write the dynamic equations as

$$
\begin{cases}
m_1 \dfrac{d^2 x_1}{dt^2} = m_1 g - k_1 x_1 + k_2 (x_2 - x_1), \\[4mm]
m_2 \dfrac{d^2 x_2}{dt^2} = m_2 g - k_2 (x_2 - x_1).
\end{cases}
$$

The equilibrium state of the system is defined by the two values x_{10} and x_{20} of the coordinates x_1, x_2 such that

$$\begin{cases} m_1 g - k_1 x_{10} + k_2 (x_{20} - x_{10}) = 0, \\[2mm] m_2 g - k_2 (x_{20} - x_{10}) = 0. \end{cases}$$

Now, taking $x_1 = x_{10} + \xi_1$ and $x_2 = x_{20} + \xi_2$, it is easy to obtain the dynamic equations of the system in terms of only the out-of-equilibrium displacement variables ξ_1 and ξ_2:

$$\frac{d^2}{dt^2} \begin{pmatrix} \xi_1 \\ \xi_2 \end{pmatrix} = \begin{pmatrix} -\dfrac{k_1 + k_2}{m_1} & \dfrac{k_2}{m_1} \\[4mm] \dfrac{k_2}{m_2} & -\dfrac{k_2}{m_2} \end{pmatrix} \begin{pmatrix} \xi_1 \\ \xi_2 \end{pmatrix} \qquad (11.74)$$

It is easy to verify that the circuit shown in Fig. 11.38b can be thought of as a model of the mechanical system. The descriptive equations for the capacitors can be written in the form

$$\begin{pmatrix} i_1 \\ i_2 \end{pmatrix} = \begin{pmatrix} -C_1 & 0 \\ 0 & -C_2 \end{pmatrix} \frac{d}{dt} \begin{pmatrix} v_1 \\ v_2 \end{pmatrix}.$$

Making use of this equation, we can now write the descriptive equations of the coupled inductors and eliminate the vector $(i_1 \ i_2)^T$:

$$\begin{pmatrix} v_1 \\ v_2 \end{pmatrix} = \begin{pmatrix} L_1 & M \\ M & L_2 \end{pmatrix} \frac{d}{dt} \begin{pmatrix} i_1 \\ i_2 \end{pmatrix} = \begin{pmatrix} L_1 & M \\ M & L_2 \end{pmatrix} \begin{pmatrix} -C_1 & 0 \\ 0 & -C_2 \end{pmatrix} \frac{d^2}{dt^2} \begin{pmatrix} v_1 \\ v_2 \end{pmatrix}.$$

Assuming $\Delta = L_1 L_2 - M^2 > 0$, the above equation can be reformulated as

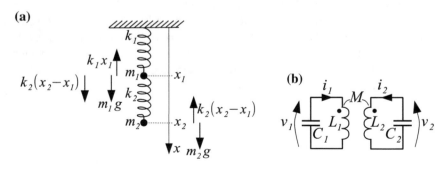

Fig. 11.38 a The double mass–spring chain; **b** its corresponding circuit

$$\frac{d^2}{dt^2}\begin{pmatrix} v_1 \\ v_2 \end{pmatrix} = \frac{1}{\Delta} \begin{pmatrix} -\dfrac{L_2}{C_1} & \dfrac{M}{C_1} \\[2ex] \dfrac{M}{C_2} & -\dfrac{L_1}{C_2} \end{pmatrix} \begin{pmatrix} v_1 \\ v_2 \end{pmatrix}. \tag{11.75}$$

This result has the same structure of Eq. 11.74, provided that $M = L_1$, which implies $\Delta = L_1L_2 - M^2 = L_1(L_2 - L_1) > 0$, that is, $L_2 > L_1$.

In the circuit, the voltages v_1, v_2 play a role corresponding to the displacements ξ_1, ξ_2 in the mechanical system. For suitable values of the circuit's physical parameters (inductances and capacitances), then, the mechanical system's behavior can be studied by observing the corresponding behavior of the circuit, which operates as a model.

For both systems it would obviously be possible to write state equations in the form $\dfrac{dx}{dt} = Ax$, in which A is a 4×4 matrix. Here, however, reference was made to Eqs. 11.74 and 11.75 to highlight the similarities between the two systems in a more compact way.

The last two sections pointed out that LTI circuits with complex conjugate natural frequencies oscillate. Case Study 2 in Sect. 11.6 and the examples analyzed in Sect. 11.7 in particular evidenced that the behavior of an oscillating system depends on a positive dimensionless damping coefficient (or damping ratio) expressed in terms of the system parameters.

The considered case studies are examples of *harmonic oscillators*, that is, systems that when displaced from their equilibrium position experience a restoring force proportional to the displacement. Besides the already considered masses connected to springs and RLC circuits, other examples of harmonic oscillators are the pendulum (with small angles of displacement) and acoustical systems such as bells and vibrating strings.

11.8 Solution for Nonstate Output Variables

We return to the general problem set up in Sect. 11.3.2 and in Fig. 11.14. Once the solution is found for the state vector $x(t)$, we can apply the substitution theorem to the components with memory (each capacitor is replaced by a voltage source and each inductor by a current source) in the circuit of Fig. 11.14, thus obtaining a memoryless circuit.

In all cases, we have a linear resistive circuit, whose solution can be found algebraically. In other words, the nonstate output vector $y(t)$ can be found as

$$y(t) = Cx(t) + D\hat{u}(t) \tag{11.76}$$

if there are no algebraic constraints ($M = 0$) and as

$$y(t) = Cx(t) + D_0\hat{u}(t) + D_1\frac{d\hat{u}}{dt} + D_2\frac{d^2\hat{u}}{dt^2} + \cdots \tag{11.77}$$

if there are $M > 0$ algebraic constraints.

This is called the output equation and is solved algebraically. Actually, this is a system of output equations; in each of them, each nonstate variable is expressed as a linear combination of state variables (already known) and inputs.

By applying the discontinuity balance to any equation of system Eq. 11.76, it is evident that in the case $M = 0$, in contrast to a state variable, every output variable *has the same maximum degree of discontinuity as the most discontinuous input* appearing in the corresponding output equation.

We remark that Eqs. 11.37 and 11.76 form an algebraic-differential system. Therefore, as stated in Sect. 11.3.1, owing to the linearity assumption, we have been able to split the general system of algebraic differential equations governing the circuit into two simpler systems: one purely differential (Eq. 11.37) and one purely algebraic (Eq. 11.76).

For instance, returning to Case Study 1 in Sect. 11.3.3, the currents i_1 and i_2 (see Fig. 11.15) are not state variables and can be expressed as linear combinations of the state variables v_1 and v_2, as was shown during the development of the case study.

11.9 Response of LTI Dynamical Circuits to Discontinuous Inputs

In this section, we provide some examples of analysis of LTI dynamical circuits subject to discontinuous inputs.

We use as a workbench the RLC series circuit shown in Fig. 11.31a and already analyzed in Sect. 11.6, Case Study 2. The I/O relationship for v is

$$LC\frac{d^2v}{dt^2} + RC\frac{dv}{dt} + v = e(t). \tag{11.78}$$

The circuit's natural frequencies are (see Eq. 11.66)

$$\lambda_{1,2} = \frac{-RC \pm \sqrt{(RC)^2 - 4LC}}{2LC}.$$

We assume that $v(t) = 0$ for $t < 0$; therefore, we immediately have $v(0^-) = 0$ and $\left.\frac{dv}{dt}\right|_{0^-} = 0$.

Case Study 1: Step Response
Find the complete response $v(t)$ for the RLC series circuit shown in Fig. 11.31a, with $e(t) = E \cdot u(t)$.

The discontinuity balance across $t = 0$ implies that $\dfrac{d^2v}{dt^2}$ has a step discontinuity; thus both v and $\dfrac{dv}{dt}$ are continuous across 0, that is, $v(0^+) = 0$ and $\dfrac{dv}{dt}\bigg|_{0^+} = 0$.

For $t > 0$, the I/O relationship becomes

$$LC\frac{d^2v}{dt^2} + RC\frac{dv}{dt} + v = E.$$

The forced response can easily be found using the similarity criterion: $v_{fo} = E$.

The complete response is

$$v(t) = v_{fr}(t) + v_{fo}(t) = K_1 e^{\lambda_1 t} + K_2 e^{\lambda_2 t} + E.$$

The coefficients K_1 and K_2 are found by imposing the initial conditions

$$v(0^+) = K_1 + K_2 + E = 0,$$
$$\frac{dv}{dt}\bigg|_{0^+} = K_1\lambda_1 + K_2\lambda_2 = 0.$$

The task of finding expressions for K_1 and K_2 by solving this linear inhomogeneous system is left to the reader.

Figure 11.39 shows the complete response for real natural frequencies (panel a) and for complex conjugate natural frequencies (panel b).

Case Study 2: Impulse Response
Find the complete response $v(t)$ for the RLC series circuit shown in Fig. 11.31a, with $e(t) = \Phi_0 \cdot \delta(t)$.

The discontinuity balance across $t = 0$ implies that across 0:

- $\dfrac{d^2v}{dt^2}$ has an impulse discontinuity;

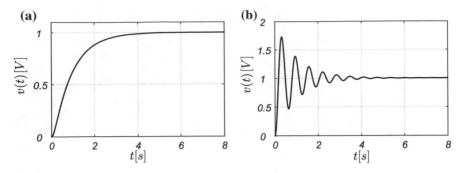

Fig. 11.39 Case Study 1. Step response $v(t)$ obtained by numerically integrating the I/O relationship for **a** $\zeta > 1$, with $RC = 1\,\mathrm{s}$ and $LC = 0.1\,\mathrm{s}^2$ (overdamped system), and **b** $\zeta < 1$, with $RC = 0.02\,\mathrm{s}$ and $LC = 0.01\,\mathrm{s}^2$ (underdamped system)

- $\dfrac{dv}{dt}$ has a step discontinuity;
- v is continuous, that is, $v(0^+) = 0$.

    To find $\dfrac{dv}{dt}\bigg|_{0^+}$, we integrate the I/O relationship for v (Eq. 11.78) between 0^- and 0^+:

$$LC\left[\frac{dv}{dt}\bigg|_{0^+} - \frac{dv}{dt}\bigg|_{0^-}\right] + 0 + 0 = \Phi_0,$$

thereby obtaining

$$\frac{dv}{dt}\bigg|_{0^+} = \frac{\Phi_0}{LC}.$$

For $t > 0$, the I/O relationship becomes

$$LC\frac{d^2v}{dt^2} + RC\frac{dv}{dt} + v = 0.$$

Therefore, there is only free response

$$v(t) = v_{fr}(t) = K_1 e^{\lambda_1 t} + K_2 e^{\lambda_2 t}.$$

The coefficients K_1 and K_2 are found by imposing the initial conditions

$$v(0^+) = K_1 + K_2 = 0,$$
$$\frac{dv}{dt}\bigg|_{0^+} = K_1\lambda_1 + K_2\lambda_2 = \frac{\Phi_0}{LC}.$$

You can check that the solution is

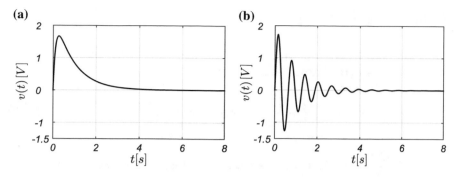

Fig. 11.40 Case Study 2. Impulse response $v(t)$ obtained by numerically integrating the I/O relationship for **a** $\zeta > 1$, with $RC = 1\,\text{s}$ and $LC = 0.1\,\text{s}^2$ (overdamped system) and $\Phi_0 = 2$ Vs, and **b**
$\zeta < 1$, with $RC = 0.02\,\text{s}$ and $LC = 0.01\,\text{s}^2$ (underdamped system) and $\Phi_0 = 0.2$ Vs

$$K_1 = \frac{\Phi_0}{LC}\frac{1}{\lambda_1 - \lambda_2},$$
$$K_2 = -K_1.$$

Figure 11.40 shows the complete response for real natural frequencies
(panel a) and for complex conjugate natural frequencies (panel b).

11.10 Generic Periodic Inputs

In this section we consider the case of inputs that are periodic but not sinusoidal. For
applications, this is a case of considerable importance in electronic circuits. The most
important physical situation concerns the case of absolutely stable linear circuits, for
which it is reasonable to expect that when the natural modes become negligible, the
response becomes periodic with the same period as for the input. Therefore, from
this point on, we consider only absolutely stable circuits.

For this kind of input (as already shown in Sect. 9.10.1.1), generally speaking it
is not possible to apply the similarity criterion, and the complete analysis is usually
quite involved and often carried out with other techniques, based on numerical or
analytical methods. In these cases, the Fourier series provides the most important
analytical tool to get some information about the periodic response of the circuit.

11.10.1 *Fourier Series*

Every function $f(t)$ periodic with period T and integrable on the interval $[0, T]$ can be *associated* with the Fourier series[3]

$$S(t) = \bar{f} + \sum_{k=1}^{+\infty} [a_k \cos(k\omega t) + b_k \sin(k\omega t)], \qquad (11.79)$$

where $\omega = \dfrac{2\pi}{T}$ is the *fundamental angular frequency* or *first harmonic*, its integer multiple $k\omega$ is the so-called kth *harmonic*,[4] and

$$\bar{f} = \frac{1}{T} \int_0^T f(t)dt = \frac{1}{T} \int_t^{t+T} f(\tau)d\tau \qquad (11.80)$$

is the **mean (or average) value** of the periodic function $f(t)$ over the period T.

We remark that the integral in Eq. 11.80 is nothing but the area under $f(t)$ over one period. Notice that the mean value over the period of any sinusoid with angular frequency $k\omega$ ($k = 1, 2, \ldots$) is null. Henceforth, we use integrals defined over the interval $[0, T]$, meaning that the integration spans an interval of length T.

The terms a_k, b_k ($k = 1, 2, \ldots$) in the Fourier series are related to $f(t)$ as follows:

$$a_k = \frac{2}{T} \int_0^T f(t) \cos(k\omega t)dt, \qquad (11.81)$$

$$b_k = \frac{2}{T} \int_0^T f(t) \sin(k\omega t)dt. \qquad (11.82)$$

Therefore, the terms $\bar{f}$, a_1, b_1, $\ldots$, a_k, b_k, $\ldots$ can be viewed as coefficients of the functions forming the *trigonometric system*:

$$1, \ \cos(\omega t), \ \sin(\omega t), \ \ldots, \ \cos(k\omega t), \ \sin(k\omega t), \ \ldots. \qquad (11.83)$$

You can check that the integral over a period T of the product of any pair of distinct elements of the trigonometric system is null. This property is called *orthogonality of the trigonometric system*. For an in-depth discussion of this aspect, see Sect. 11.10.2.

---

[3] Fourier series are named after Jean-Baptiste Joseph Fourier (1768–1830), a French mathematician and physicist who made important contributions to the study of trigonometric series. He introduced these series for solving problems of heat transfer and vibrations.

[4] This name derives from the concept of overtones or harmonics in musical instruments: the wavelengths of the overtones of a vibrating string or a column of air are derived from the string's (or air column's) fundamental wavelength.

For compactness of notation, it is useful to introduce the *inner product* between two functions f and g defined in the closed domain $[0, T]$,

$$(f, g) \doteq \frac{2}{T} \int_0^T f(t)g(t)dt, \tag{11.84}$$

which allows us to write, for instance, $a_k = (f, \cos(k\omega t))$, according to Eq. 11.81. The notion of inner product between two functions corresponds to the well-known dot product between two vectors. By analogy, if $(f, g) = 0$ (with f, g not identically zero), we say that the functions f and g are *orthogonal*.

For the trigonometric system, in addition to orthogonality, we have

$$(\cos(k\omega t), \cos(k\omega t)) = 1; \quad (\sin(k\omega t), \sin(k\omega t)) = 1; \quad (1, 1) = 2. \tag{11.85}$$

These results and their correspondence with the concepts known for vectors allow us to interpret the terms of the Fourier series $S(t)$ in a particularly effective way. First, however, we need to deepen the relationship between $f(t)$ and its Fourier series $S(t)$.

In general, when $f(t)$ satisfies only the condition of being integrable on $[0, T]$, we *cannot* write $f(t) = S(t)$. A sufficient condition for the relation $f(t) = S(t)$ to be valid is that $f(t)$ be a *piecewise-smooth* function. This category of functions (which includes, as a particular case, the continuous functions) is sufficiently general to cover all applications of interest to us.

This equivalence holds within the limits set by the following theorem, which is reported here for the sake of completeness. As outlined in Fig. 11.41, when $f(t)$ is a piecewise-smooth function, we can break up the interval $[0, T]$ into a finite number of subintervals

$$[0, t_1], \ldots, [t_i, t_{i+1}], \ldots, [t_N, T]$$

such that both $f(t)$ and $\dfrac{df(t)}{dt}$ are continuous within each subinterval and are bounded on the whole interval $[0, T]$.

Theorem 11.1 (Convergence of Fourier series [2]) *If $f(t)$ is a piecewise-smooth function over the interval $[0, T]$, then its Fourier series*

$$S(t) = \bar{f} + \sum_{k=1}^{+\infty} [a_k \cos(k\omega t) + b_k \sin(k\omega t)]$$

converges at each point t of the interval, and we have:

1. *$S(t) = f(t)$ if $t \in (0, T)$ and $f(t)$ is continuous at the point t;*
2. *$S(t) = \dfrac{f(t^+) + f(t^-)}{2}$ if $t \in (0, T)$ and $f(t)$ has a jump at the point t;*
3. *$S(0) = S(T) = \dfrac{f(0^+) + f(T^-)}{2}$.*

Fig. 11.41 Structure of a piecewise-smooth periodic function $f(t)$. Within the period, the instants t_i indicate points where f is continuous but its time derivative has a jump (t_1) or where also f has a jump (t_2, t_3)

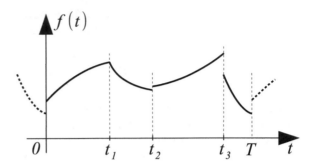

Notice that because a function continuous at a point t is such that $\dfrac{f(t^+) + f(t^-)}{2} = \dfrac{2f(t)}{2} \equiv f(t)$, the second equality of the theorem includes the first one. Therefore, this "average value" for $S(t)$ applies to every point $t \in (0, T)$, without the need to distinguish where $f(t)$ is continuous and where discontinuous.

Henceforth we assume that f is a piecewise-smooth function, so we can write $f(t) = S(t)$ for every point t where f is continuous.

11.10.2   Some Supplementary Notes About Fourier Series

✂ **Shortcut.** This section can be skipped without compromising the comprehension of the next sections.

11.10.2.1   Analogies with Vectors

Every element of a linear vector space can be expressed as a linear combination of linearly independent elements forming a basis; see Sect. 2.2.3 in Vol. 1. The number of these elements is the dimension of the space. When the elements are mutually orthogonal, the coefficients defining a vector v in terms of the basis are particularly easy to obtain. For instance, a vector v in three-dimensional space (see Fig. 11.42) can be expressed as the weighted sum of the vector elements q_1, q_2, q_3 forming an orthogonal basis, that is, such that $(q_i, q_k) = q_i^T q_k = 0$ for $i \neq k$ and $i, k = 1, 2, 3$:

$$v = \sum_{k=1}^{3} v_k q_k. \tag{11.86}$$

Vectors q_i may have different lengths. Forming the dot product between the vector q_i and each side of Eq. 11.86, we obtain

Fig. 11.42 A vector in three-dimensional space and its components along three orthogonal directions

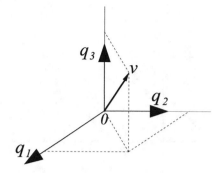

$$(q_i, v) = \left(q_i, \sum_{k=1}^{3} v_k q_k \right) = \sum_{k=1}^{3} v_k (q_i, q_k) = v_i (q_i, q_i),$$

because $(q_i, q_k) = q_i^T q_k = 0$ for any $k \neq i$. Therefore, the coefficient v_i is

$$v_i = \frac{(q_i, v)}{(q_i, q_i)}, \qquad i = 1, 2, 3.$$

In the function space identified by the trigonometric system Eq. 11.83, $f(t)$ plays the same role as v in the vector space discussed above. The coefficients of the Fourier series' elements follow by a completely similar procedure by taking the inner products between any element of the trigonometric system and both members of the expression $f(t) = S(t)$. Owing to the orthogonality of the trigonometric system and to Eq. 11.85, we obtain

$$\begin{cases} (S, 1) = (\bar{f}, 1) = \bar{f}(1, 1) = 2\bar{f}, \\ (f, 1) = (S, 1) \end{cases} \Rightarrow \bar{f} = \frac{1}{2}(f, 1).$$

We remark that this expression for $\bar{f}$ recasts Eq. 11.80 in terms of the inner product defined by Eq. 11.84.

Similarly, the results for a_k and b_k confirm Eqs. 11.81 and 11.82:

$$(f, \cos(k\omega t)) = a_k \underbrace{(\cos(k\omega t), \cos(k\omega t))}_{= 1} \Rightarrow a_k = (f, \cos(k\omega t)),$$

$$(f, \sin(k\omega t)) = b_k \underbrace{(\sin(k\omega t), \sin(k\omega t))}_{= 1} \Rightarrow b_k = (f, \sin(k\omega t)).$$

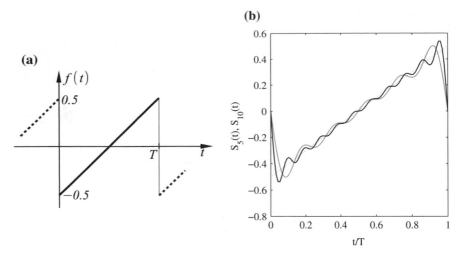

Fig. 11.43 Case Study 1. **a** Sawtooth function $f(t)$; **b** functions $S_5(t)$ (gray line) and $S_{10}(t)$ (black line), containing the first 5 and 10 terms of its Fourier series, respectively

11.10.2.2   Truncated Series

The series $S(t)$ reconstructs the function f through the contribution of its harmonics. The way in which this happens can be studied by considering the function obtained from the truncated series

$$S_n(t) = \bar{f} + \sum_{k=1}^{n} [a_k \cos(k\omega t) + b_k \sin(k\omega t)],$$

which contains only the first $(2n + 1)$ terms of S.

Case Study 1: Sawtooth Wave
Consider the function

$$f(t) = \frac{t}{T} - \frac{1}{2}, \qquad t \in [0, T],$$

which is the core of the periodic function shown in Fig. 11.43a, and determine the coefficients of its Fourier series.

We obviously have $\bar{f} = 0$. You can check that the coefficients a_k and b_k are

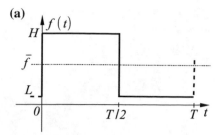

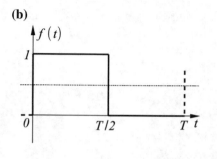

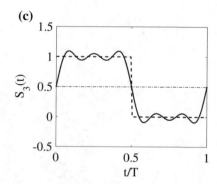

Fig. 11.44 Case Study 2. **a** Piecewise-constant function $f(t)$; **b** piecewise-constant function $f(t)$ with $H = 1$ and $L = 0$; **c** function $S_3(t)$, containing the first 3 terms of its Fourier series

$$a_k = (f, \cos(k\omega t)) = \frac{2}{T} \int_0^T \left(\frac{t}{T} - \frac{1}{2}\right) \cos(k\omega t) dt = 0,$$

$$b_k = (f, \sin(k\omega t)) = -\frac{1}{k\pi}.$$

The function $f(t)$ is continuous over the interval $(0, T)$, and we have $f(0^+) = -\frac{1}{2}$ and $f(T^-) = \frac{1}{2}$. Therefore, we have $S(t) = f(t)$ for all $t \in (0, T)$ and $S(0) = S(T) = \frac{1}{2}\left(f(0^+) + f(T^-)\right) = 0$.

A very simple way to show how the Fourier series converges to $f(t)$ is to consider the truncated series $S_n(t)$:

$$S_n(t) = -\frac{1}{\pi} \sum_{k=1}^{n} \frac{1}{k} \sin(k\omega t); \qquad \lim_{n \to \infty} S_n(t) = S(t).$$

For instance, by setting $n = 5$ and $n = 10$, Fig. 11.43b clearly shows how, as n grows, S_n approaches f inside the interval. Moreover, the plot points out that S_n converges to 0 as $t \to 0$ with $t > 0$ and as $t \to T$ with $t < T$.

Case Study 2: Square Wave
Consider the piecewise-constant function f defined as

$$f(t) = \begin{cases} H \ \ \text{for} \ t \in \left(0, \dfrac{T}{2}\right), \\ L \ \ \text{for} \ t \in \left(\dfrac{T}{2}, T\right). \end{cases}$$

This function is the core of the periodic square wave shown in Fig. 11.44a. Determine the coefficients of its Fourier series. Then set $H = 1$ and $L = 0$ as in Fig. 11.44b and find the truncated series $S_3(t)$.

The mean value of f is $\bar{f} = \dfrac{H + L}{2}$. The other terms in the Fourier series are

$$a_k = (f, \cos(k\omega t)) = 0,$$
$$b_k = (f, \sin(k\omega t)) = \begin{cases} 0 \ \ \text{for} \ k \ \text{even}, \\ \dfrac{2}{\pi k}(H - L) \ \ \text{for} \ k \ \text{odd}. \end{cases}$$

Therefore, the Fourier series is

$$S(t) = \frac{H + L}{2} + \frac{2(H - L)}{\pi} + \sum_{k=1,3,\dots}^{\infty} \frac{\sin(k\omega t)}{k}.$$

Now, taking $H = 1$ and $L = 0$, we obtain

$$S_3(t) = \frac{1}{2} + \frac{2}{\pi}\left(\sin(\omega t) + \frac{\sin(3\omega t)}{3} + \frac{\sin(5\omega t)}{5}\right),$$

which corresponds to the curve shown in Fig. 11.44c. The overshoots of $S_3(t)$ close to the step discontinuities of $f(t)$ are a consequence of the so-called Gibbs phenomenon. In addition to the comments already made for Case Study 1, we notice that $S_3\left(\dfrac{T}{2}\right) = \bar{f}$.

11.10.3   Mean Value of Circuit Variables

In a circuit, $f(t)$ is an electrical variable representing a periodic voltage or current. In particular, when $f(t)$ represents a voltage (current) source, its Fourier series $S(t)$ points out that this source can be represented through the series (parallel) connection of sources, each of which corresponds to a term of $S(t)$. In general, the number of these sources is infinite, but in some particular cases only a finite number of terms $\bar{f}$, a_k, b_k ($k = 1, 2, \ldots$) are different from zero. Periodic inputs of this type have already been considered in Sect. 11.5.2. For instance, a voltage $v(t) = E + E_1 \sin(\omega t) + E_2 \cos(2\omega t)$ is a periodic signal with mean value E and with $b_1 = E_1$ and $a_2 = E_2$. All the other coefficients are null.

After the input is represented through its Fourier series, the periodic response of each circuit variable is also represented by a similar series (from this point on called an output series), the terms of which can be easily obtained from the circuit equations. By virtue of linearity and the superposition principle, each term of the output series represents the forced response induced by a single input, either constant or sinusoidal. Therefore, we predict what the output is for a nonsinusoidal periodic input $f(t)$ by first expanding f in a Fourier series, then computing the particular integral due to each term in the series, and finally summing all these outputs to obtain the result. In addition, some analytic properties of the periodic response (including its mean value) can be obtained from the expression of the series, without having to carry out expensive calculations.

The mean value of any electric variable in an absolutely stable LTI circuit working in periodic steady state can be found by applying the superposition principle. Each periodic input $f(t)$ can be expanded in a Fourier series:

- the constant term $\bar{f}$ in the series is the mean value of $f(t)$ and induces a constant forced response contribution, owing to the similarity criterion;
- each sinusoid in the series corresponds to a sinusoidal forced response contribution with the same frequency, owing to the similarity criterion.

Inasmuch as the mean value over the period of any sinusoid with frequency equal to the fundamental one or a multiple thereof is null, *the (infinite) sinusoidal terms do not contribute at all to the mean value of any circuit variable*. Therefore, we can study an equivalent circuit in which each circuit variable (inputs included) is replaced by the corresponding mean value. Moreover, if the circuit is absolutely stable (as assumed), each capacitor can be replaced by an open circuit, and each inductor by a short circuit. This means that **to find the mean value of any electric variable, we have to study a memoryless circuit.**

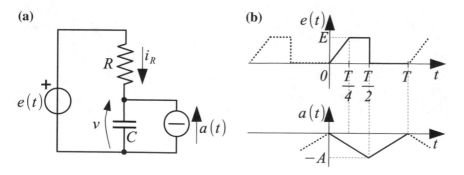

Fig. 11.45 Case Study 1. **a** Circuit to be analyzed; **b** periodic input signals (with period T)

Fig. 11.46 Memoryless
circuit to be studied to solve
Case Study 1

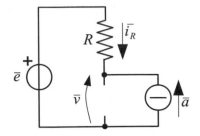

Case Study 1
*Find the mean values $\bar{v}$ and $\bar{i}_R$ for the circuit shown in Fig. 11.45a, which
works in periodic steady state with the T-periodic inputs $e(t)$ and $a(t)$ shown
in Fig. 11.45b. Determine the value of A (if it exists) such that for a fixed value
of E, $\bar{v} = 0$.*

To find the requested mean values, we have to study the memoryless circuit
shown in Fig. 11.46. It is straightforward to find $\bar{i}_R = -\bar{a}$ and $\bar{v} = \bar{e} + R\bar{a}$.

The mean value is nothing but the area under one period of the corresponding
input signal, divided by T. For $e(t)$ we have to compute the area of the trapezoid,
thereby obtaining $\bar{e} = \dfrac{3}{8}E$. For $a(t)$ we have to compute the (negative) area of
the triangle, which yields $\bar{a} = -\dfrac{A}{2}$.

Finally, we obtain $\bar{i}_R = \dfrac{A}{2}$ and $\bar{v} = \dfrac{3}{8}E - \dfrac{RA}{2}$.

The value of A such that $\bar{v} = 0$ is $A = \dfrac{3E}{4R}$.

Case Study 2
*For the circuit analyzed in Case Study 1 of Sect. 11.5.2, find $\bar{i}$ and $\bar{v}$ at steady
state.*

Owing to the analysis already carried out, we immediately obtain $\bar{v} =
v_{DC} = 0$ and $\bar{i} = i_{DC} = A_0$.

11.10.4 Root Mean Square Value

There are two alternative formulations of Eq. 11.79 for a function $f(t)$ that can be associated with the Fourier series. The first one is

$$f(t) = \bar{f} + \sum_{k=1}^{+\infty} A_k \cos(k\omega t + \phi_k), \qquad (11.87)$$

where $A_k = \sqrt{a_k^2 + b_k^2}$ and ϕ_k is the phase of the complex number $a_k - jb_k$. Indeed,

$$\cos(k\omega t + \phi_k) = \cos(k\omega t)\cos(\phi_k) - \sin(k\omega t)\sin(\phi_k),$$

and therefore

$$a_k = A_k \cos(\phi_k) \quad \text{and} \quad b_k = -A_k \sin(\phi_k),$$

that is,

$$a_k - jb_k = A_k e^{j\phi_k}.$$

The second one is

$$f(t) = \frac{1}{2} \sum_{k=-\infty}^{+\infty} \dot{A}_k e^{jk\omega t}, \qquad (11.88)$$

where $\dot{A}_k = A_k e^{j\phi_k}$, $\dot{A}_k^* = \dot{A}_{-k}$, and $\dot{A}_0 = 2\bar{f}$.

> The **root mean square** (RMS) or **effective** value of a periodic waveform $f(t)$ of period T is
>
> $$f_{eff} = \sqrt{\frac{1}{T} \int_0^T f^2(t)dt}. \qquad (11.89)$$

For instance, the RMS value of a pure sinusoid $A_k \cos(k\omega t + \phi_k)$ is its amplitude divided by $\sqrt{2}$. You can easily deduce that by rewriting the expression for the sinusoid's RMS value in terms of Eq. 11.84 (inner product):

$$\sqrt{\frac{1}{T} \int_0^T (A_k \cos(k\omega t + \phi_k))^2 \, dt} = \sqrt{\frac{1}{2} A_k^2 \underbrace{(\cos(k\omega t + \phi_k), \cos(k\omega t + \phi_k))}_{= 1}} = \frac{A_k}{\sqrt{2}}.$$

For a constant signal (which can be viewed as a limiting case of a periodic signal), the mean value coincides with the constant itself, whereas the RMS value coincides with its absolute value.

Remark: The mean power $\bar{p}$ absorbed by a resistor in periodic steady state is

$$\bar{p} = \frac{1}{T} \int_0^T p(t)dt = \frac{1}{T} \int_0^T Ri^2(t)dt = \frac{1}{T} \int_0^T \frac{v^2(t)}{R}dt, \qquad (11.90)$$

that is,

$$\bar{p} = Ri_{eff}^2 = \frac{v_{eff}^2}{R}. \qquad (11.91)$$

Theorem 11.2 (Root mean square value theorem) *the root mean square value of a piecewise-smooth periodic function $f(t)$ of period T can be computed in one of the following alternative ways:*

$$f_{eff} = \sqrt{\bar{f}^2 + \frac{1}{2} \sum_{k=1}^{+\infty} [a_k^2 + b_k^2]}, \qquad (11.92)$$

$$f_{eff} = \sqrt{\bar{f}^2 + \sum_{k=1}^{+\infty} \frac{A_k^2}{2}}, \qquad (11.93)$$

$$f_{eff} = \sqrt{\bar{f}^2 + \sum_{k=1}^{+\infty} \frac{|\dot{A}_k|^2}{2}}. \qquad (11.94)$$

Proof The proof is given for Eq. 11.92 by making reference to the Fourier series expression for Eq. 11.79. The other equivalences derive from the alternative formulations of Eqs. 11.87 and 11.88.

Since $f(t)$ is piecewise-smooth, it can be represented through its Fourier series, which is abbreviated by denoting the sum of the trigonometric terms by $g(t)$:

$$f(t) = \bar{f} + \underbrace{\sum_{k=1}^{+\infty} [a_k \cos(k\omega t) + b_k \sin(k\omega t)]}_{g(t)}.$$

Owing to Eq. 11.89, we have

$$f_{eff}^2 = \frac{1}{T} \int_0^T f^2(t)dt = \frac{1}{2}(f,f). \qquad (11.95)$$

The inner product can now be expanded as

$$(f,f) = (\bar{f} + g, \bar{f} + g) = (\bar{f}, \bar{f}) + 2(\bar{f}, g) + (g, g) =$$

$$= \bar{f}^2 \underbrace{(1, 1)}_{= 2} + 2\bar{f} \underbrace{(1, g)}_{= 0} + (g, g) = 2\bar{f}^2 + (g, g).$$

Owing to orthogonality, the product $(1, g)$ is null, and the product (g, g) can be further simplified:

$$(g, g) = \left(\sum_{p=1}^{+\infty} [a_p \cos(p\omega t) + b_p \sin(p\omega t)], \sum_{q=1}^{+\infty} [a_q \cos(q\omega t) + b_q \sin(q\omega t)] \right) =$$

$$= \sum_{p=1}^{+\infty} \sum_{q=1}^{+\infty} \left(a_p \cos(p\omega t) + b_p \sin(p\omega t), a_q \cos(q\omega t) + b_q \sin(q\omega t) \right) =$$

$$= \sum_{p=1}^{+\infty} a_p^2 \underbrace{(\cos(p\omega t), \cos(p\omega t))}_{= 1} + b_p^2 \underbrace{(\sin(p\omega t), \sin(p\omega t))}_{= 1} = \sum_{p=1}^{+\infty} \left[a_p^2 + b_p^2 \right].$$

Therefore, recalling Eq. 11.95, we obtain the final result

$$f_{eff} = \sqrt{\frac{1}{2}(f,f)} = \sqrt{\bar{f}^2 + \frac{1}{2} \sum_{p=1}^{+\infty} \left[a_p^2 + b_p^2 \right]}.$$

□

This result, also known as Parseval's theorem,[5] implies that the average power absorbed by a resistor in periodic steady state (Eq. 11.91) is the sum of the average powers for the separate harmonics. This is a further reason to carry out the analysis of a periodic waveform in terms of its harmonics. Parseval's formula (Eq. 11.92) can be interpreted as the infinite-dimensional analogue of the Pythagorean theorem in two-dimensional space and of the Euclidean distance in finite-dimensional spaces.

As a final remark, Eq. 11.92 obviously holds when only finitely many components of the Fourier series differ from zero, for example, when a periodic voltage (current) source $f(t)$ in a circuit can be represented by a series (parallel) connection of a finite number of sources.

---

[5]The theorem is named after Marc-Antoine Parseval des Chênes (1755–1836), a French mathematician. His theorem presaged the unitarity of the Fourier transform.

Fig. 11.47 Case Study 1

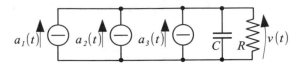

Case Study 1

For the circuit shown in Fig. 11.47, where $a_1(t) = A_1 \sin(\omega t)$, $a_2(t) = A_2 \cos(2\omega t)$, $a_3(t) = A_3 \sin(3\omega t)$, find the RMS value of $v(t)$ over the period T.

In this case, it is convenient to reason in the phasor domain, by applying the superposition principle:

$$\dot{V}_1 = \frac{R\dot{A}_1}{1 + j\omega RC}; \qquad \dot{V}_2 = \frac{R\dot{A}_2}{1 + j2\omega RC}; \qquad \dot{V}_3 = \frac{R\dot{A}_3}{1 + j3\omega RC}.$$

Phasors $\dot{V}_1$, $\dot{V}_2$, and $\dot{V}_3$ are related to sinusoidal responses at frequencies ω, 2ω, and 3ω, respectively. Using Eq. 11.94, we obtain

$$v_{eff} = \sqrt{\frac{|\dot{V}_1|^2}{2} + \frac{|\dot{V}_2|^2}{2} + \frac{|\dot{V}_3|^2}{2}} = \frac{1}{\sqrt{2}} \sqrt{\frac{(RA_1)^2}{1 + (\omega RC)^2} + \frac{(RA_2)^2}{1 + (2\omega RC)^2} + \frac{(RA_3)^2}{1 + (3\omega RC)^2}}.$$

Case Study 2

For the memoryless circuit shown in Fig. 11.48, where the descriptive equations of the three-terminal are $v_1 = Ri_1$ and $i_2 = \alpha i_1 + g v_2$ and the inputs are $e_1(t) = E_1 \cos(\omega t)$ and $e_2(t) = E_2 \cos(2\omega t)$, find:

1. *the RMS values of $i_1(t)$ and $i_2(t)$ over the period $T = \dfrac{2\pi}{\omega}$;*
2. *the mean value of the power absorbed by the three-terminal.*

Fig. 11.48 Case Study 2

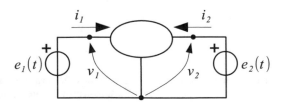

1. From the first descriptive equation of the three-terminal, we immediately obtain $i_1 = \dfrac{e_1}{R} = \dfrac{E_1}{R}\cos(\omega t)$. This is a pure cosine, whose RMS value is $i_{1eff} = \dfrac{E_1}{\sqrt{2}R}$.

   From the second descriptive equation we obtain

   $$i_2 = \alpha i_1 + g e_2 = \alpha \frac{E_1}{R}\cos(\omega t) + g E_2 \cos(2\omega t).$$

   On applying Eq. 11.93 (or Eq. 11.92), we obtain

   $$i_{2eff} = \sqrt{\left(\frac{\alpha E_1}{\sqrt{2}R}\right)^2 + \left(\frac{g E_2}{\sqrt{2}}\right)^2}.$$

2. The instantaneous power absorbed by the three-terminal is

   $$p(t) = v_1 i_1 + v_2 i_2 = \frac{e_1^2}{R} + \alpha \frac{e_1}{R} e_2 + g e_2^2.$$

   Therefore,

   $$\bar{p} = \frac{1}{R}\underbrace{\frac{1}{T}\int_0^T e_1^2 dt}_{= e_{1eff}^2} + \frac{\alpha}{R}\frac{1}{T}\int_0^T e_1 e_2 dt + g\underbrace{\frac{1}{T}\int_0^T e_2^2 dt}_{= e_{2eff}^2}.$$

   Given that due to orthonormality we have

   $$\int_0^T e_1 e_2 dt = E_1 E_2 \int_0^T \cos(\omega t)\cos(2\omega t)dt = 0,$$

   we finally obtain

   $$\bar{p} = \frac{E_1^2}{2R} + g\frac{E_2^2}{2}.$$

Case Study 3

The circuit shown in Fig. 11.49a is subject to periodic inputs. Assume that it is working under periodic steady-state conditions and answer the following questions.

1. Let $a(t) = A_0 + A\cos(\omega t)$ and $e(t) = f(t)$, where $f(t)$ is the periodic signal shown in Fig. 11.49b. Find the mean value of $i(t)$ and $v(t)$ over the common period $T = \dfrac{2\pi}{\omega}$.

2. Let $a(t) = A_0 + A\cos(\omega t)$ and $e(t) = E_0 + E\cos(2\omega t)$; to simplify the expressions, assume that the inductors are closely coupled, by setting $M = L$. Find the mean value of the power absorbed by the resistor with resistance R.

3. Connect the capacitor C in parallel to the resistor $2R$ and consider the modified circuit, again subject to the first set of inputs and working under periodic steady-state conditions. Find the new mean values of $i(t)$ and $v(t)$ over the period T.

1. The circuit to be studied to find the mean values of $i(t)$ and $v(t)$ is shown in Fig. 11.50a. By solving this memoryless circuit, we obtain

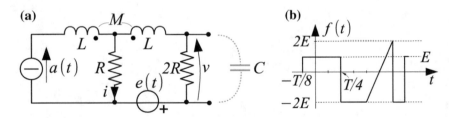

Fig. 11.49 Case Study 3. **a** Circuit to be analyzed; **b** periodic input signal (with period T)

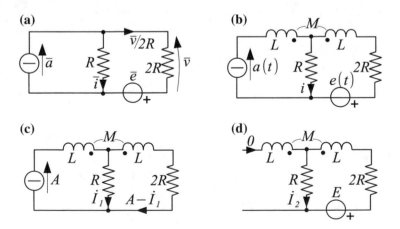

Fig. 11.50 Case Study 3. Auxiliary circuits to solve **a** point 1, and **b** point 2 and to find phasors **c** $\dot{I}_1$, and **d** $\dot{I}_2$

$$\bar{v} = \frac{2}{3}R\bar{a} - \frac{2}{3}\bar{e},$$

$$\bar{i} = \frac{2}{3}\bar{a} + \frac{\bar{e}}{3R}.$$

From the expression for $a(t)$, it is immediate that $\bar{a} = A_0$. Computing the area under $f(t)$ (shown in Fig. 11.49b), we obtain

$$\bar{e} = \frac{1}{T}\left(\frac{3}{8}TE - \frac{T}{4}2E - \frac{T}{8}2E\right) = -\frac{3}{8}E.$$

Therefore,

$$\bar{v} = \frac{2}{3}RA_0 + \frac{E}{4},$$

$$\bar{i} = \frac{2}{3}A_0 - \frac{E}{8R}.$$

2. The circuit to be studied under the given assumptions is shown in Fig. 11.50b, with $M = L$. The instantaneous power absorbed by the resistor with resistance R is $p(t) = Ri^2(t)$. As a consequence, by applying the definition of mean value and Eq. 11.94, we obtain

$$\bar{p} = Ri_{eff}^2 = R\left(\bar{i}^2 + \frac{|\dot{I}_1|^2}{2} + \frac{|\dot{I}_2|^2}{2}\right),$$

where each term within parentheses corresponds to a different steady state: DC steady state (mean value $\bar{i}$), AC steady state with angular frequency ω (phasor $\dot{I}_1$), and AC steady state with angular frequency 2ω (phasor $\dot{I}_2$). First of all, we derive $\bar{i}$. The circuit to be studied is once again the one shown in Fig. 11.50a; thus (see point 1)

$$\bar{i} = \frac{2}{3}\bar{a} + \frac{\bar{e}}{3R} = \frac{2}{3}A_0 + \frac{E_0}{3R}.$$

Now we find the phasor $\dot{I}_1$, which regards the AC steady state at angular frequency ω, corresponding to the current input term $A\cos(\omega t)$. Therefore, we turn off the voltage source $e(t)$ and the DC component of $a(t)$. We can either transform the I/O relationship for i_1 into the phasor domain or write directly the circuit equations in the phasor domain, according to Fig. 11.50c. In both cases, you can check that the phasor is

$$\dot{I}_1 = \frac{2RA}{3R + j\omega L}.$$

Therefore,

$$|\dot{I}_1|^2 = \frac{(2RA)^2}{(3R)^2 + (\omega L)^2}.$$

Phasor $\dot{I}_2$ regards the AC particular integral at angular frequency 2ω induced by the voltage input term $E\cos(2\omega t)$ of $e(t)$. Therefore, we must turn off the current source $a(t)$ and the DC component of $e(t)$. We can either transform the I/O relationship for i_1 into the phasor domain or write the circuit equations directly in the phasor domain, according to Fig. 11.50d. In both cases, you can check that the phasor is

$$\dot{I}_2 = \frac{E}{3R + j2\omega L}.$$

Therefore,

$$|\dot{I}_2|^2 = \frac{E^2}{(3R)^2 + (2\omega L)^2}.$$

In conclusion,

$$\bar{p} = R\left[\left(\frac{2}{3}A_0 + \frac{E_0}{3R}\right)^2 + \frac{1}{2}\frac{(2RA)^2}{(3R)^2 + (\omega L)^2} + \frac{1}{2}\frac{E^2}{(3R)^2 + (2\omega L)^2}\right].$$

3. By connecting C in parallel to the $2R$ resistor, the voltages and currents for the periodic steady state in the circuit obviously change with respect to those of point 1. However, the circuit to find the mean values of $i(t)$ and $v(t)$ is the same as shown in Fig. 11.50a, because C is replaced by an open circuit. Consequently, the mean values $\bar{i}$ and $\bar{v}$ do not change with respect to point 1.

Case Study 4

For the circuit shown in Fig. 11.51a, find:

1. *the mean values of $i(t)$ and $v(t)$ over the period $T = \dfrac{2\pi}{\omega}$, by assuming that $a(t) = A_0 + A_1 \sin(\omega t) + A_2 \cos(\omega t)$, $e(t)$ is the T-periodic signal shown in Fig. 11.51b, and the two-port is the one shown in Fig. 11.51c;*

2. *the mean value of the power absorbed by the two-port, by assuming that $a(t) = A_0$, $e(t) = E_0 + E\cos(\omega t)$, $C = 0$, and two-port (memoryless) described by the resistance matrix $\begin{pmatrix} 0 & 0 \\ R & 0 \end{pmatrix}$.*

1. The circuit to be studied to find the mean values of $i(t)$ and $v(t)$ is shown in Fig. 11.52a. By solving this memoryless circuit, we obtain $\bar{i} = \dfrac{\bar{a}}{3}$ (current divider) and $\bar{v} = 2R\bar{i} = \dfrac{2}{3}R\bar{a}$. Thus $e(t)$ does not influence the mean values.

   From the expression for $a(t)$, it is immediate that $\bar{a} = A_0$. Therefore,

   $$\bar{v} = \frac{2}{3}RA_0,$$

   $$\bar{i} = \frac{A_0}{3}.$$

2. The circuit to be studied under the given assumptions is shown in Fig. 11.52b. Notice that the Norton equivalent on the right of the original circuit is replaced by the corresponding Thévenin equivalent.

   The instantaneous power absorbed by the two-port is $p(t) = v_1 i_1 + v_2 i_2$. From the two-port descriptive equations ($v_1 = 0$ and $v_2 = Ri_1$) and from the circuit equations, we obtain

   $$i_1 = \frac{e}{R},$$

   $$i_2 = \frac{a}{3} - \frac{e}{3R}.$$

As a consequence, by applying the definition of mean value, we obtain

$$\bar{p} = \frac{1}{T}\int_0^T e\left(\frac{a}{3} - \frac{e}{3R}\right) dt,$$

that is,

$$\bar{p} = -\frac{1}{3R}\underbrace{\frac{1}{T}\int_0^T e^2 dt}_{= e_{\text{eff}}^2} + \frac{1}{3}\underbrace{\frac{1}{T}\int_0^T E_0 A_0 + A_0 E \cos(\omega t) dt}_{= E_0 A_0}.$$

Given that

$$e_{\text{eff}}^2 = E_0^2 + \frac{E^2}{2},$$

we finally obtain

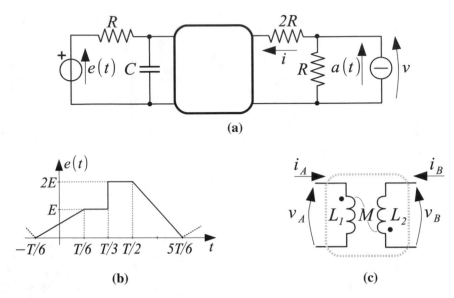

Fig. 11.51 Case Study 4. Circuit to be analyzed (**a**), periodic input signal (with period T) (**b**), two-port for question 1(**c**)

$$\bar{p} = -\frac{1}{3R}\left(E_0^2 + \frac{E^2}{2}\right) + \frac{E_0 A_0}{3}.$$

Notice that depending on the values of E_0, A_0, and E, we can have either positive or negative $\bar{p}$. This is compatible with the descriptive equations of the two-port ($v_1 = 0$ and $v_2 = Ri_1$), which is in fact a CCVS, that is, an active component.

11.11   Multi-input Example

In this section we propose a summarizing example.

Case Study

For the circuit shown in Fig. 11.53 (already analyzed in Sect. 11.5.2, Case Study 1), with $a(t) = A_0 + A_1 \sin(\omega t)u(-t) + Q_0\delta(t)$ and with the 2-port described by its hybrid matrix $H = \begin{pmatrix} R & 0 \\ \alpha & g \end{pmatrix}$, with $\alpha \neq 0$, find:
1. *the I/O relationship for $i(t)$;*

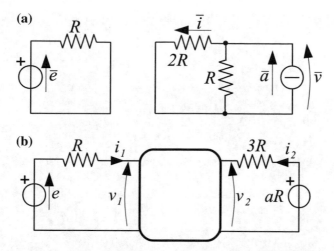

Fig. 11.52 Case Study 4. Auxiliary circuits to solve **a** point 1 and **b** point 2

2. *the I/O relationship for* $v(t)$;
3. *the natural frequencies and the absolute stability condition(s) (henceforth, assume that they are satisfied);*
4. $i(t)$ *for* $t < 0$ *and the conditions in* $t = 0^-$ *needed to proceed with the analysis;*
5. *the mean value and RMS value of* $i(t)$ *for* $t < 0$;
6. $v(t)$ *for* $t < 0$ *and the conditions in* $t = 0^-$ *needed to proceed with the analysis;*
7. *conditions in* $t = 0^+$ *needed to find* $i(t)$ *for* $t > 0$;
8. $i(t)$ *for* $t > 0$.

Some answers were already provided in Sect. 11.5.2, Case Study 1, and are reported here for ease of reference.

1. I/O relationship for $i(t)$:

$$L\frac{di}{dt} + Ri = Ra. \tag{11.96}$$

2. I/O relationship for $v(t)$:

Fig. 11.53 Case Study (summarizing example)

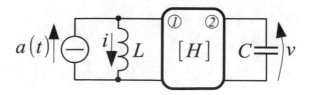

$$LC\frac{d^2v}{dt^2} + \tau\frac{dv}{dt} + Rgv = -\alpha L\frac{da}{dt}, \tag{11.97}$$

with $\tau = Lg + RC$.

3. natural frequencies: $\lambda_1 = -\dfrac{R}{L}$ and $\lambda_2 = -\dfrac{g}{C}$. The circuit is absolutely stable, provided that $R > 0$ and $g > 0$.
4. For $t < 0$, the input is $a(t) = A_0 + A_1 \sin(\omega t)$; thus $i(t)$ (coinciding with the forced response) is the sum of two terms:

(DC) the part of solution due to the DC term is (by substitution in Eq. 11.96)
$i_{DC} = A_0$;

(AC) the part of the solution due to the sinusoidal term can be found using phasors. Equation 11.96 in the phasor domain becomes

$$(R + j\omega L)\dot{I} = R\dot{A} = R(-jA_1),$$

from which we obtain

$$\dot{I} = -\frac{RA_1}{R^2 + (\omega L)^2}(\omega L + jR).$$

Therefore, the AC part of the solution is

$$i_{AC}(t) = -\frac{RA_1}{R^2 + (\omega L)^2}[\omega L \cos(\omega t) - R\sin(\omega t)].$$

On the whole,

$$i(t) = i_{DC} + i_{AC} = A_0 - \frac{RA_1}{R^2 + (\omega L)^2}[\omega L \cos(\omega t) - R\sin(\omega t)]. \tag{11.98}$$

Given that the I/O relationship (Eq. 11.96) is of the first order, we just need one condition ($i(0^-)$), which can be directly computed from Eq. 11.98:

$$i(0^-) = A_0 - \frac{RA_1\omega L}{R^2 + (\omega L)^2}.$$

5. The mean value of $i(t)$ for $t < 0$ coincides with i_{DC}; thus $\bar{i} = A_0$. The RMS value can be computed by applying either Eq. 11.94, exploiting the knowledge of the phasor, such that

$$|\dot{I}|^2 = \frac{(RA_1)^2}{R^2 + (\omega L)^2},$$

or Eq. 11.92, exploiting Eq. 11.98. In both cases, we obtain

$$i_{eff} = \sqrt{A_0^2 + \frac{1}{2}\frac{(RA_1)^2}{R^2 + (\omega L)^2}}.$$

6. Owing to Eq. 11.97, only the sinusoidal term of $a(t)$ influences $v(t)$ for $t < 0$. The solution was already obtained in Sect. 11.5.2, Case Study 1 (AC1 term):

$$v(t) = v_{AC}(t) = \frac{-\alpha\omega LA_1}{\beta^2 + (\omega\tau)^2}[\beta\cos(\omega t) + \omega\tau\sin(\omega t)], \qquad (11.99)$$

with $\beta = Rg - \omega^2 LC$. The I/O relationship for v is of second order; we therefore need two conditions in 0^-, which we compute using Eq. 11.99:

$$v(0^-) = \frac{-\alpha\omega LA_1\beta}{\beta^2 + (\omega\tau)^2}$$
$$\left.\frac{dv}{dt}\right|_{0^-} = \frac{-\alpha\omega^3 LA_1\tau}{\beta^2 + (\omega\tau)^2}.$$

7. From the discontinuity balance applied to Eq. 11.96, we deduce that $i(t)$ has a step discontinuity across $t = 0$. By integrating Eq. 11.96 between 0^- and 0^+, we obtain

$$L[i(0^+) - i(0^-)] = RQ_0,$$

that is,

$$i(0^+) = \frac{RQ_0}{L} + i(0^-).$$

8. For $t > 0$, we have $a(t) = A_0$. Therefore, given that the forced response is the same as for $t < 0$, the structure of the solution is

$$i(t) = i_{fr}(t) + i_{fo}(t) = Ke^{\lambda_1 t} + A_0,$$

and K is determined by the initial condition

$$i(0^+) = K + A_0.$$

Therefore,

$$i(t) = \underbrace{\left(\frac{RQ_0}{L} - \frac{RA_1\omega L}{R^2 + (\omega L)^2}\right)e^{\lambda_1 t}}_{i_{fr}(t)} + \underbrace{A_0}_{i_{fo}(t)}.$$

11.12 Problems

11.1 For the circuit shown in Fig. 11.54a, where the two-port is described by the transmission matrix $T = \begin{pmatrix} \alpha & R \\ g & \beta \end{pmatrix}$, with $\alpha, \beta, R, g \neq 0$ and $e(t) = \Phi_0 \delta(t) + [E_0 + E_1 \sin(\omega t)]u(-t)$, find:

1. the I/O relationship (state equation) for the state variable $i(t)$;
2. the circuit's natural frequency;
3. the absolute stability conditions;
4. $i(t)$ for $t < 0$ (by assuming that the circuit works in steady state) and $i(0^-)$;
5. $i(0^+)$;
6. $i(t)$ for $t > 0$;
7. the energy w dissipated by the circuit for $t > 0$.

11.2 For the circuit shown in Fig. 11.54b, working in steady state for $t < 0$, where $e_1(t) = \Phi_0 \delta(t) + E_1 \cos(\omega t)u(t)$ and $e_2(t) = E_0 u(-t) + E_2 \sin(2\omega t)u(t)$, find:

1. the I/O relationship for the state variable $v_1(t)$;
2. the I/O relationship for the state variable $v_2(t)$ and order of discontinuity of $v_2(t)$ at $t = 0$;
3. the circuit's natural frequencies;
4. $v_1(t)$ for $t < 0$ and $v_1(0^-)$;
5. $v_1(0^+)$;
6. $v_1(t)$ for $t > 0$;
7. the mean value $\bar{v}_1$ and RMS value v_{1eff} of $v_1(t)$ at steady state for $t > 0$.

11.3 For the circuit shown in Fig. 11.55a, working in steady state for $t < 0$, where $\alpha \neq 1$ and $e(t) = \Phi_0 \delta(t) + E_0 + E_1 \cos(\omega t)u(t)$, find:

1. the I/O relationship for $i_2(t)$;
2. the circuit's natural frequencies;
3. the absolute stability condition;
4. $i_2(t)$ for $t < 0$ and $i_2(0^-)$;

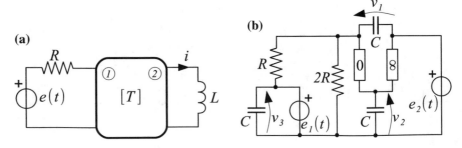

Fig. 11.54 a Problem 11.1; **b** Problem 11.2

(a)   **(b)**

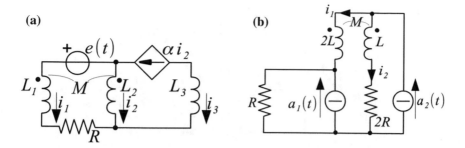

Fig. 11.55 a Problem 11.3; **b** Problem 11.4

5. $i_2(0^+)$;
6. $i_2(t)$ for $t > 0$.

11.4 For the circuit shown in Fig. 11.55b, working in steady state for $t < 0$, where
$a_1(t) = Q_0\delta(t) + A_0u(-t)$ and $a_2(t) = A_0u(t) + A_1\cos(\omega t)u(-t)$, find:

1. the I/O relationship for $i_2(t)$;
2. the circuit's natural frequencies;
3. the absolute stability condition;
4. $i_2(t)$ for $t < 0$ and $i_2(0^-)$;
5. $i_2(0^+)$;
6. $i_2(t)$ for $t > 0$.

11.5 For the circuit shown in Fig. 11.56, working in steady state for $t < 0$, where
$e(t) = \Phi_0\delta(t) + E_0u(t) + E_1\cos(\omega t)u(-t)$ and $a(t) = A_0 + A_1\sin(2\omega t)u(-t)$, find:

1. the I/O relationship for $i_1(t)$;
2. the circuit's natural frequencies;
3. the absolute stability condition;

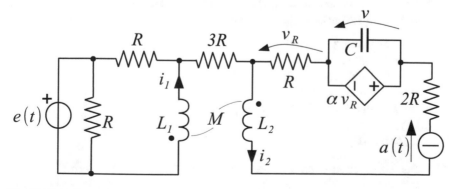

Fig. 11.56 Problem 11.5

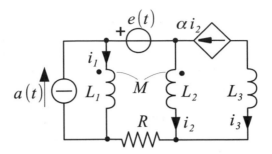

Fig. 11.57 Problem 11.6

4. $i_1(t)$ for $t < 0$ and $i_1(0^-)$;
5. the mean power $\bar{p}$ absorbed by resistor $2R$ over the period for $t < 0$;
6. $i_1(0^+)$;
7. $i_1(t)$ for $t > 0$.

11.6 For the circuit shown in Fig. 11.57, working in steady state for $t < 0$, where $\alpha \neq 1$, $e(t) = \Phi_0\delta(t) + E_0 + E_1\cos(\omega t)u(t)$ and $a(t) = A_0 u(t) + A_1\sin(2\omega t)u(t)$, find:

1. the I/O relationship for $i_2(t)$;
2. the circuit's natural frequencies;
3. choose the conditions ensuring the circuit absolute stability:

   - the circuit is always absolutely stable (no conditions are required);
   - $\alpha < 1$ and $M < 0$;
   - $1 < \alpha < 2$ and $M > 0$;

4. $i_2(t)$ for $t < 0$ and $i_2(0^-)$;
5. $i_2(0^+)$;
6. $i_2(t)$ for $t > 0$.

11.7 For the circuit shown in Fig. 11.58, working in steady state for $t < 0$, where $a_1(t) = A_0 + A_1\sin(\omega t)u(-t)$ and $a_2(t) = Q_0\delta(t) + Au(-t) + A_2 u(t)$, find:

1. the I/O relationship for $i_2(t)$;
2. the circuit's natural frequencies;
3. the absolute stability condition;
4. $i_2(t)$ for $t < 0$ and $i_2(0^-)$;
5. the mean value $\bar{i}_2$ and RMS value i_{2eff} of $i_2(t)$ for $t < 0$.
6. $i_2(0^+)$;
7. $i_2(t)$ for $t > 0$.

11.8 For the circuit shown in Fig. 11.59, where $e(t) = \Phi_0\delta(t) + E_2\cos(2\omega t)u(t)$, $a(t) = Q_0\delta(t)$ and all memory components are uncharged at $t = 0^-$, find:

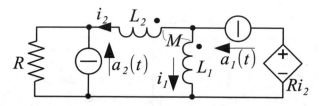

Fig. 11.58 Problem 11.7

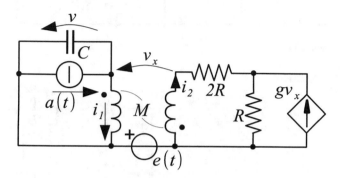

Fig. 11.59 Problem 11.8

1. the I/O relationship for $i_2(t)$;
2. the number of SSVs;
3. the circuit's natural frequencies;
4. the stability condition; henceforth, assume that it is satisfied;
5. the conditions at $t = 0^+$ needed for the subsequent analysis for $t > 0$;
6. $i_2(t)$ for $t > 0$.

11.9 For the circuit shown in Fig. 11.60, where $e(t) = E_0 + E_1 \sin(\omega t)u(-t)$ and $a(t) = Au(-t)$, find:

1. the I/O relationship for $v(t)$;
2. the number of SSVs;
3. the circuit's natural frequencies;
4. the stability condition; henceforth, assume that it is satisfied;
5. $v(t)$ for $t < 0$ and $v(0^-)$, assuming that $v(t)$ has reached a steady state;
6. the RMS value of $v(t)$ for $t < 0$;
7. $v(0^+)$;
8. $v(t)$ for $t > 0$.

11.10 For the circuit shown in Fig. 11.61, where $e(t) = Eu(t) + \Phi_0\delta(t)$ and $i(0^-) = 0$, find:

1. the I/O relationship for $i(t)$;
2. the number of SSVs;

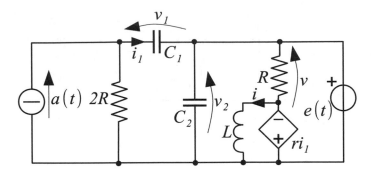

Fig. 11.60   Problem 11.9

Fig. 11.61   Problem 11.10

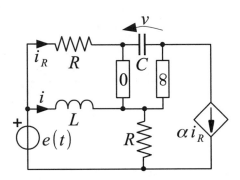

3. the circuit's natural frequencies;
4. the stability condition; henceforth, assume that it is satisfied;
5. $i(0^+)$;
6. $i(t)$ for $t > 0$;
7. the energy dissipated by the circuit for $t > 0$, by assuming that $E = 0$ and $v(0^-)$ is known.

11.11 For the circuit shown in Fig. 11.62, where $e(t) = E + \Phi_0\delta(t)$, $a_1(t) = A_0u(t)$ $+ A_1 \sin(\omega t)u(-t)$, and $a_2(t) = A_0$, find:

1. the I/O relationship for $i_2(t)$;
2. the number of SSVs;
3. the circuit's natural frequencies;
4. the stability condition;
5. $i_2(t)$ for $t < 0$ and $i_2(0^-)$, assuming that $i_2(t)$ has reached a steady state;
6. the mean value $\bar{i}_2$ and RMS value i_{2eff} of $i_2(t)$ for $t < 0$;
7. $i_2(0^+)$;
8. $i_2(t)$ for $t > 0$; provide a qualitative graphical description of the complete solution;
9. how would the answer to question 8 change if the I/O relationship were of order 2, with complex conjugate natural frequencies (say, λ and λ^*) and with the same

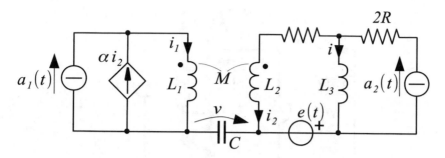

Fig. 11.62 Problem 11.11

Fig. 11.63 Problem 11.12

inputs? Which missing data would be needed to answer the latter question? Provide a qualitative graphical description of the complete solution.

11.12 For the circuit shown in Fig. 11.63, where $e(t) = \Phi_0\delta(t) - [E_0 + E_1 \cos(\omega t)]$ $u(t)$ and all the inductors are uncharged at $t = 0^-$, find:

1. the I/O relationship for $i_1(t)$;
2. the number of SSVs;
3. the circuit's natural frequencies;
4. the stability condition; henceforth, assume that it is satisfied;
5. $i_1(0^+)$;
6. $i_1(t)$ for $t > 0$.

11.13 For the circuit shown in Fig. 11.64, working in steady state for $t < 0$, where $e(t) = Eu(t) + \Phi_0\delta(t)$, $a_1(t) = A_0 + A_1 \cos(\omega t)u(-t)$, and $a_2(t) = A_2 \sin(2\omega t)$ $u(-t)$, find:

1. the I/O relationship for $i(t)$;
2. the I/O relationship for $v(t)$;
3. the number of SSVs;
4. the circuit's natural frequencies;
5. the stability condition;
6. $i(t)$ for $t < 0$ and conditions at $t = 0^-$ needed for the subsequent analysis;
7. the mean value $\bar{i}$ and RMS value i_{eff} of $i(t)$ for $t < 0$;
8. the conditions at $t = 0^+$ needed for the subsequent analysis;
9. $i(t)$ for $t > 0$,
10. $v(t)$ for $t > 0$, by assuming that the conditions at $t = 0^-$ are known.

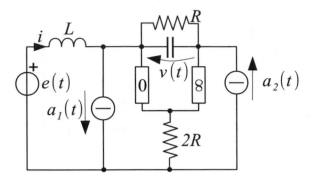

Fig. 11.64 Problem 11.13

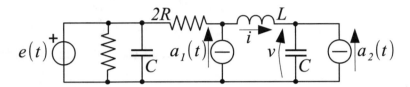

Fig. 11.65 Problem 11.14

11.14 For the circuit shown in Fig. 11.65, working in steady state for $t < 0$, where $e(t) = E_0 u(t) + E_1 \cos(2\omega t)u(-t)$, $a_1(t) = A_1 u(t) + A \sin(\omega t)u(-t)$, and $a_2(t) = A_2 u(-t)$, find:

1. the I/O relationship for $i(t)$;
2. the number of SSVs;
3. the circuit's natural frequencies;
4. the condition to have at least two complex conjugate natural frequencies;
5. the stability condition by assuming that at least two natural frequencies are complex conjugates;
6. $i(t)$ for $t < 0$ and conditions at $t = 0^-$ needed for the subsequent analysis;
7. the conditions at $t = 0^+$ needed for the subsequent analysis;
8. $i(t)$ for $t > 0$, by assuming that at least two natural frequencies are complex conjugates.

11.15 Assuming that the I/O relationship in the phasor domain for a second-order circuit working in AC steady state with a sinusoidal input $a(t)$ with angular frequency ω is

$$(\alpha - \omega^2 LC)\dot{V} = -(R - j2\omega L)\dot{A}$$

with $R, L, C > 0$, find:

1. the corresponding I/O relationship in the time domain;
2. the circuit's natural frequencies;

3. the condition to have complex conjugate natural frequencies;
4. the stability condition;
5. conditions at $t = 0^+$ needed for the analysis for $t > 0$, by assuming that $a(t) = A_0 u(t)$ and that the conditions at $t = 0^-$ are known.

11.16 For the circuit shown in Fig. 11.66, working in steady state for $t < 0$, where
$e(t) = E_0\, u(-t) + E_1 \cos(\omega t)\, u(t) + E_2 \sin(2\omega t)\, u(t) + \Phi_0 \delta(t)$, find:

1. the I/O relationship for $i(t)$;
2. the I/O relationship for $v(t)$;
3. the circuit's natural frequencies;
4. the stability condition;
5. $i(t)$ for $t < 0$ and conditions at $t = 0^-$ needed for the subsequent analysis;
6. $v(t)$ for $t < 0$;
7. $i(0^+)$;
8. $i(t)$ for $t > 0$;
9. the mean value $\bar{p}$ of the power absorbed by the nullor, assuming that $e(t) = E \cos(\omega t)$ and the circuit is working in AC steady state.

11.17 For the circuit shown in Fig. 11.68, working in AC steady state, where $e(t) = E \sin(\omega t)$ and $a(t) = A \sin(\omega t)$, find:

1. the I/O relationship for $i(t)$;
2. the circuit's natural frequencies;
3. the stability condition;
4. $i(t)$.

11.18 For the circuit shown in Fig. 11.67, working in steady state for $t < 0$, where
$a(t) = A_0\, u(t) + A_1 \sin(\omega t) u(-t)$ and $R, C > 0$, find:

1. the I/O relationship for $v(t)$;
2. the I/O relationship for $i(t)$;
3. the circuit's natural frequencies;
4. the stability condition;

Fig. 11.66 Problem 11.16

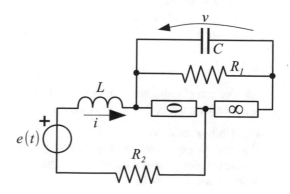

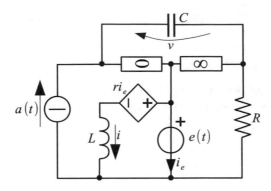

Fig. 11.67 Problem 11.18

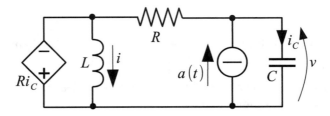

Fig. 11.68 Problem 11.17

Fig. 11.69 Problem 11.19

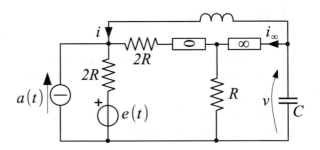

5. $v(t)$ for $t < 0$ and $v(0^-)$;
6. $i(t)$ for $t < 0$ and $i(0^-)$;
7. the conditions at $t = 0^+$ for both $v(t)$ and $i(t)$ needed for the subsequent analysis;
8. $v(t)$ for $t > 0$;
9. $i(t)$ for $t > 0$.

11.19 For the circuit shown in Fig. 11.69, working in steady state, where $a(t) = A_0 + A_1 \sin(\omega t)$ and $e(t) = E_0 + E_1 \cos(\omega t)$, find:

1. the stability condition;
2. the mean value $\bar{i}_\infty$ over the period $T = 2\pi/\omega$ of the current i_∞;
3. the mean value $\bar{p}_1$ over the period $T = 2\pi/\omega$ of the power absorbed by the resistor R, by assuming $C = 0$.

Fig. 11.70 Problem 11.20

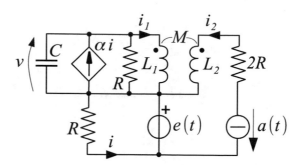

11.20 For the circuit shown in Fig. 11.70, working in steady state for $t < 0$, where $a(t) = A_0 \, u(t)$ and $e(t) = [E_0 + E_1 \sin(\omega t)] \, u(-t)$, find:

1. the I/O relationship for $v(t)$;
2. the circuit's natural frequencies;
3. the stability condition;
4. $v(t)$ for $t < 0$ and the conditions at $t = 0^-$ needed for the subsequent analysis;
5. the conditions at $t = 0^+$ for $v(t)$ needed for the subsequent analysis;
6. $v(t)$ for $t > 0$.

11.21 For the circuit shown in Fig. 11.71, working in steady state for $t < 0$, where $a(t) = A \, u(-t) + Q_0 \delta(t)$, find:

1. the I/O relationship for $v_2(t)$;
2. the circuit's natural frequencies;
3. the condition to have two complex conjugate natural frequencies and, assuming that this condition is satisfied, the stability condition;
4. $v_2(t)$ for $t < 0$ and the conditions at $t = 0^-$ needed for the subsequent analysis;
5. the conditions at $t = 0^+$ for $v_2(t)$ needed for the subsequent analysis;
6. $v_2(t)$ for $t > 0$.

11.22 For the circuit shown in Fig. 11.72a, working in steady state, find:

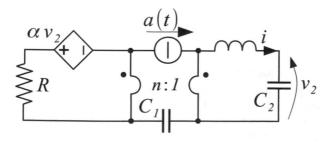

Fig. 11.71 Problem 11.21

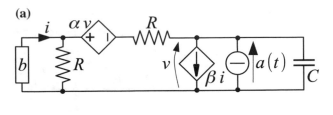

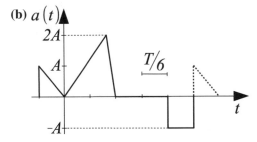

Fig. 11.72 Problem 11.22: **a** circuit; **b** T-periodic input

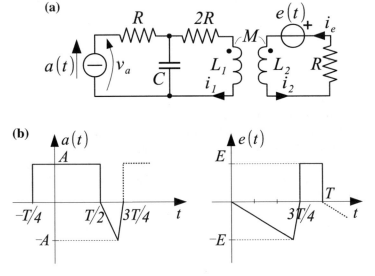

Fig. 11.73 Problem 11.23: **a** circuit; **b** T-periodic inputs

1. the mean values $\bar{v}$ and $\bar{i}$ when the input is T-periodic with the shape shown in
   Fig. 11.72b, $\alpha = 1$, $\beta = 2$, and the two-terminal b is a resistor of resistance $2R$;
2. the mean power $\bar{p}$ absorbed over the period $T = 2\pi/\omega$ by the resistor connected
   in parallel to b when $a(t) = A_0 + A_1 \cos(\omega t)$, $\alpha = \beta = 0$, and the two-terminal
   b is an inductor of inductance L.

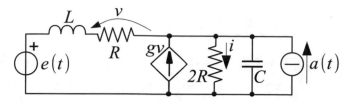

Fig. 11.74 Problem 11.24

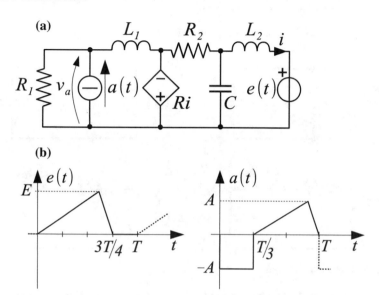

Fig. 11.75 Problem 11.25: **a** circuit; **b** T-periodic inputs

11.23 For the circuit shown in Fig. 11.73a, working in steady state, find:

1. the mean values $\bar{v}_a$ and $\bar{i}_e$ when the inputs are T-periodic with the shapes shown in Fig. 11.73b;
2. the RMS value $i_{1\text{eff}}$ of the current $i_1(t)$ over the period $T = 2\pi/\omega$ when $R = 0$, $a(t) = A$, $e(t) = E_0 + E_1 \cos(\omega t)$, and the inductors are closely coupled.

11.24 For the circuit shown in Fig. 11.74, working in steady state with $a(t) = A_0 + A_1 \sin(2\omega t)$ and $e(t) = E_0 + E_1 \cos(\omega t)$, find:

1. the mean values $\bar{v}$ and $\bar{i}$ over the period $T = 2\pi/\omega$;
2. the mean power $\bar{p}$ absorbed over the period $T = 2\pi/\omega$ by the resistor $2R$ when $L = 0$ and $g = 0$.

11.25 For the circuit shown in Fig. 11.75a, working in steady state, find:

1. the mean value $\bar{v}_a$ when the inputs are T-periodic with the shapes shown in Fig. 11.75b;

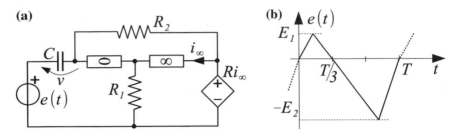

Fig. 11.76 Problem 11.26: **a** circuit; **b** T-periodic input

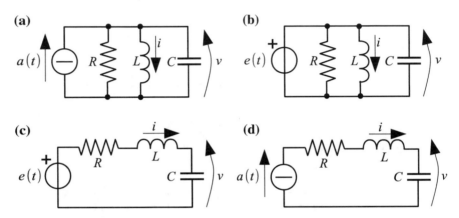

Fig. 11.77 Problem 11.27

2. the mean power $\bar{p}$ absorbed over the period $T = 2\pi/\omega$ by the resistor R_2 when $C = 0$, $a(t) = A_0 + A_1 \cos(2\omega t)$ and $e(t) = E_0 + E_1 \sin(\omega t)$.

11.26 For the circuit shown in Fig. 11.76a, working in steady state with the T-periodic input shown in Fig. 11.76b, find:

1. the condition for the existence of a steady state;
2. the mean value $\bar{v}$.

11.27 The four circuits shown in Fig. 11.77 are obtained by connecting a composite two-terminal made by R, L, C in parallel (series) to a current or to a voltage source for $t \geq 0$. The initial state of the circuits can be nonzero.
Indicate which of these circuits, if any, cannot reach a steady state and justify your answer.

References

1. Apostol T (1967) Calculus, vol 1. Wiley, New York
2. Budak BM, Fomin SV (1973) Multiple integrals, field theory and series. MIR Publishers, Moscow
3. Golub G, Van Loan CF (2013) Matrix computations. The Johns Hopkins University Press, Baltimore

Chapter 12
Advanced Concepts: Higher-Order Nonlinear Circuits—State Equations and Equilibrium Points

> *Science may set limits to knowledge, but should not set limits to imagination.*
> Bertrand Russell

Abstract In this chapter, some of the concepts introduced in Chap. 11 are extended to nonlinear higher-order circuits. In particular, we generalize the complementary component method to find the state equations of a given circuit and propose a method for analyzing and classifying steady-state DC solutions, corresponding to equilibrium points. The most common bifurcations of equilibria are also introduced and described. Finally, we propose some remarks about small-signal analysis of nonlinear circuits.

12.1 Nonlinear State Equations

Consider a circuit that contains, in addition to linear time-invariant components (both dynamic and memoryless) and input sources, also nonlinear memoryless components with two or more terminals.

Inasmuch as the dynamic components are linear, we can apply to them the same procedure followed in Sect. 11.3.2. According to this procedure, coupled inductors and coupled capacitors are replaced by equivalent models (see Sect. 11.2.5) containing ideal transformers plus two-terminal inductors and capacitors, respectively. Moreover, for every linear algebraic constraint on WSVs, one memory component is replaced by a memoryless two-terminal consisting of controlled and independent sources, as shown in Sect. 9.3.2; see Figs. 9.13 and 9.15. For each of these independent sources, the impressed voltage/current is proportional to the time derivative of the original sources.

After these replacements, the circuit contains only two-terminal capacitors and inductors (N_C and N_L of them, respectively) whose WSVs are also SSVs. Therefore, the original circuit can be represented as the connection of a set of $n = N_C + N_L$

© Springer Nature Switzerland AG 2020

M. Parodi and M. Storace, *Linear and Nonlinear Circuits: Basic and Advanced Concepts*,
Lecture Notes in Electrical Engineering 620,
https://doi.org/10.1007/978-3-030-35044-4_12

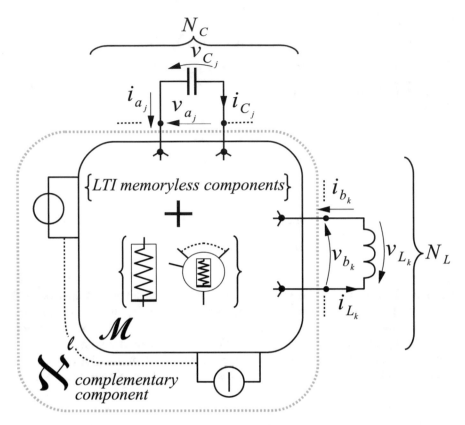

Fig. 12.1 Circuit representation as a memoryless complementary component ℵ connected to linear inductors and capacitors

linear memory components to a *complementary component* ℵ, which is a memoryless n-port, as shown in Fig. 12.1. The vectors v_C (N_C entries) and i_L (N_L entries) are the sets of SSVs for capacitors and inductors, respectively.

The state vector

$$x = \begin{pmatrix} v_C \\ i_L \end{pmatrix}$$

has n entries. The ℓ independent sources (including those coming from the equivalent circuit representation of the algebraic constraints) are conceived as internal to ℵ and represented as inputs to the $(n + \ell)$-port $\mathcal{M}$. They are collected in an ℓ-size vector $w(t)$. In particular, $\mathcal{M}$ includes the controlled sources deriving from the equivalent circuit representation of the algebraic constraints, as shown in Fig. 12.2.

Inasmuch as the nonlinear memoryless elements are confined within $\mathcal{M}$, the multiport's descriptive equations are nonlinear and of an algebraic nature. The number of descriptive variables of $\mathcal{M}$ is $2(n + \ell)$. Those concerning the N_C ports are denoted by the vectors v_a and i_a; similarly, vectors v_b, i_b denote the descriptive variables for

Fig. 12.2 For every algebraic constraint transformation, the resulting controlled source is included within $\mathcal{M}$, whereas the independent source is represented as an input for $\mathcal{M}$

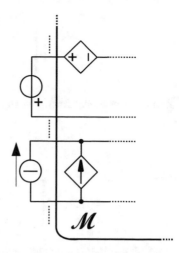

the N_L ports. We assume that v_a, i_b and the input vector $w(t)$ constitute a (mixed) basis of definition for $\mathcal{M}$ [1]. This amounts to stating that the descriptive equations for $\mathcal{M}$ can be written in the hybrid form

$$\begin{cases} i_a = h_a \left(v_a, i_b, w(t) \right), \\ v_b = h_b \left(v_a, i_b, w(t) \right). \end{cases}$$

Inasmuch as in the circuit satisfies $v_a = v_C$, $i_b = i_L$, $i_a = -i_C$, and $v_b = -v_L$, the previous equations can be reformulated as

$$\begin{cases} i_C = -h_a \left(v_C, i_L, w(t) \right), \\ v_L = -h_b \left(v_C, i_L, w(t) \right). \end{cases} \tag{12.1}$$

Each element of the vectors h_a and h_b is a nonlinear function of its arguments. A similar set of equations could be written for the ℓ variables on the input ports.

Now, by defining the diagonal matrices

$$[C] = \begin{bmatrix} C_1 & 0 & \cdots & 0 \\ 0 & C_2 & 0 & \vdots \\ \vdots & \cdots\cdots & \vdots \\ 0 & \cdots\cdots & C_{N_C} \end{bmatrix}; \quad [L] = \begin{bmatrix} L_1 & 0 & \cdots & 0 \\ 0 & L_2 & 0 & \vdots \\ \vdots & \cdots\cdots & \vdots \\ 0 & \cdots\cdots & L_{N_L} \end{bmatrix},$$

we can write

$$i_C = [C] \frac{dv_C}{dt}, \quad v_L = [L] \frac{di_L}{dt}.$$

By substituting these equations in Eq. 12.1, we easily obtain the following result:

$$\frac{dx}{dt} = \begin{pmatrix} \dfrac{dv_C}{dt} \\[2ex] \dfrac{di_L}{dt} \end{pmatrix} = \begin{pmatrix} -[C]^{-1} h_a\,(x,\ w(t)) \\[1ex] -[L]^{-1} h_b\,(x,\ w(t)) \end{pmatrix}. \tag{12.2}$$

In a more compact form, the state equations (Eq. 12.2) can be recast as

$$\frac{dx}{dt} = -[D]^{-1} h(x, w(t)), \tag{12.3}$$

with

$$[D] = \begin{bmatrix} [C] & 0_{N_C \times N_L} \\[2ex] 0_{N_L \times N_C} & [L] \end{bmatrix}; \quad h(\cdot) = \begin{pmatrix} h_a(\cdot) \\[1ex] h_b(\cdot) \end{pmatrix}. \tag{12.4}$$

As a final remark, consider the case in which the original circuit has only constant input sources. In this case, the voltage/current independent sources generated by algebraic constraints disappear, because they include a d/dt operator. The remaining sources can be viewed as nonlinear, time-invariant, resistive elements, and as such, they can be incorporated into $\mathcal{M}$. After this modification, we have $w(t) = 0$ and $\mathcal{M} \equiv \aleph$. Therefore, the nonlinear state equations (Eq. 12.3) can be expressed in the *autonomous form*

$$\frac{dx}{dt} = \underbrace{-[D]^{-1} h(x)}_{f(x)} = f(x). \tag{12.5}$$

12.1.1   Existence of the State Equation

The process of formulating the state equations is not as simple as it might appear from the procedure described above, which is based on the assumption that the hybrid vector function $h\,(\cdot)$ in Eq. 12.3 is known a priori. Actually, $h\,(\cdot)$ can be determined, provided that some conditions are met within the circuit. In particular, every resistor inside $\mathcal{M}$ must admit a basis that can be expressed in terms of x and w, that is, the basis admitted by $\mathcal{M}$.

To show this, consider a multiterminal (or multiport) resistor inside $\mathcal{M}$ and suppose that its (known) descriptive equations are

$$\begin{cases} i_a = h_a(v_a, i_b), \\ v_b = h_b(v_a, i_b). \end{cases} \tag{12.6}$$

This explicit form implies that the resistor admits the basis (v_a, i_b). In other words, for known values of v_a and i_b, the complementary variables i_a and v_b follow directly from Eq. 12.6. In the particular case of a two-terminal resistor, Eq. 12.6 reduces to a

Fig. 12.3 Descriptive
variables at the jth terminal
(or port) of a nonlinear
resistor

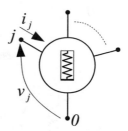

Fig. 12.4 Circuit constraint
to be satisfied when v_j
belongs to a basis admitted
by the resistor. **a** General
case; **b** particular case

(a)　　　　　　　　**(b)**

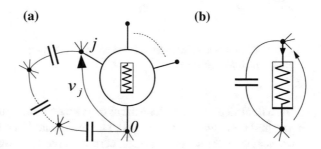

single equation, either $i = h(v)$ or $v = h(i)$, which identifies a voltage-controlled or
a current-controlled element, respectively.

With regard to $\mathcal{M}$, for simplicity we set the components of $w(t)$ to zero, and we
focus our attention on the remaining terms (x components) of the basis.

Let us consider the jth terminal (or port) of the resistor, together with its descriptive variables v_j, i_j, as shown in Fig. 12.3. Let h_j be the function element of h_a or
h_b in Eq. 12.6 related to v_j, i_j. We have two alternatives:

- If the basis (v_a, i_b) contains v_j, that is, if one of the descriptive equations (Eq. 12.6)
  can be written as

$$i_j = h_j(\ldots, v_j, \ldots),$$

  then a necessary condition to determine h explicitly is that v_j be expressible in
  terms of the v_C components, that is, the terminal pair $(j, 0)$ (or the jth port)
  belongs to a loop containing only capacitors, in addition to itself. This is shown
  in Fig. 12.4a and, as a particular case, in Fig. 12.4b.
- If the basis variable is i_j,

$$v_j = h_j(\ldots, i_j, \ldots),$$

  the circuit must allow us to express i_j in terms of the i_L components. Therefore,
  the jth terminal belongs to a cut-set that, in addition to i_j, involves only inductor
  currents. This corresponds to the general case shown in Fig. 12.5a. Figure 12.5b
  shows a particular case.

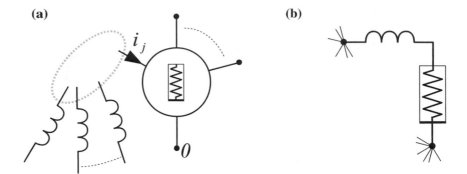

Fig. 12.5 Circuit constraint to be satisfied when i_j belongs to a basis admitted by the resistor. **a** general case; **b** particular case

As a final step, we can extend the above two conditions to the case in which $w \neq 0$. In the first case, the voltage v_j has to be formulated in terms of v_C and the voltage sources in w; in the second case, the current i_j has to be expressed in terms of i_L and the current sources in w. Correspondingly, the path in Fig. 12.4 includes branches with voltage sources, whereas the cut-set in Fig. 12.5 includes branches with current sources.

The violation of the above conditions in at least one case makes it impossible to formulate state equations in the canonical form of Eq. 12.3.

Case Study 1

The circuit shown in Fig. 12.6a includes two nonlinear resistive elements. The three-terminal resistor is voltage-controlled with descriptive equations

$$\begin{cases} i_a = h_a \left(v_a, v_b \right), \\ i_b = h_b \left(v_a, v_b \right). \end{cases} \tag{12.7}$$

The two-terminal resistor is current-controlled. Its descriptive equation is $v_R = h(i_R)$. Following the general procedure, find:

- *the resistive multiport $\mathcal{M}$;*
- *the bases admitted by $\mathcal{M}$;*
- *the nonlinear state equations for the circuit.*

  We first consider the algebraic constraint between the WSVs v_1, v_2, v_3 and the input voltage $e(t)$. The capacitor C_3 can be replaced by the algebraic model α shown in Fig. 12.6b, because

$$i_3 = \frac{C_3}{C_1} i_1 - \frac{C_3}{C_2} i_2 + C_3 \frac{de}{dt}, \tag{12.8}$$

where i_1, i_2 are the capacitor currents through C_1, C_2, respectively, and the last term is viewed as an independent current source.

This transformation allows us to represent the circuit (see Fig. 12.7a) as a resistive multiport connected to C_1, C_2, and L and to the independent sources $e(t)$ and $C_3 \frac{de}{dt}$, which represent the components of the input vector $w(t)$. Observe that the voltages v_a and v_b, which are a basis for the three-terminal resistor, are closed on C_1 and on a loop containing only C_2 and C_1, respectively; moreover, the current-controlled resistor is in series with the inductor L.

Now let us verify that v_1, v_2, i_L plus the elements of w are a basis for the multiport $\mathcal{M}$. To this end, consider the circuit shown in Fig. 12.7b, where the two capacitors and the inductor connected to $\mathcal{M}$ are replaced by sources with impressed voltages v_1, v_2 and current i_L, respectively. A direct inspection of this circuit confirms that all the variables belonging to a basis for the elements inside $\mathcal{M}$ can be expressed in terms of v_1, v_2, and i_L plus e and a, which are a basis for $\mathcal{M}$.

Taking into account Eqs. 12.7 and 12.8, we obtain

$$\begin{cases} i_1 + i_2 = h_a\left(v_1, v_1 - v_2\right), \\ i_b = h_b\left(v_1, v_1 - v_2\right), \\ i_b = -i_2\left(1 + \dfrac{C_3}{C_2}\right) + i_L + \dfrac{C_3}{C_1}i_1 + \underbrace{C_3\dfrac{de}{dt}}_{a}, \\ v_1 - v_2 = v_L + h(-i_L). \end{cases} \qquad (12.9)$$

The state equations can be recast in the canonical form of Eq. 12.3. To this end, we use the first three equations to eliminate i_b and then to find algebraically i_1 and i_2; after this step, setting $\Delta = -\left(1 + \dfrac{C_3}{C_2} + \dfrac{C_3}{C_1}\right)$ for brevity, we have

$$\begin{cases} \dfrac{dv_1}{dt} = -\dfrac{i_1}{C_1} = -\dfrac{1}{\Delta C_1}\left\{-\left(1 + \dfrac{C_3}{C_2}\right)h_a(v_1, v_1 - v_2) + i_L - h_b(v_1, v_1 - v_2) + C_3\dfrac{de}{dt}\right\}, \\ \dfrac{dv_2}{dt} = -\dfrac{i_2}{C_2} = -\dfrac{1}{\Delta C_2}\left\{-i_L + h_b(v_1, v_1 - v_2) - C_3\dfrac{de}{dt} - \dfrac{C_3}{C_1}h_a(v_1, v_1 - v_2),\right\} \\ \dfrac{di_L}{dt} = -\dfrac{1}{L}\left(v_1 - v_2 - h(-i_L)\right). \end{cases}$$

$$(12.10)$$

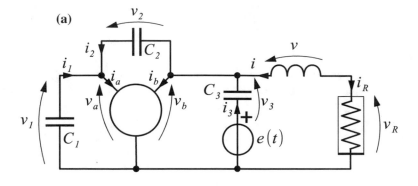

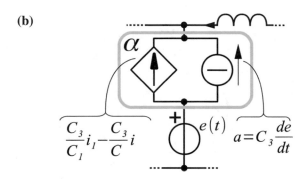

Fig. 12.6 **a** Case Study 1; **b** replacement of C_3 with the memoryless equivalent two-terminal α

Case Study 2
The circuit shown in Fig. 12.8a includes two nonlinear resistive elements. The three-terminal has a mixed basis v_a, i_b and descriptive equations

$$\begin{cases} i_a = h_a\,(v_a, i_b)\,, \\ v_b = h_b\,(v_a, i_b)\,. \end{cases} \tag{12.11}$$

The two-terminal resistor is voltage-controlled. Its descriptive equation is $i_R = h(v_R)$. Verify that the capacitor voltages v_1, v_2 and the inductor current i_L form a basis of definition for $\mathcal{M}$ and find the nonlinear state equations for the circuit.

This circuit can be represented as a resistive multiport $\mathcal{M}$ connected to the linear memory elements C_1, C_2, and L. No independent sources are present. Observe that v_a is the voltage across C_1, whereas the i_b-terminal is in series with the inductor L. Therefore, all the variables v_a, i_b, i_R can be expressed in terms of v_1, v_2, and i_L. In order to verify that v_1, v_2, and i_L are a basis for

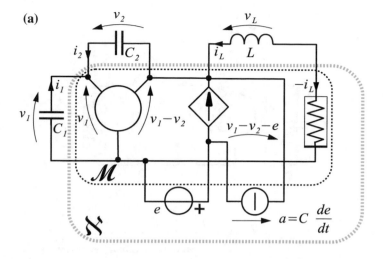

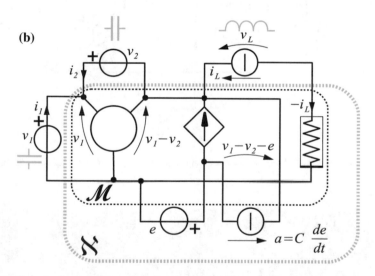

Fig. 12.7 Case Study 1: **a** equivalent representation of the circuit after the transformation of Fig. 12.6b; **b** (v_1, v_2, i_L, e, a) is a basis for the multiport $\mathcal{M}$

$\mathcal{M}$ ($\equiv \aleph$), we can assign them by replacing C_1, C_2, and L with voltage and current sources, as shown in Fig. 12.8b, and verify that the complementary port variables can always be calculated.

(a)

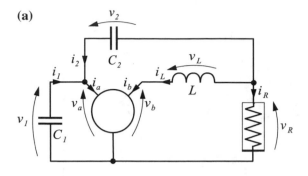

(b)

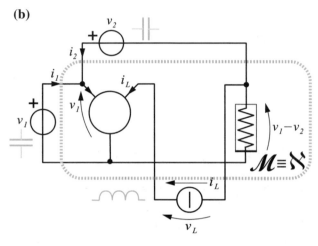

Fig. 12.8 **a** Case Study 2; **b** equivalent circuit showing that (v_1, v_2, i_L) is a basis for the multiport $\mathcal{M} \equiv \aleph$

Taking into account Eq. 12.11, the corresponding circuit equations are

$$
\begin{cases}
i_1 + i_2 = h_a(v_1, i_L), \\
v_b = h_b(v_1, i_L), \\
i_2 + i_L + h(v_1 - v_2) = 0, \\
v_b - v_L = v_1 - v_2,
\end{cases}
\Rightarrow
\begin{cases}
i_1 = i_L + h(v_1 - v_2) + h_a(v_1, i_L), \\
i_2 = -i_L - h(v_1 - v_2), \\
v_L = -v_1 + v_2 + h_b(v_1, i_L).
\end{cases}
$$

$$(12.12)$$

These equations give the port variables i_1, i_2, and v_L in terms of v_1, v_2, and i_L, as expected.

Finally, taking the above equations and writing i_1, i_2, v_L in terms of the time derivatives of v_1, v_2, and i_L, we obtain the state equations in canonical form:

$$
\begin{cases}
\dfrac{dv_1}{dt} = -\dfrac{1}{C_1}\{i_L + h(v_1 - v_2) + h_a(v_1, i_L)\}, \\[2ex]
\dfrac{dv_2}{dt} = \dfrac{1}{C_2}\{i_L + h(v_1 - v_2)\}, \\[2ex]
\dfrac{di_L}{dt} = -\dfrac{1}{L}\{-v_1 + v_2 + h_b(v_1, i_L)\}.
\end{cases}
\tag{12.13}
$$

These autonomous state equations can be trivially recast as Eq. 12.5, with $x = (v_1 \ v_2 \ i_L)^T$ and the right-hand-side terms of Eq. 12.13 as components of the vector f.

Case Study 3

The circuits shown in Figs. 12.9 and 12.10a are a modified version of those considered in Case Study 2 and Case Study 1, respectively. The only variant consists in the addition, in both circuits, of a linear resistor R, marked in gray in the figures. The remaining components are defined as in the case study they come from.

Show that for both circuits, the state equations in canonical form can no longer be obtained.

- Owing to the presence of R, the voltage-controlled nonlinear resistor in Fig. 12.9 is no longer connected to a path formed by capacitors only. This can be easily verified by writing the circuit equations

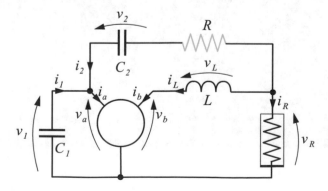

Fig. 12.9 Case study 3: circuit similar to that of Case Study 2

(a)

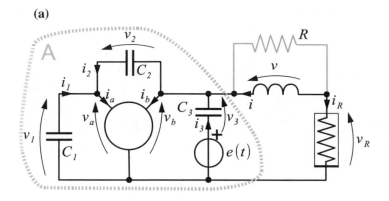

(b)

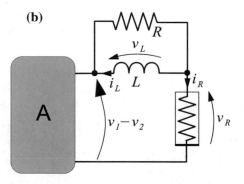

Fig. 12.10 Case Study 3: circuit similar to that of Case Study 1

$$\begin{cases} i_1 + i_2 = h_a\,(v_1, i_L)\,, \\ v_b = h_b\,(v_1, i_L)\,, \\ i_2 + i_L + h(v_R) = 0, \\ v_b - v_L = v_R. \end{cases} \tag{12.14}$$

The first two equations are the same as for Case Study 2 (Eq. 12.12), but the last two show that in this case, v_R can no longer be expressed in terms of v_1 and v_2, and we have $i_2 + i_L + h(v_b - v_L) = 0$. Now we can write v_b in terms of v_1 and i_L using Eq. 12.11 and the variables i_2 and v_L as the time derivatives of v_2 and i_L, respectively. The resulting equation is

$$-C_2\frac{dv_2}{dt} + i_L + h\left(h_b\,(v_1, i_L) + L\frac{di_L}{dt}\right) = 0.$$

Since the term $\dfrac{di_L}{dt}$ appears within the argument of the nonlinear function $h(\cdot)$, the state equations in canonical form cannot be obtained.

- For $R \to \infty$, the whole circuit reduces to that of Case Study 1. The region A of the circuit in Fig. 12.10 is the same as in Case Study 1, and its capacitor C_3 can be replaced accordingly. Therefore, A is thought of as a two-terminal element with voltage $v_1 - v_2$. The branches associated with R, L and with the current-controlled nonlinear resistor form a cut-set. Making reference to the compact representation shown in Fig. 12.10b, we have

$$\begin{cases} i_L - \dfrac{v_L}{R} + i_R = 0, \\ v_1 - v_2 = v_L + h(i_R), . \end{cases} \Rightarrow \quad v_1 - v_2 = v_L + h\left(\dfrac{v_L}{R} - i_L\right). \quad (12.15)$$

Since $v_L = -L\dfrac{di_L}{dt}$, the time derivative of i_L could be written in terms of v_1, v_2, i_L by solving the above nonlinear implicit equation, which rules out the possibility to obtain the state equations in canonical form.

12.2 Linear Autonomous Circuits Revisited

Before analyzing nonlinear circuits, we briefly summarize the concepts already introduced in Chap. 11 for linear circuits, but from a dynamical systems standpoint. This is useful also for nonlinear circuits, as pointed out in Sect. 12.3. As usual (see Chap. 10), we restrict our analysis to a specific class of linear circuits, where:

(1) all components are time-invariant, including the sources, which can only be constant;
(2) the solution exists and is unique, for an assigned initial condition.

Our main goal is to find their equilibrium points and their state portraits.

Owing to the assumptions listed above, after a proper change of variables, the circuit equations can always be recast in canonical form as $\dot{x} = f(x) = Ax$, where x is the (normalized) state vector and $f(x) = Ax$ is the (linear and autonomous) **vector field**, which defines the law of the state evolution in time. The DC solutions of this dynamical system must satisfy the equilibrium condition $\dot{x} = 0$.

The origin is an equilibrium point, inasmuch as $x = 0$ always satisfies the equilibrium condition, but it is not necessarily either unique or stable. This will become clearer in the next subsections. We just notice that when $det(A) = 0$, the equilibrium condition $\dot{x} = 0$ has infinitely many solutions, each one corresponding to a different initial condition. The existence of nontrivial solutions for $Ax = 0$ means that the null space of A is not empty: $\dim(\mathcal{N}(A)) \neq 0$.

The circuit's natural frequencies are the eigenvalues of state matrix A and can be counted (taking into account the corresponding multiplicity) and grouped into three sets (notice that some of them can be empty):

- S^- contains the n^- eigenvalues with $\Re\{\lambda_j\} < 0$, corresponding to a set of eigenvectors $\{\Lambda^-\}$;
- S^+ contains the n^+ eigenvalues with $\Re\{\lambda_j\} > 0$, corresponding to a set of eigenvectors $\{\Lambda^+\}$;
- S^0 contains the n^0 eigenvalues with $\Re\{\lambda_j\} = 0$, corresponding to a set of eigenvectors $\{\Lambda^0\}$.

The jth eigenvector is a solution of the equation $A\Lambda_j = \lambda_j \Lambda_j$. The direction (straight line) spanned by an eigenvector is also called an eigendirection.

Of course, for a circuit with n SSVs, $n^- + n^+ + n^0 = n$. Moreover, the circuit is absolutely stable if and only if $n^+ = n^0 = 0$.

If at least two among n^-, n^+, and n^0 are not null, the linear subspace of the state space generated by each set of eigenvectors is called a *manifold*:

- $W^s = span\{\Lambda^-\}$ (with dimension n^-) is the **stable manifold**;
- $W^u = span\{\Lambda^+\}$ (with dimension n^+) is the **unstable manifold**;
- $W^0 = span\{\Lambda^0\}$ (with dimension n^0) is the **center manifold**.

Within S^-, the eigenvalue(s) closest to the imaginary axis is called the *leading eigenvalue*. It corresponds to the vanishing natural mode that lasts the longest, given that it corresponds to the maximum time constant (see Sect. 11.6).

Similarly, within S^+, the eigenvalue(s) closest to the imaginary axis is called the *leading eigenvalue*. It corresponds to the slowest increasing natural mode.

Examples related to the above definitions are provided in the following subsections.

12.2.1   Second-Order Circuits

In order to reiterate and reinforce the concepts introduced above, we initially focus on second-order circuits.

Case Study
Consider the linear circuit shown in Fig. 12.11 and find its state equations in the canonical form $\dot{x} = Ax$, through proper normalization and change of variables. Then find eigenvalues and eigenvectors of A.

Fig. 12.11 Case Study

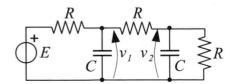

The circuit's state equations are

$$C\frac{dv_1}{dt} = \frac{E - v_1}{R} + \frac{v_2 - v_1}{R},$$

$$C\frac{dv_2}{dt} = \frac{v_1 - v_2}{R} - \frac{v_2}{R},$$

The DC solution is found by imposing the equilibrium condition $\frac{dv_1}{dt} = \frac{dv_2}{dt} = 0$, which provides $v_1^* = \frac{2}{3}E$ and $v_2^* = \frac{E}{3}$.

The "perturbation" with respect to the equilibrium solution is given by the shifted variables $\hat{v}_1 = v_1 - v_1^*$ and $\hat{v}_2 = v_2 - v_2^*$. See Sect. 11.7.1 for a similar change of variables in a mechanical example. The state equations can be recast in terms of these variables:

$$\frac{d}{dt}\begin{pmatrix} \hat{v}_1 \\ \hat{v}_2 \end{pmatrix} = \underbrace{\begin{pmatrix} -\dfrac{2}{RC} & \dfrac{1}{RC} \\ \dfrac{1}{RC} & -\dfrac{2}{RC} \end{pmatrix}}_{\hat{A}} \begin{pmatrix} \hat{v}_1 \\ \hat{v}_2 \end{pmatrix}. \tag{12.16}$$

We now introduce the normalized variables x_1, x_2:

$$\begin{pmatrix} \hat{v}_1 \\ \hat{v}_2 \end{pmatrix} = \underbrace{\begin{pmatrix} V_0 & 0 \\ 0 & V_0 \end{pmatrix}}_{Q} \begin{pmatrix} x_1 \\ x_2 \end{pmatrix},$$

where V_0 is an arbitrary reference voltage. Matrix Q is nonsingular (therefore, invertible) and allows defining a *similarity transformation* (or conjugation) of matrix $\hat{A}$. Indeed, Eq. 12.16 can be recast as follows in terms of the normalized variables:

$$Q\frac{d}{dt}\begin{pmatrix} x_1 \\ x_2 \end{pmatrix} = \hat{A}Q\begin{pmatrix} x_1 \\ x_2 \end{pmatrix},$$

that is,

$$\frac{d}{dt}\begin{pmatrix} x_1 \\ x_2 \end{pmatrix} = Q^{-1}\hat{A}Q\begin{pmatrix} x_1 \\ x_2 \end{pmatrix}.$$

For any invertible matrix Q, $Q^{-1}\hat{A}Q$ has the same eigenvalues, trace, and determinant as $\hat{A}$ [2]. In linear algebra, the trace of a square matrix is defined to be the sum of the elements on the main diagonal, that is, the diagonal from the upper left to the lower right corner.

We now define a normalized time $\tau = \dfrac{t}{RC}$, thus obtaining the normalized state equations:

$$\begin{pmatrix} \dot{x}_1 \\ \dot{x}_2 \end{pmatrix} = \underbrace{\begin{pmatrix} -2 & 1 \\ 1 & -2 \end{pmatrix}}_{= A} \begin{pmatrix} x_1 \\ x_2 \end{pmatrix}, \tag{12.17}$$

that is, $\dot{x} = Ax$.

The circuit's natural frequencies can be found by looking for the eigenvalues of A, that is, by solving the characteristic equation $\lambda^2 + 4\lambda + 3 = 0$. The solutions are $\lambda_1 = -1$ (leading eigenvalue) and $\lambda_2 = -3$. Notice that the eigenvalues of $\hat{A}$ are $\dfrac{\lambda_1}{RC}$ and $\dfrac{\lambda_2}{RC}$.

The corresponding eigenvectors are found by solving the equation $A\Lambda_j = \lambda_j \Lambda_j$, thus obtaining $\Lambda_1 = \begin{pmatrix} 1 \\ 1 \end{pmatrix}$ and $\Lambda_2 = \begin{pmatrix} -1 \\ 1 \end{pmatrix}$.

The complete solution starting in $\tau = 0$ from a given initial condition $(x_1(0), x_2(0))$ is

$$x(\tau) = \alpha_1 \Lambda_1 e^{\lambda_1 \tau} + \alpha_2 \Lambda_2 e^{\lambda_2 \tau}, \tag{12.18}$$

where the coefficients α_1 and α_2 can be found by imposing the initial conditions

$$x(0) = \begin{pmatrix} x_1(0) \\ x_2(0) \end{pmatrix} = \alpha_1 \begin{pmatrix} 1 \\ 1 \end{pmatrix} + \alpha_2 \begin{pmatrix} -1 \\ 1 \end{pmatrix}.$$

Therefore,

$$\alpha_1 = \frac{x_1(0) + x_2(0)}{2},$$
$$\alpha_2 = \frac{x_2(0) - x_1(0)}{2}.$$

Figure 12.12a shows the system state portrait, obtained by plotting the parametric equations

$$\begin{pmatrix} x_1(\tau) \\ x_2(\tau) \end{pmatrix} = \begin{pmatrix} \alpha_1 & -\alpha_2 \\ \alpha_1 & \alpha_2 \end{pmatrix} \begin{pmatrix} e^{\lambda_1 \tau} \\ e^{\lambda_2 \tau} \end{pmatrix}$$

starting from eight different initial conditions. The two straight lines correspond to the eigenvectors Λ_1 (slow eigendirection, denoted by a single arrow and corresponding to the principal eigenvalue) and Λ_2 (fast eigendirection, denoted by a double arrow).

The figure points out that all trajectories initially evolve parallel to the fast eigendirection, whereas they asymptotically converge to the origin along the slow eigendirection.

Fig. 12.12 Case Study: **a** state portrait; **b** time evolution of the state variables x_1 (dashed black line) and x_2 (solid gray line)

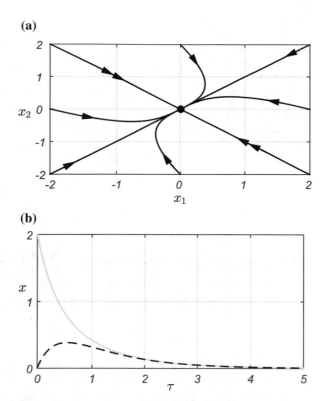

Figure 12.12b shows the time evolution (starting from the initial condition $(0, 2)$) of the normalized state variables x_1 (dashed black line) and x_2 (solid gray line).

Finally, Fig. 12.13 shows a 3D plot of the solution starting from the initial condition $(0, 2)$ in the space (τ, x_1, x_2) (thick black line). Moreover, for the sake of comparison with Fig. 12.12, it shows also:

- the solution projection on the state plane (x_1, x_2), that is, the trajectory (dark gray line), which can be compared with the corresponding trajectory in Fig. 12.12a;
- the solution projections on the planes (τ, x_1) (thin black line) and (τ, x_2) (light gray line), which can be compared with the corresponding plots in Fig. 12.12b.

The above case study allowed us to exemplify some definitions provided at the beginning of Sect. 12.2. In general, we would like to analyze all possible behaviors of second-order circuits, corresponding to different features of the natural frequencies: real, complex conjugate, null, with positive/negative real part, and so forth. To this

Fig. 12.13 Case Study: 3D plot of the solution starting from the initial condition $(0, 2)$ in the space (τ, x_1, x_2) (thick black line) and its projections on the state plane (x_1, x_2) (dark gray line), on the plane (τ, x_1) (thin black line), and on the plane (τ, x_2) (light gray line)

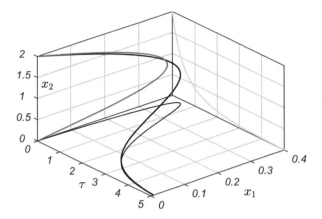

end, we focus on a generic state matrix $A = \begin{pmatrix} A_{11} & A_{12} \\ A_{21} & A_{22} \end{pmatrix}$ and study its eigenvalues. We initially assume that the eigenvalues are distinct.

The characteristic equation is $\det(\lambda I - A) = 0$, that is,

$$\lambda^2 - \tau\lambda + \Delta = 0, \tag{12.19}$$

where $\tau = A_{11} + A_{22}$ is the trace[1] of A, whereas $\Delta = A_{11}A_{22} - A_{12}A_{21}$ is the determinant of A. The trace of a matrix is also the sum of its eigenvalues; therefore $\tau = \lambda_1 + \lambda_2$. The determinant of a matrix is also the product of its eigenvalues; therefore $\Delta = \lambda_1\lambda_2$.

The expressions for λ_1 and λ_2 are

$$\lambda_1 = \frac{\tau + \sqrt{\tau^2 - 4\Delta}}{2}, \qquad \lambda_2 = \frac{\tau - \sqrt{\tau^2 - 4\Delta}}{2}. \tag{12.20}$$

Therefore, the eigenvalues are complex conjugates if and only if $\tau^2 - 4\Delta < 0$.

Making reference to Fig. 12.14, it is immediate that the two eigenvalues are complex conjugates when the pair (Δ, τ) lies inside the parabola $\tau^2 = 4\Delta$. Indeed, for every point on the positive half-axis Δ, we have $\tau = 0$ and $\Delta > 0$, and therefore $\tau^2 - 4\Delta < 0$.

We begin by focusing on real eigenvalues; thus we analyze the plane (Δ, τ) outside the parabola.

- In region ①, we have positive Δ and negative τ, that is, $\lambda_1\lambda_2 > 0$ and $\lambda_1 + \lambda_2 < 0$. This means that both λ_1 and λ_2 are negative, as well as in the case study, where $\tau = -4$ and $\Delta = 2$.

---

[1] We use the symbol τ for the trace, following the notation used in [3]. We remark that τ is dimensionless, as well as the elements of A, and has nothing to do with the normalized variable introduced in the last case study.

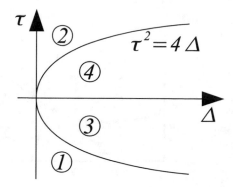

Fig. 12.14 Regions in the plane (Δ, τ) corresponding to different features of the eigenvalues λ_1 and λ_2

(a)   **(b)**

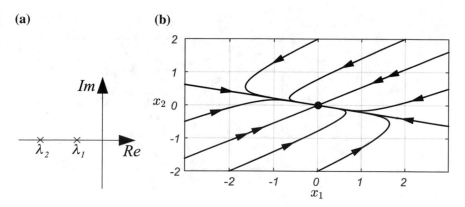

Fig. 12.15 Stable node: **a** eigenvalues; **b** state portrait

In this case, Λ_1 and Λ_2 span the whole state space ($n = n^-$), and therefore there are no manifolds. The origin is *globally stable* and is called a *stable node*.

Figure 12.15 shows an example for $A = \begin{pmatrix} -1 & -3 \\ -\dfrac{1}{3} & -2 \end{pmatrix}$, corresponding to $\lambda_1 \approx -0.38$ (leading eigenvalue) and $\lambda_2 \approx -2.62$.

- In region ②, Δ and τ are both positive, meaning that $\lambda_1 \lambda_2 > 0$ and $\lambda_1 + \lambda_2 > 0$. Therefore, both λ_1 and λ_2 are positive.

  Also in this case, Λ_1 and Λ_2 span the whole state space ($n = n^+$), and therefore there are no manifolds. The origin is (strongly) *unstable* and is called an *unstable node*.

  Figure 12.16 shows an example for $A = \begin{pmatrix} 3 & -1 \\ -2 & 4 \end{pmatrix}$, corresponding to $\lambda_1 = 5$ and $\lambda_2 = 2$. The two straight lines correspond to the eigenvectors Λ_1 (slow eigendirection, denoted by a single arrow and corresponding to the principal eigenvalue) and Λ_2 (fast eigendirection, denoted by a double arrow).

(a)   **(b)**

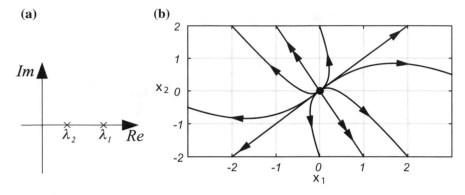

Fig. 12.16 Unstable node: **a** eigenvalues; **b** state portrait

(a)   **(b)**

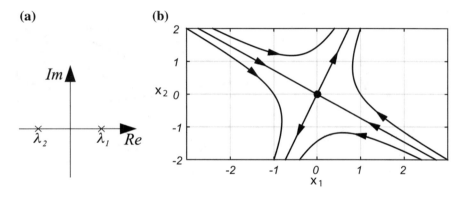

Fig. 12.17 Saddle: **a** eigenvalues; **b** state portrait

The figure points out that all trajectories initially evolve parallel to the slow eigendirection, whereas they asymptotically diverge along the fast eigendirection.

- In the half-plane where $\Delta < 0$, we have $\lambda_1 \lambda_2 < 0$. Therefore, λ_1 and λ_2 have opposite signs. According to Eq. 12.20, we assume that $\lambda_1 > 0$ and $\lambda_2 < 0$.
  The origin is (strongly) *unstable*, the stable $W^- = span\{\Lambda_2\}$ and the unstable $W^+ = span\{\Lambda_1\}$ manifolds are straight lines, and the origin is called a *saddle*.
  The trajectories initially evolve parallel to the stable manifold and asymptotically tend to the unstable manifold, thus describing hyperbolic curves in the state plane, as shown in Fig. 12.17, obtained with $A = \begin{pmatrix} -1 & -1 \\ 2 & 1 \end{pmatrix}$, corresponding to $\lambda_1 \approx 1.73$ and $\lambda_2 \approx -1.73$.

- On the τ-axis, we have $\Delta = 0$, and thus at least one eigenvalue is null, and the matrix A is singular. This implies that the origin is no longer the unique equilibrium point, since we have infinite solutions of the equilibrium condition. In particular, if $\tau < 0$, we have $\lambda_1 = 0$ and $\lambda_2 < 0$.

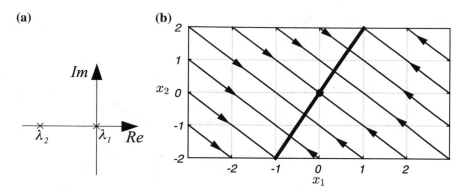

Fig. 12.18 Infinite Lyapunov-stable equilibrium points: eigenvalues (**a**) and state portrait (**b**)

The origin is (simply) *stable*, and the stable $W^- = span\{\Lambda_2\}$ and the center $W^0 = span\{\Lambda_1\}$ manifolds are straight lines.

The center manifold contains infinite Lyapunov-stable (that is, nonattractive; see Sect. 10.1) equilibrium points. As shown in Fig. 12.18, the trajectories evolve parallel to the stable manifold and asymptotically end on the center manifold (thick black line), thus describing straight lines in the state plane. In this example, $A = \begin{pmatrix} -2 & 1 \\ 2 & -1 \end{pmatrix}$, corresponding to $\lambda_1 = 0$ and $\lambda_2 = -3$.

Similarly, if $\tau > 0$, we have $\lambda_1 > 0$ and $\lambda_2 = 0$, the origin is *unstable*, and the unstable $W^+ = span\{\Lambda_1\}$ and the center $W^0 = span\{\Lambda_2\}$ manifolds are straight lines.

The center manifold contains infinite unstable equilibrium points. The trajectories evolve parallel to the unstable manifold and asymptotically diverge, describing straight lines in the state plane.

The trajectories in the state plane are completely similar to those shown in Fig. 12.18, but with reversed arrows.

We now consider the case of complex conjugate eigenvalues, analyzing the plane (Δ, τ) inside the parabola, where $\Delta > 0$.

- In region ③, we have negative τ, that is $\Re\{\lambda_1\} = \Re\{\lambda_2\} < 0$.

  In this case there are no manifolds; the origin is *globally stable* and is called a *stable focus*.

  The solutions are sinusoids with decaying amplitude (see Sect. 11.5.1), and thus the corresponding trajectories in the state plane describe spirals converging to the origin, as shown in Fig. 12.19b, obtained with $A = \begin{pmatrix} 0 & -1 \\ 5 & -1 \end{pmatrix}$. Figure 12.19c shows the time evolution of the state variables x_1 (black curve) and x_2 (gray curve). Panels b and c are nothing but projections of the solution curve in 3-dimensional space (τ, x_1, x_2), which is shown in Fig. 12.20.

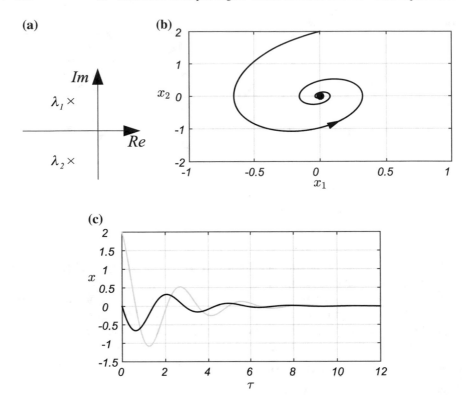

Fig. 12.19 Stable focus: **a** eigenvalues; **b** state portrait; **c** time evolution of the state variables x_1 (black curve) and x_2 (gray curve)

- In region ④, we have positive τ, that is, $\Re\{\lambda_1\} = \Re\{\lambda_2\} > 0$.

  Also in this case there are no manifolds; the origin is *unstable* and is called an *unstable focus*.

  The solutions are sinusoids with increasing amplitude, and thus the corresponding trajectories in the state plane describe spirals diverging from the origin.

  The trajectories in the state plane are completely similar to the one shown in Fig. 12.19b, but with reversed arrows.

- Finally, on the $\Delta > 0$ half-axis, we have $\tau = 0$, that is, $\Re\{\lambda_1, \lambda_2\} = 0$.

  Also in this case there are no manifolds; the origin is *Lyapunov-stable* and is called a *center*.

  The solutions are sinusoids with constant amplitude, and thus the corresponding trajectories in the state plane describe closed (ellipsoidal) curves around the origin, as shown in Fig. 12.21.

For instance, the RLC series circuit shown in Fig. 11.31 (Sect. 11.6), with constant input and assigned initial condition, corresponds to a normalized system in which the origin (after a proper change of variables) is either a stable node (overdamped

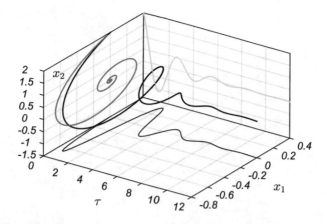

Fig. 12.20 Case Study: 3D plot of the solution starting from the initial condition $(0, 2)$ in the space (τ, x_1, x_2) (thick black line) and its projections on the state plane (x_1, x_2) (dark gray line), on the plane (τ, x_1) (thin black line), and on the plane (τ, x_2) (light gray line)

(a) **(b)**

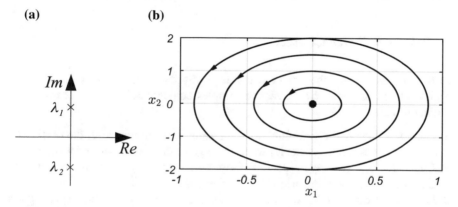

Fig. 12.21 Center: **a** eigenvalues; **b** state portrait

system) or a stable focus (underdamped system). In the limit case for $R = 0$, it is a center.

Finally, we should consider matrices A with two identical eigenvalues $\lambda_1 = \lambda_2 = \frac{\tau}{2}$, corresponding to the parabola $\tau^2 = 4\Delta$. Actually, this is a limit case not particularly significant from a circuit viewpoint, and we address the interested reader to specific textbooks such as [3].

Figure 12.22 summarizes the main behaviors described above.

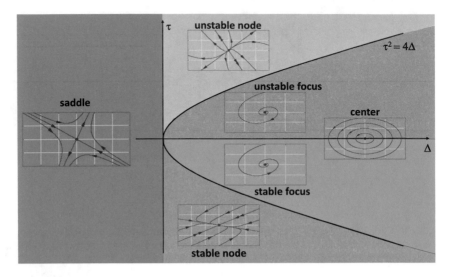

Fig. 12.22 Summary of the main possible behaviors for a linear 2D system

12.2.2   nth-Order Circuits

In third-order circuits, the matrix A has three eigenvalues, one of which must be real, while the other two can be real or complex conjugates. Here are the four most common cases:

- There are three real eigenvalues with the same sign. In this case, the origin is a **node**, which is stable (unstable) when the eigenvalues are negative (positive).
- There are three real eigenvalues, at least one of them positive and at least one negative. In this case, the origin is a **saddle** and is unstable.
- There are one real and two complex conjugate eigenvalues, all having real parts with the same sign. In this case, the origin is a **node-focus** and is stable (unstable) when the real part of the eigenvalues is negative (positive).
- There are two complex conjugate eigenvalues and one real eigenvalue with sign opposite that of the real part of the other two. In this case, the origin is a **saddle-focus** and is unstable.

Figure 12.23 shows an example of a stable node, with $A = \begin{pmatrix} -1 & 0 & 0 \\ 0 & -2 & 0 \\ 0 & 0 & -3 \end{pmatrix}$.

Figure 12.24 shows an example of a stable node-focus, with $A = \begin{pmatrix} 0 & -1 & 0 \\ 5 & -1 & 0 \\ 0 & 0 & -2 \end{pmatrix}$.

In this case, the leading eigenvalue is a pair of complex conjugate eigenvalues; therefore, the gray trajectory initially evolves along the fast direction (coinciding

(a)

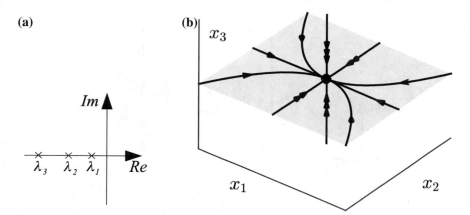

Fig. 12.23 3D stable node: **a** eigenvalues; **b** state portrait

(a)

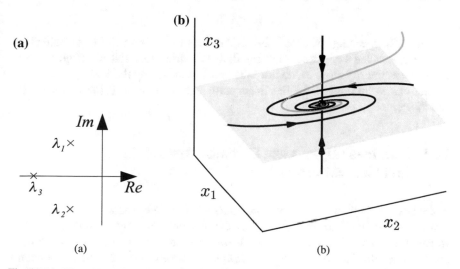

(a)   (b)

Fig. 12.24 3D stable node: **a** eigenvalues; **b** state portrait

with the axis x_3) and progressively rotates around the origin as it approaches the plane $x_3 = 0$.

Figure 12.25 shows another example of a stable node-focus, with $A = \begin{pmatrix} 0 & -1 & 0 \\ 18 & -1 & 0 \\ 0 & 0 & -0.3 \end{pmatrix}$. In this case, the leading eigenvalue is real, and therefore, the gray trajectory shrinks as it rotates around the slow direction (coinciding with the x_3-axis) and progressively approaches the origin.

For higher-order cases, one usually simply checks the position of the A eigenvalues in the complex plane to evaluate the origin's stability.

(a)   **(b)**

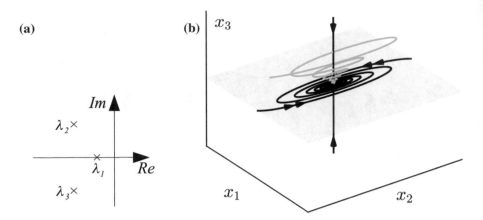

Fig. 12.25 3D stable node: **a** eigenvalues; **b** state portrait

Generally speaking, when all the eigenvalues of A lie in the right half-plane, the origin is called a **repellor**, meaning that every trajectory starting from an initial condition arbitrarily close to the origin is "pushed away" by the flow.

The names introduced for the linear case will also be used for nonlinear circuits/systems.

12.3 Nonlinear Dynamical Circuits: Assumptions and General Properties

In the next sections, we generalize the concepts already introduced in Chap. 10 for first-order circuits. Also for higher-order circuits, described by more than one state equation, we can either find a *numerical solution* through integration algorithms or obtain some information about the *qualitative behavior* of the solution and about the presence and properties of equilibrium points (in this section) or periodic solutions (in Chap. 14), through both analytic and geometric tools [3].

Also in this case we restrict our analysis to a specific class of nth-order nonlinear circuits for which a unique solution exists for an assigned initial condition, and

the **state equation** can always be written in the following **canonical form**:

$$
\begin{aligned}
\dot{x}_1 &= f_1(x_1, x_2, \ldots, x_n), \\
\dot{x}_2 &= f_2(x_1, x_2, \ldots, x_n),
\end{aligned}
\tag{12.21}
$$

$$\cdots$$
$$\dot{x}_n = f_n(x_1, x_2, \ldots, x_n), \tag{12.22}$$

or more compactly,

$$\dot{x} = f(x), \tag{12.23}$$

where $x = \begin{pmatrix} x_1 \\ x_2 \\ \vdots \\ x_n \end{pmatrix}$ is the state vector and $f = \begin{pmatrix} f_1 \\ f_2 \\ \vdots \\ f_n \end{pmatrix}$ is called a **vector field**

and defines the law of the state evolution in time, thus inducing a **flow** of the state. For a given initial condition, the state trajectory follows the flow.

The **state space** is the set of all possible states x of the considered circuit, that is, the domain of the function f.

According to our assumptions,

- inasmuch as we have n SSVs, the state space is n-dimensional;
- we consider only autonomous circuits (see the definition given in Sect. 10.1) whose state evolution is described by Eq. 12.5, which is equivalent to Eq. 12.23;
- each component of f is a Lipschitz-continuous function. This ensures that the solution of Eq. 12.23 for a given initial condition $x(t_0)$ exists and is unique [3].

The class of circuits described in Sect. 12.1 satisfies the condition about the canonical form.

The jth **nullcline** of the ODE system described by Eq. 12.21 is the geometric shape for which $\dot{x}_j = 0$, that is, the locus of points of the state space satisfying the constraint $f_j = 0$.

For instance, in a two-dimensional linear system, the nullclines can be represented by two lines in a plane; in a nonlinear two-dimensional system they are arbitrary curves. In an n-dimensional system, they are lower-dimensional manifolds of the state space.

Every equilibrium solution of Eq. 12.21 can be found by imposing the condition $\dot{x} = 0$; such a solution is also called an **equilibrium point** (or simply *equilibrium*) of the circuit (or system, more generally) described by Eq. 12.21.

The equilibrium points of a nth-order dynamical system $\dot{x} = f(x)$ can be found algebraically by imposing

(a)   **(b)**

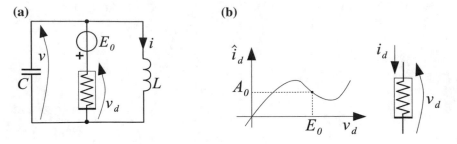

Fig. 12.26 a Van der Pol oscillator and **b** tunnel diode and its DP characteristic

$$f(x) = 0, \qquad\qquad (12.24)$$

which is called the **equilibrium condition**.

Thus the equilibrium points of the circuit can be found either by analytically solving Eq. 12.24 (*analytic method*) or graphically by finding (at least for $n = 2$) the intersections of all the nullclines (*graphical method*).

The following case study is concerned with one of the most widely used models of nonlinear autonomous oscillators, called a **van der Pol oscillator**. It is named after Balthasar van der Pol (1889–1959), a Dutch physicist whose main interests were in radio wave propagation, theory of electrical circuits, and mathematical physics. We remark that the example provides not only an application of the concepts introduced above, but also a road map for finding proper normalizations.

Case Study: Van der Pol oscillator
Consider the van der Pol oscillator shown in Fig. 12.26. By assuming that $i_d = \hat{i}_d(v_d) = A_0 + \gamma(v_d - E_0)^3 - \alpha(v_d - E_0)$, *with positive-dimensional coefficients* γ *[A/V³] and* α *[Ω^{-1}], find its state equations in the canonical form* $\dot{x} = f(x)$ *through proper normalization and change of variables. Then find the equilibrium solutions, both analytically and graphically.*
    The circuit equations are

$$\begin{cases} C\dfrac{dv}{dt} + \hat{i}_d(v_d) + i = 0, \\ v + E_0 = v_d, \\ v = L\dfrac{di}{dt}, \end{cases}$$

corresponding to the state equations (in canonical form)

$$
\begin{cases}
\dfrac{dv}{dt} = \dfrac{-\hat{\imath}_d(v + E_0) - i}{C} = \dfrac{-A_0 - \gamma v^3 + \alpha v - i}{C}, \\[3mm]
\dfrac{di}{dt} = \dfrac{v}{L}.
\end{cases}
$$

By imposing $\dfrac{dv}{dt} = 0$ and $\dfrac{di}{dt} = 0$ (equilibrium condition), we immediately obtain $v^* = 0$ and $i^* = -A_0$. Now we introduce the shifted variable $\tilde{\imath} = i - i^* = i + A_0$, so that the origin of the plane $(v, \tilde{\imath})$ is the only equilibrium:

$$
\begin{cases}
\dfrac{dv}{dt} = \dfrac{-\gamma v^3 + \alpha v - \tilde{\imath}}{C}, \\[3mm]
\dfrac{d\tilde{\imath}}{dt} = \dfrac{v}{L}.
\end{cases}
$$

By introducing the normalized time variable

$$
\tau = \frac{t}{\sqrt{LC}},
$$

the state equations can be recast as follows:

$$
\begin{cases}
\dfrac{dv}{d\tau} = \sqrt{\dfrac{L}{C}}\left(-\gamma v^3 + \alpha v\right) - \sqrt{\dfrac{L}{C}}\tilde{\imath} = \alpha\sqrt{\dfrac{L}{C}}\left(1 - \dfrac{\gamma}{\alpha}v^2\right)v - \sqrt{\dfrac{L}{C}}\tilde{\imath}, \\[3mm]
\dfrac{d\tilde{\imath}}{d\tau} = \sqrt{\dfrac{C}{L}}v.
\end{cases}
$$

Now we introduce the normalized parameter $\varepsilon = \alpha\sqrt{\dfrac{L}{C}} > 0$, thus obtaining

$$
\begin{cases}
\dfrac{dv}{d\tau} = \varepsilon\left(1 - \dfrac{\gamma}{\alpha}v^2\right)v - \sqrt{\dfrac{L}{C}}\tilde{\imath}, \\[3mm]
\dfrac{d\tilde{\imath}}{d\tau} = \sqrt{\dfrac{C}{L}}v.
\end{cases}
$$

Finally, we normalize the state variables

$$
x = \frac{v}{V_0}, \qquad y = \frac{\tilde{\imath}}{I_0},
$$

with V_0 and I_0 arbitrary. Their expressions can be determined with the aim of obtaining the simplest form for the equations to be handled.

After a few manipulations, the (normalized) state equations can be recast as follows:

$$\begin{cases} \dfrac{dx}{d\tau} = \dot{x} = \varepsilon \left(1 - \dfrac{\gamma}{\alpha}V_0^2 x^2\right) x - \sqrt{\dfrac{L}{C}}\dfrac{I_0}{V_0}y, \\ \dfrac{dy}{d\tau} = \dot{y} = \sqrt{\dfrac{C}{L}}\dfrac{V_0}{I_0}x. \end{cases}$$

Therefore, we can choose $\sqrt{\dfrac{C}{L}\dfrac{V_0}{I_0}} = 1$, thus obtaining

$$\begin{cases} \dot{x} = \varepsilon \left(1 - \dfrac{\gamma}{\alpha}V_0^2 x^2\right) x - y, \\ \dot{y} = x. \end{cases}$$

Now we have two possibilities:

(a) If we reason from the state equations, we can set $\dfrac{\gamma}{\alpha}V_0^2 = 1$, that is, $V_0 = \sqrt{\dfrac{\alpha}{\gamma}}$ and $I_0 = \sqrt{\dfrac{C}{L}}\sqrt{\dfrac{\alpha}{\gamma}}$. In this case, the normalized state equations are

$$\begin{cases} \dot{x} = \varepsilon \left(1 - x^2\right) x - y, \\ \dot{y} = x. \end{cases} \tag{12.25}$$

With this choice, the I/O relationship for x is $\ddot{x} = \varepsilon \left(1 - 3x^2\right) \dot{x} - x$.

(b) If we reason from the I/O relationship for x, we have $\ddot{x} = \varepsilon \left(1 - \dfrac{3\gamma}{\alpha}V_0^2 x^2\right) \dot{x} - x$, and thus we can set $\dfrac{3\gamma}{\alpha}V_0^2 = 1$, that is $V_0 = \sqrt{\dfrac{\alpha}{3\gamma}}$ and $I_0 = \sqrt{\dfrac{C}{L}}\sqrt{\dfrac{\alpha}{3\gamma}}$. In this case, the normalized state equations are

$$\begin{cases} \dot{x} = \varepsilon \left(1 - \dfrac{x^2}{3}\right) x - y, \\ \dot{y} = x, \end{cases} \tag{12.26}$$

and the I/O relationship for x is $\ddot{x} = \varepsilon \left(1 - x^2\right) \dot{x} - x$.

In both cases, the only equilibrium is the origin, as stated at the beginning. This is confirmed also graphically: Fig. 12.27 shows the nullclines $y = \varepsilon \left(1 - x^2\right) x$ (solid black curve) and $x = 0$ (gray curve)—whose point of intersection is indeed the origin—corresponding to Eq. 12.25 for $\varepsilon = 2$.

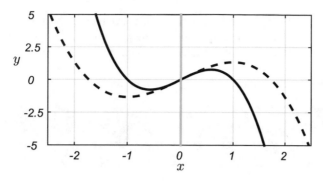

Fig. 12.27 Nullclines for the normalized state equations (Eq. 12.25) describing the van der Pol oscillator (for $\varepsilon = 2$): $y = \varepsilon \left(1 - x^2\right) x$ (solid black curve) and $x = 0$ (gray curve). Using Eq. 12.26, the solid black curve is replaced by the dashed black curve $y = \varepsilon \left(1 - \dfrac{x^2}{3}\right) x$

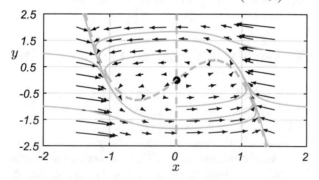

Fig. 12.28 Equation 12.25: vector field (black arrows), examples of trajectories starting from different initial conditions (solid gray lines), and nullclines (dashed gray lines)

The dashed black curve is the nullcline $y = \varepsilon \left(1 - \dfrac{x^2}{3}\right) x$ corresponding to Eq. 12.26, for $\varepsilon = 2$.

For the case of Eq. 12.25, Fig. 12.28 shows the vector field (black arrows), some examples of trajectories starting from different initial conditions (solid gray lines), and the nullclines (dashed gray lines). All trajectories converge to a periodic solution, corresponding to a closed curve in the state plane.

Figure 12.28 points out some important concepts. First of all, the phase portrait contains some loci of points that are invariant under the action of the vector field and for this reason are called *invariant sets*.

An **invariant set** of a dynamical system $\dot{x} = f(x)$ is a subset S of the state space such that the trajectory starting from any initial condition $x(t_0) \in S$ evolves within S for all $t > t_0$.

For instance, examples of invariant sets are any equilibrium point, any asymptotic solution of the dynamical system (see the periodic solution in Fig. 12.28), any trajectory (by definition), any stable/unstable/center manifold (see the examples in Sect. 12.2).

A second element evidenced by Fig. 12.28 is that

> in the **nonlinear case**, the **stability is no longer a property of the whole system**, as in the linear case, **but of each invariant set**.

For instance, it is apparent that the equilibrium point (origin of the state plane) is unstable, whereas the periodic solution to which the trajectories tend under the action of the vector field is stable.

Stability analysis is the subject of the next two sections.

12.4 Equilibrium Stability Analysis Through Linearization

Similarly to the first-order case analyzed in Sect. 10.1.3, also in higher-order circuits the stability of an equilibrium point x^* (satisfying the condition $f(x^*) = 0$) can be simply checked, both analytically and graphically.

Graphically, it is sufficient to analyze the flow directions around the equilibrium point, as stated above. For instance, in Fig. 12.28, the orientations of the vector field arrows show that the flow brings the state far from the origin, which is the unique equilibrium point.

Analytically, we can draw the same conclusions by analyzing the evolution of a small perturbation $w = x - x^*$ away from x^*. Since $\dot{w} = \dot{x} = f(x) = f(w + x^*)$, we can use a Taylor expansion, thereby obtaining

$$\dot{w}_j = f_j(x^*) + \left[\nabla f_j(x)\big|_{x=x^*}\right]^T w + \mathcal{O}(w_j^2), \qquad j = 1, \ldots, n, \qquad (12.27)$$

where

$$\left[\nabla f_j(x)\right]^T = \left[\frac{\partial f_j}{\partial x_1}, \frac{\partial f_j}{\partial x_2}, \ldots, \frac{\partial f_j}{\partial x_n}\right]$$

is the *gradient* of f_j with respect to the vector of Cartesian coordinates x.

On the whole, in matrix form we have

$$\dot{w} = f(x^*) + J(x)\big|_{x=x^*} w + \mathcal{O}(||w||^2), \qquad (12.28)$$

where

$$
J(x) = \begin{pmatrix}
\dfrac{\partial f_1}{\partial x_1} & \dfrac{\partial f_1}{\partial x_2} & \cdots & \dfrac{\partial f_1}{\partial x_n} \\[2ex]
\dfrac{\partial f_2}{\partial x_1} & \dfrac{\partial f_2}{\partial x_2} & \cdots & \dfrac{\partial f_2}{\partial x_n} \\[2ex]
\vdots & \vdots & \ddots & \vdots \\[2ex]
\dfrac{\partial f_n}{\partial x_1} & \dfrac{\partial f_n}{\partial x_2} & \cdots & \dfrac{\partial f_n}{\partial x_n}
\end{pmatrix}
$$

is the **Jacobian matrix** of f with respect to x.

But $f(x^*) = 0$ (by the definition of equilibrium point), so

$$
\dot{w} = J(x)|_{x=x^*} \, w + \mathcal{O}(||w||^2) \tag{12.29}
$$

An equilibrium point x^* of a system $\dot{x} = f(x)$ is **hyperbolic** if $\det[\, J(x)|_{x=x^*}] \neq 0$. Otherwise, it is nonhyperbolic. In other words, the Jacobian matrix of a system computed in a hyperbolic equilibrium point does not have any eigenvalue lying on the imaginary axis of the complex plane, and the corresponding linearized system has no center manifold.

The higher-order terms $O(||w||^2)$ can be neglected only if x^* is hyperbolic and for low perturbation values. Under these assumptions, the perturbation evolution is approximately governed by the linear nth-order system

$$
\dot{w} = J(x)|_{x=x^*} \, w. \tag{12.30}
$$

Therefore, the perturbation either grows or decays in time depending on the eigenvalues of $J(x)|_{x=x^*}$. Returning to the definitions provided in Sect. 12.2, we can assume that $J(x)|_{x=x^*}$ has n^- eigenvalues with $\Re\{\lambda_j\} < 0$ and n^+ eigenvalues with $\Re\{\lambda_j\} > 0$. Since x^* is hyperbolic by assumption, we have no eigenvalues with $\Re\{\lambda_j\} = 0$, that is, $n^0 = 0$.

In particular, the equilibrium point x^* is:

- (locally) stable if and only if $n^+ = 0$; this property holds locally, within the limits imposed by approximation 12.30;
- unstable if $n^+ > 0$.

In contrast, if x^* is nonhyperbolic we cannot draw any conclusion from Eq. 12.30 about the stability of x^*, because the perturbation evolution is ruled by the higher-order terms $O(||w||^2)$, neglected by the linearization. In this case, we have to resort to other methods (typically graphical or based on numerical integration) to evaluate the equilibrium's stability.

According to their local behavior, the equilibrium points are classified using the same terminology as for linear systems.

An unstable equilibrium point with $n^+ = n$ (all the eigenvalues of $J(x)|_{x=x^*}$ lie in the right half-plane) is called a **repellor**, since it locally pushes away any trajectory starting in its neighborhood.

Case Study: stability analysis of the van der Pol oscillator
Analyze the stability of the equilibrium solution of the van der Pol oscillator by applying the linearization method.

We already found that the only equilibrium point is the origin. We refer to Eq. 12.25, whose Jacobian matrix is

$$J(x) = \begin{pmatrix} \varepsilon(1 - 3x^2) & -1 \\ 1 & 0 \end{pmatrix}$$

Therefore,

$$J(x)|_{x=x^*} = \begin{pmatrix} \varepsilon & -1 \\ 1 & 0 \end{pmatrix}.$$

Inasmuch as $n = 2$, we can evaluate the stability through τ and Δ, as shown in Sect. 12.2.1: $\tau = \varepsilon > 0$ and $\Delta = 1$. This point lies in the first quadrant of the (Δ, τ)-plane, and the equilibrium is unstable, either a node or a focus, according to the value of ε. In any case, it is a repellor. For instance, with the value used to obtain Fig. 12.28 ($\varepsilon = 2$), we have $\tau^2 - 4\Delta = 0$; therefore, we are exactly at the edge between node and focus, with two coincident real and positive eigenvalues $\lambda_1 = \lambda_2 = 1$.

In a neighborhood of the generic equilibrium point x^*, if at least two among n^-, n^+, and n^0 are not null, there exist:

- a **stable manifold** W^s, of dimension n^-;
- an **unstable manifold** W^u, of dimension n^+;
- a **center manifold** W^0, of dimension n^0.

These manifolds are invariant sets, and:

- at x^* they are tangent to the corresponding manifolds of the linearized system (Eq. 12.30);
- the dynamics on W^s and W^u are equivalent to that of the linearized system;
- the dynamics on W^0, instead, depends on the higher-order terms $O(||w||^2)$, neglected by the linearization.

12.5 Bifurcations

In Sect. 10.4 we introduced the concept of *bifurcation* as a qualitative change in the dynamics of a given circuit/system due to a change in one or more parameters, often called *control parameters*. Moreover, for first-order circuits we introduced the *fold bifurcation of equilibria*, through which two equilibrium points are created or destroyed. This type of bifurcation occurs when an eigenvalue of the circuit state equation linearized around one of the two equilibrium points crosses the edge of stability, that is, when $f'(x^*) = 0$.

Also for circuits of order greater than one, the simplest way to have a bifurcation for an equilibrium x^* is that one or more eigenvalues of the Jacobian matrix $J(x)|_{x=x^*}$ cross the edge of stability, that is, when one or more eigenvalues have real part equal to 0. In other words, this happens when $n^0 > 0$, meaning that x^* is nonhyperbolic.

In the simplest cases, in the complex plane we have either only one real eigenvalue that becomes null (see Fig. 12.29a) or a pair of complex conjugate eigenvalues that cross the imaginary axis, as sketched in Fig. 12.29b.

12.5.1 Fold Bifurcation of Equilibria

The first case is a fold bifurcation of equilibria; it corresponds to the collision of two equilibrium points in the control space $(x; p)$, as described in Sect. 10.4.2 for first-order circuits.

For instance, for a second-order system with a parameter p, Figs. 12.30, 12.31 and 12.32 show, for three values of p, a qualitative example pointing out:

- the equilibrium points E_1 (stable node) and E_2 (saddle point) and two sections (Σ_1 corresponding to $p = p_1$ and Σ_2 corresponding to $p = p_2 > p_1$) in the control space $(x; p)$ (Fig. 12.30);

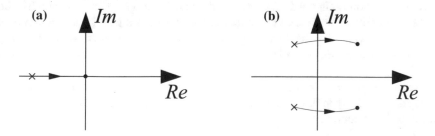

Fig. 12.29 Changes of eigenvalues that determine change of stability of equilibria: **a** fold bifurcation; **b** Hopf bifurcation. The crosses (dots) mark the initial (final) positions of the eigenvalues. The transition from initial to final position is due to a continuous change in a bifurcation parameter

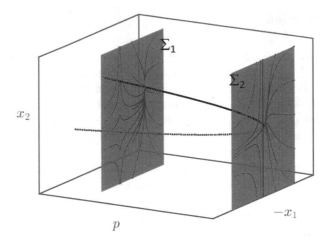

Fig. 12.30 Fold bifurcation of equilibria for circuits/systems of order greater than one: by changing (increasing in the figure) the control parameter p, two equilibrium points E_1 (stable node, solid line) and E_2 (saddle point, dotted line) collide in the control space (p, x_1, x_2) at $p = p_2$

- the state portraits in the (x_1, x_2)-plane corresponding to three sections $p = p_1$ (see Σ_1 in Fig. 12.30), $p = p_2$ (Σ_2), and $p = p_3 > p_2$, respectively in the control space (Fig. 12.31);
- the position of the eigenvalues of E_1 and E_2 in the complex plane for $p = p_1$ and $p = p_2$ (Fig. 12.32).

12.5.2   Hopf Bifurcation

The second case is the so-called Hopf (or Andronov–Hopf or Poincaré–Andronov–Hopf) bifurcation,[2] which corresponds to the collision of a focus and a self-sustained periodic solution (called a *limit cycle*) in the control space.[3] Correspondingly, the Jacobian matrix of the system computed at the focus has a simple pair of complex eigenvalues on the imaginary axis, $\lambda_{1,2} = \pm j\omega$, and these eigenvalues are the only ones with null real part. A Hopf bifurcation can occur in state spaces of any dimension $n \geq 2$.

A bifurcation can occur in one of two possible ways. In both cases, the equilibrium changes its stability in the same way. What distinguish the two cases is the behavior of the periodic solution.

---

[2]It is named after Eberhard Frederich Ferdinand Hopf (1902–1983), an Austrian mathematician and astronomer, Aleksandr Aleksandrovich Andronov (1901–1952), a Soviet physicist, and Henri Poincaré (1854–1912), a French mathematician.

[3]A limit cycle (as detailed in Chap. 14) is isolated in the state space, unlike the Lyapunov-stable periodic solutions around a center encountered in Sect. 12.2.1.

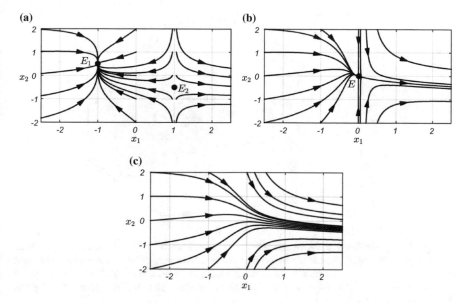

Fig. 12.31 Fold bifurcation of equilibria for circuits/systems of order greater than one: state portraits of the complete circuit/system on three Poincaré sections **a** Σ_1, **b** Σ_2 (where E_1 and E_2 collide, collapsing in a unique equilibrium point E), **c** Σ_3 (not shown in Fig. 12.30), corresponding to parameter values $p_1 < p_2 < p_3$, respectively

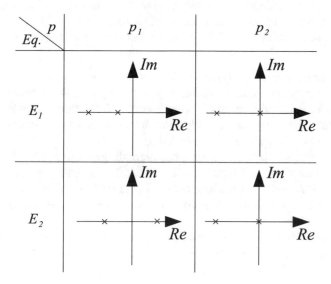

Fig. 12.32 Fold bifurcation of equilibria for circuits/systems of order greater than one: eigenvalues of the system computed at E_1 and E_2 for $p = p_1$ and $p = p_2$

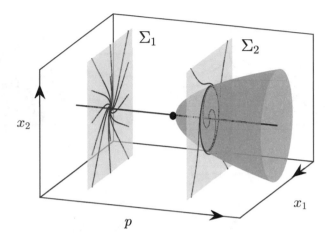

Fig. 12.33 Supercritical Hopf bifurcation for a circuit/system of order two in the control space (p, x_1, x_2): by changing (increasing in the figure) the control parameter p, the equilibrium point (focus, solid line) changes its stability from stable (dark green) to unstable (red) and collides in the control space (p, x_1, x_2) at $p = p*$ (black dot) with a periodic solution (light green), which exists for $p > p*$ and is stable

In the first case (*supercritical* Hopf bifurcation), the periodic solution is stable and, under the usual assumption of smooth vector field f, branches with small amplitude around the stationary solution when the equilibrium point becomes unstable. For instance, for a second-order system with a parameter p, for two values of p, Figs. 12.33, 12.34, 12.35 and 12.36 illustrate a qualitative example pointing out the following:

- the equilibrium point (focus) and the limit cycle and two sections (Σ_1 corresponding to $p = p_1$ and Σ_2 corresponding to $p = p_2 > p_1$) in the control space $(x; p)$ (Fig. 12.33); it is apparent that by changing the bifurcation parameter, there is a continuous transition from a stable invariant set to another one; the bifurcation point (at $p = p^*$) is marked by a black dot;
- the state portraits in the (x_1, x_2)-plane corresponding to $p = p_1$ (Fig. 12.34a, which displays in detail what happens on Σ_1 in Fig. 12.33) and $p = p_2$ (Fig. 12.34b, which displays in detail what happens on Σ_2) in the control space;
- the time plots in the (t, x_1)-plane corresponding to $p = p_1$ (Fig. 12.35a) and $p = p_2$ (Fig. 12.35b);
- the position in the complex plane of the eigenvalues of the Jacobian matrix computed in the equilibrium, for $p = p_1$ (Fig. 12.36a), $p = p^*$ (Fig. 12.36b), and $p = p_2$ (Fig. 12.36c).

We remark that *the supercritical Hopf bifurcation is one of the most common ways to generate persistent and self-sustained stable oscillations in smooth systems.*

In the second case (*subcritical* Hopf bifurcation), the periodic solution is unstable and branches with small amplitude from the equilibrium point when it becomes unstable, as shown in Fig. 12.37. In this case, unlike the supercritical one, the system does not have a continuous transition from a stable invariant set to another one; when

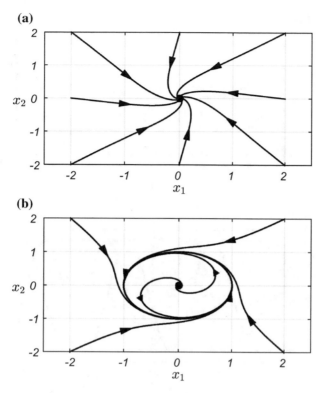

Fig. 12.34 State portraits for **a** $p = p_1$, corresponding to Σ_1 in Fig. 12.33, and for **b** $p = p_2$, corresponding to Σ_2 in Fig. 12.33

the equilibrium loses its stability, the solution must diverge or jump to another stable invariant set.

In summary, the Hopf bifurcation is one of the most common ways to obtain stable and unstable *periodic solutions*, which are treated in Chap. 14.

Case Study: Hopf bifurcation in the van der Pol oscillator

Find the value of the bifurcation parameter ε such that the van der Pol oscillator undergoes a Hopf bifurcation.

We know that for this oscillator, $\tau = \varepsilon > 0$ and $\Delta = 1$. A Hopf bifurcation occurs when a pair of complex conjugate eigenvalues cross the imaginary axis. This means that at the bifurcation, the eigenvalues are purely imaginary, and consequently, the Hopf bifurcation condition is $\tau = \varepsilon = 0$. Inasmuch as ε is strictly positive by assumption, we cannot have a Hopf bifurcation in this oscillator.

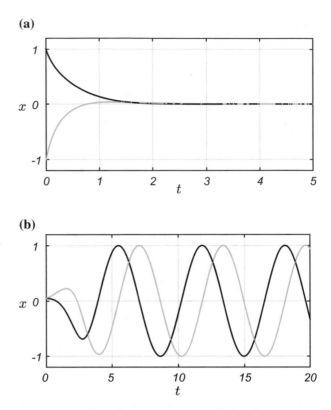

Fig. 12.35 Time plots $x_1(t)$ (black line) and $x_2(t)$ (gray line). **a** For $p = p_1$, starting from the initial condition $(1, -1)$, the solution converges to the stable focus; **b** for $p = p_2$, starting from the initial condition $(0.05, 0.05)$, the solution converges to the stable periodic solution

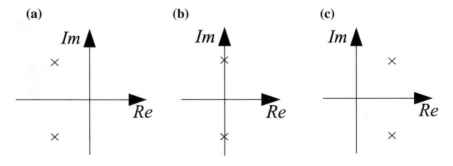

Fig. 12.36 Eigenvalues of the focus E for **a** $p = p_1$; **b** $p = p^*$; and **c** $p = p_2$

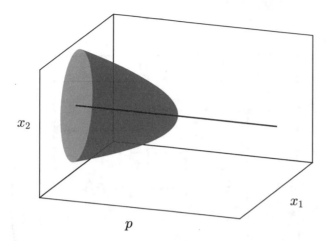

Fig. 12.37 Subcritical Hopf bifurcation for a circuit/system of order two in the control space (p, x_1, x_2): by changing (increasing in the figure) the control parameter p, the equilibrium point (focus, solid line) changes its stability from stable (dark green) to unstable (red) and collides in the control space (p, x_1, x_2) at $p = p*$ with a periodic solution (red paraboloid), which exists for $p < p*$ and is unstable

In the following sections we analyze two well-known oscillators using the conceptual tools introduced in this chapter.

12.6 Wien Bridge Oscillator

A Wien bridge oscillator is a type of electronic oscillator based on a bridge circuit originally developed by Max Wien.[4] It produces an almost-sinusoidal oscillation at a certain frequency. We analyze the configuration shown in Fig. 12.38.

First of all, we determine the characteristic $v_o = f(v_2)$, in order to analyze the equivalent (more compact) circuit shown in Fig. 12.39, where the op-amp, R_S, and R_F are replaced by a nonlinear VCVS.[5] From the KVL for the central mesh, we obtain

$$v_2 = v_x + v_o \frac{R_S}{R_S + R_F}.$$

---

[4]Max Karl Werner Wien (1866–1938) was a German physicist known for the invention of a circuit (the Wien bridge) used to measure the impedance of capacitors and inductors.

[5]Notice that the nonlinear VCVS could be obtained by properly combining linear controlled sources and a nonlinear resistor (see Sect. 5.2.1.5 in Vol. 1). This is proposed as an exercise in the problems at the end of this chapter.

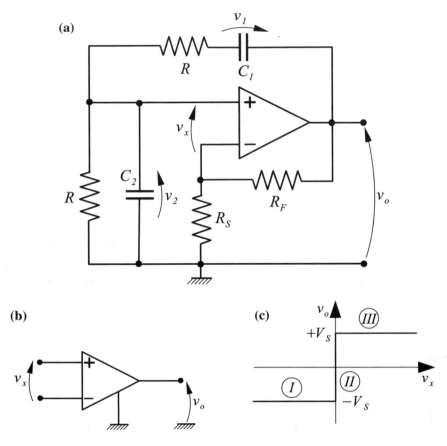

Fig. 12.38 a Wien bridge oscillator, whose **b** op-amp has **c** an idealized piecewise-constant I/O relationship

Fig. 12.39 Equivalent representation of the Wien bridge oscillator

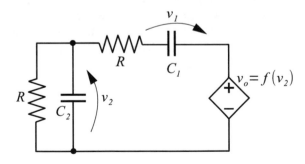

- In region ⑪, $v_x = 0$ and $v_o \in (-V_S, V_S)$. Therefore,

$$v_2 = v_o \frac{R_S}{R_S + R_F},$$

that is,

$$v_o = v_2 \frac{R_S + R_F}{R_S} = \alpha v_2,$$

where $\alpha = \dfrac{R_S + R_F}{R_S}$.

- In region ⑪, $v_o = V_S$ and $v_x > 0$. Therefore,

$$v_2 - V_S \frac{R_S}{R_S + R_F} > 0,$$

that is,

$$v_2 > \frac{V_S}{\alpha}.$$

- Similarly, in region ①, $v_o = -V_S$ and $v_x < 0$. Therefore,

$$v_2 < -\frac{V_S}{\alpha}.$$

As a consequence, the characteristic $v_o = f(v_2)$ has the shape displayed in Fig. 12.40 (gray thick curve).

The circuit equations

$$\begin{cases} v_1 + RC_1 \dfrac{dv_1}{dt} + v_2 = f(v_2), \\ C_1 \dfrac{dv_1}{dt} = \dfrac{v_2}{R} + C_2 \dfrac{dv_2}{dt}, \end{cases}$$

can be easily recast as state equations as follows:

$$\begin{cases} C_1 \dfrac{dv_1}{dt} = \dfrac{1}{R}[f(v_2) - v_1 - v_2], \\ \\ C_2 \dfrac{dv_2}{dt} = \dfrac{1}{R}[f(v_2) - 2v_2 - v_1]. \end{cases} \tag{12.31}$$

Figure 12.41 evidences the complementary-component equivalent representation for the Wien bridge oscillator. You can check that the descriptive equations for the memoryless two-port ℵ (shown in Fig. 12.41) are

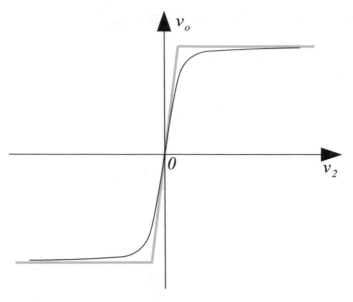

Fig. 12.40 Characteristic $v_o = f(v_2)$ corresponding to the op-amp piecewise-constant I/O relationship of Fig. 12.38c (gray thick curve) and its smoothed version (black thin curve)

Fig. 12.41 Memoryless two-port ℵ for the Wien bridge oscillator

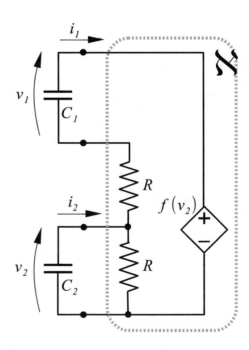

$$\begin{cases} i_1 = \dfrac{1}{R}[v_1 + v_2 - f(v_2)], \\[4mm] i_2 = \dfrac{1}{R}[v_1 + 2v_2 - f(v_2)]. \end{cases}$$

12.6.1 Normalized State Equations

By imposing the equilibrium condition, from Eq. 12.31 we obtain

$$\begin{cases} v_2^* = 0, \\ v_1^* = f(0) = 0. \end{cases}$$

Therefore, the only equilibrium point is the origin.

Henceforth we assume that $C_1 = C_2 = C$. Under this assumption, the state equations can be recast as

$$\begin{cases} \dfrac{dv_1}{dt} = \dfrac{1}{RC}[f(v_2) - v_1 - v_2], \\[4mm] \dfrac{dv_2}{dt} = \dfrac{1}{RC}[f(v_2) - 2v_2 - v_1]. \end{cases} \tag{12.32}$$

Therefore, we can define a normalized time $\tau = \dfrac{t}{RC}$. The state variables, in turn, can be normalized as follows:

$$x_1 = \frac{v_1}{V_S} \qquad x_2 = \frac{v_2}{V_S}.$$

At this point, we replace the piecewise-linear function $f(v_2)$ shown in Fig. 12.40 (gray thick curve) with a smooth function (black thin curve) [4]:

$$f(v_2) = \frac{2}{\pi} V_S \arctan\left(\frac{\pi}{2}\frac{\alpha}{V_S}v_2\right).$$

Since $f(v_2) = f(V_S x_2) = \dfrac{2}{\pi} V_S \arctan\left(\dfrac{\pi}{2}\alpha x_2\right)$, the normalized state equations can be written as

$$\begin{cases} \dot{x}_1 = \dfrac{2}{\pi} \arctan\left(\dfrac{\pi}{2}\alpha x_2\right) - x_1 - x_2, \\[4mm] \dot{x}_2 = \dfrac{2}{\pi} \arctan\left(\dfrac{\pi}{2}\alpha x_2\right) - x_1 - 2x_2, \end{cases} \tag{12.33}$$

where $\dot{x}_1 = \dfrac{dx_1}{d\tau}$, and so forth.

12.6.2 Hopf Bifurcation of the Equilibrium Point

First of all, we compute the Jacobian matrix for the system described by Eq. 12.33:

$$J(x) = \begin{pmatrix} -1 & \dfrac{\alpha}{1 + \left(\dfrac{\pi}{2}\alpha x_2\right)^2} - 1, \\ -1 & \dfrac{\alpha}{1 + \left(\dfrac{\pi}{2}\alpha x_2\right)^2} - 2. \end{pmatrix} \qquad (12.34)$$

Now, computing the Jacobian matrix at the only equilibrium point, we can analyze its stability:

$$J(0, 0) = \begin{pmatrix} -1 & \alpha - 1 \\ -1 & \alpha - 2 \end{pmatrix}.$$

The characteristic equation is

$$\lambda^2 - \lambda(\alpha - 3) + 1 = 0,$$

whose solutions are

$$\lambda_\pm = \frac{\alpha - 3 \pm \sqrt{\alpha^2 - 6\alpha + 5}}{2}.$$

The origin is stable for $\alpha < 3$ and unstable for $\alpha > 3$. At $\alpha = 3$, the matrix has two eigenvalues that cross the imaginary axis: $\lambda_\pm = \pm j$, which correspond to a unitary normalized angular frequency. Therefore, the system undergoes a Hopf bifurcation.

As an alternative, we could follow the method described in Sect. 12.2.1. The characteristic equation for $J(0, 0)$ can be also written as

$$\lambda^2 - \lambda Tr[J(0, 0)] + \det[J(0, 0)] = 0,$$

where $Tr[J(0, 0)] = \alpha - 3$ is the trace of the matrix $J(0, 0)$ and $\det[J(0, 0)] = 1$ is its determinant. Therefore, according to Fig. 12.14, we can easily study the stability of the equilibrium point, drawing the same conclusions as above.

The simplest way to check whether the bifurcation is supercritical is to analyze the state portraits of the system across the bifurcation.

Figure 12.42 shows the state portraits for $\alpha = 2.5$ (panel a), $\alpha = 3.2$ (panel b), and $\alpha = 8$ (panel c). It is apparent that below the bifurcation value, the trajectories converge to a stable focus, whereas for $\alpha > 3$, the trajectories converge to a limit cycle, which grows with α.

Figure 12.43 shows the corresponding time plots. You can see that just after the bifurcation (Fig. 12.43b), the oscillation is nearly sinusoidal, whereas for higher values of α (Fig. 12.43c), the waveform is highly distorted due to the nonlinearity.

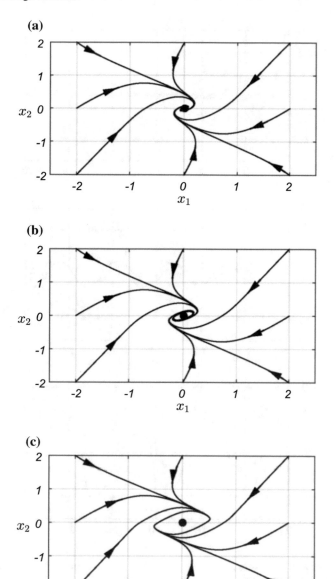

Fig. 12.42 State portraits for **a** $\alpha = 2.5$, **b** $\alpha = 3.2$, and **c** $\alpha = 8$

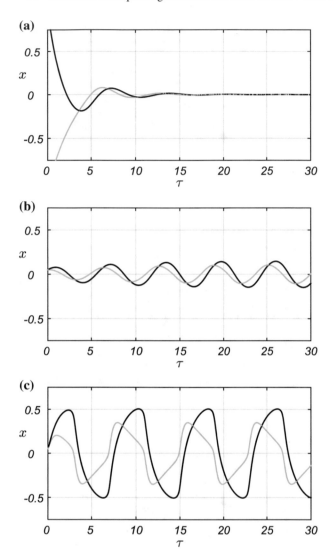

Fig. 12.43 Time plots of the normalized state variables $x_1(\tau)$ (black line) and $x_2(\tau)$ (gray line) for **a** $\alpha = 2.5$, **b** $\alpha = 3.2$, and **c** $\alpha = 8$. Initial conditions: **a** $[x_1(0), x_2(0)] = (1, 1)$; **b**, **c** $[x_1(0), x_2(0)] = (0.05, 0.05)$

12.7 Colpitts Oscillator

A Colpitts oscillator, invented in 1918 by the American engineer Edwin H. Colpitts, is one of a number of LC oscillators that use a combination of electronic devices, inductors, and capacitors to produce an oscillation at a certain frequency. It consists of a gain device (such as a BJT, field effect transistor, operational amplifier, or vacuum tube) properly connected to a parallel LC circuit.

We consider the classical configuration of the Colpitts oscillator [5–7], containing a BJT as the gain element and a resonant network consisting of an inductor and a pair of capacitors, as shown in Fig. 12.44a, where the bias current source I_0 represents a proper electronic circuit.

The BJT is in common-base configuration and can be replaced by the equivalent simplified model shown in Fig. 12.44b. The influence of the parasitic capacitances at the BC and BE junctions is neglected, thus yielding the circuit shown in Fig. 12.45.

Taking into account that the current i_E depends on $v_{BE} = -v_2$, that is,

$$i_E = \hat{i}_E(v_2) = I_S \left(e^{-\frac{v_2}{V_T}} - 1 \right),$$

the circuit equations

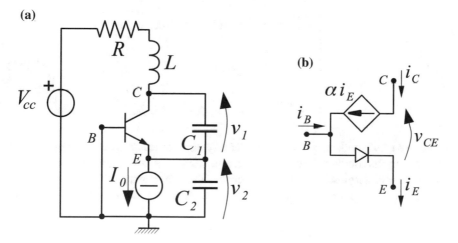

Fig. 12.44 **a** The considered Colpitts circuit and **b** model of common-base BJT

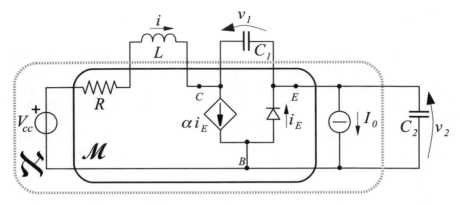

Fig. 12.45 Equivalent circuit of the circuit shown in Fig. 12.44a

$$\begin{cases} V_{CC} = L\dfrac{di}{dt} + Ri + v_1 + v_2, \\[2mm] i = \alpha\hat{i}_E(v_2) + C_1\dfrac{dv_1}{dt}, \\[2mm] C_1\dfrac{dv_1}{dt} + \hat{i}_E(v_2) = C_2\dfrac{dv_2}{dt} + I_0, \end{cases}$$

can be easily recast as state equations as follows:

$$\begin{cases} L\dfrac{di}{dt} = -Ri - v_1 - v_2 + V_{CC}, \\[3mm] C_1\dfrac{dv_1}{dt} = i - \alpha\hat{i}_E(v_2), \\[3mm] C_2\dfrac{dv_2}{dt} = (1-\alpha)\hat{i}_E(v_2) + i - I_0. \end{cases} \qquad (12.35)$$

Figure 12.45 evidences the complementary-component equivalent representation for the Colpitts oscillator. Notice that the sources V_{CC} and I_0 could also be included within the memoryless multiport $\mathcal{M}$, because they are constant. You can check that the descriptive equations for the memoryless three-port $\aleph$ (shown in Fig. 12.46) are

$$\begin{cases} i_1 = \alpha\hat{i}_E(v_2) + i_3, \\[2mm] i_2 = i_3 + I_0 + (\alpha - 1)\hat{i}_E(v_2), \\[2mm] v_3 = Ri_3 - v_1 - v_2 + V_{CC}. \end{cases}$$

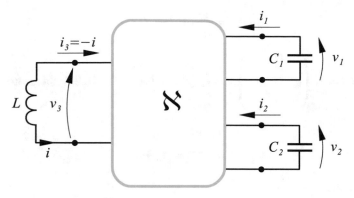

Fig. 12.46 Memoryless three-port $\aleph$ for the Colpitts oscillator

12.7.1 *Normalized State Equations*

We assume that $\alpha = 1$, which implies $I_B = 0$.

By imposing the equilibrium condition and taking into account that the function $\hat{i}_E(v_2)$ is monotonically decreasing (and therefore invertible), from Eq. 12.35 we obtain

$$\begin{cases} i^* = I_0, \\ v_2^* = \hat{i}_E^{-1}(I_0), \\ v_1^* = V_{CC} - RI_0 - \hat{i}_E^{-1}(I_0). \end{cases} \tag{12.36}$$

The "perturbation" with respect to the equilibrium solution is given by the shifted variables $\tilde{v}_1 = v_1 - v_1^*$, $\tilde{v}_2 = v_2 - v_2^*$, and $\tilde{i} = i - i^*$. The state equations can be recast in terms of these variables:

$$\begin{cases} C_1 \dfrac{d\tilde{v}_1}{dt} = \tilde{i} + \left[I_0 - \hat{i}_E(v_2^* + \tilde{v}_2) \right] = \tilde{i} + I_0 \left[1 - \dfrac{1}{I_0}\hat{i}_E(v_2^* + \tilde{v}_2) \right], \\[4mm] C_2 \dfrac{d\tilde{v}_2}{dt} = \tilde{i}, \\[4mm] L \dfrac{d\tilde{i}}{dt} = -R\tilde{i} - \tilde{v}_1 - \tilde{v}_2. \end{cases} \tag{12.37}$$

The state variables can now be normalized as follows:

$$x_1 = \frac{\tilde{v}_1}{V_T}, \qquad x_2 = \frac{\tilde{v}_2}{V_T}, \qquad x_3 = \frac{\tilde{i}}{I_0}.$$

For the sake of compactness, we define

$$f(x_2) = \frac{1}{I_0}\hat{i}_E(v_2^* + V_T x_2) - 1.$$

But $\hat{i}_E(v_2^*) = I_S\left(e^{-\frac{v_2^*}{V_T}} - 1\right)$ and $\hat{i}_E(v_2^*) = I_0$ (see Eq. 12.36); therefore,

$$I_S e^{-\frac{v_2^*}{V_T}} = I_0 + I_S.$$

Consequently,

$$f(x_2) = \frac{1}{I_0}\left[(I_0 + I_S)e^{-x_2} - (I_0 + I_S)\right].$$

Inasmuch as $I_S \ll I_0$, we can finally approximate $f(x_2)$ as

$$f(x_2) \approx e^{-x_2} - 1.$$

The state equations can thus be written as

$$\begin{cases} \dfrac{dx_1}{dt} = \dfrac{I_0}{C_1 V_T}\left[x_3 - f(x_2)\right], \\[2ex] \dfrac{dx_2}{dt} = \dfrac{I_0}{C_2 V_T}x_3, \\[2ex] \dfrac{dx_3}{dt} = -\dfrac{R}{L}x_3 - \dfrac{V_T}{L I_0}(x_1 + x_2). \end{cases} \tag{12.38}$$

The only nonlinear element in these equations is $f(x_2)$. In the absence of this term, the above equations would describe (in normalized form) the linear RLC circuit shown in Fig. 12.47.

Fig. 12.47 Linear equivalent circuit corresponding to the normalized Eq. 12.38 for $f(x_2) = 0$

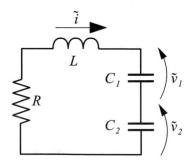

With this in mind, we can assume as a physical reference term the angular frequency $\omega_0 = \dfrac{1}{\sqrt{L\dfrac{C_1 C_2}{C_1 + C_2}}}$. Moreover, we define dimensionless parameters[6]

$$Q = \frac{\omega_0 L}{R} = \frac{1}{R}\sqrt{\frac{L}{\dfrac{C_1 C_2}{C_1 + C_2}}}, \qquad k = \frac{C_2}{C_1 + C_2}, \qquad g = \frac{L I_0}{R V_T (C_1 + C_2)}.$$

Using these parameters in the nonlinear system, we can first of all define a normalized time $\tau = \omega_0 t$, so that the state equations can be recast as

$$\begin{cases} \dot{x}_1 = \dfrac{I_0}{\omega_0 C_1 V_T} [x_3 - f(x_2)], \\[3mm] \dot{x}_2 = \dfrac{I_0}{\omega_0 C_2 V_T} x_3, \\[3mm] \dot{x}_3 = -\dfrac{R}{\omega_0 L} x_3 - \dfrac{V_T}{\omega_0 L I_0}(x_1 + x_2), \end{cases} \qquad (12.39)$$

where $\dot{x}_1 = \dfrac{dx_1}{d\tau}$, and so forth.

Using the other dimensionless parameters defined above, we finally obtain the normalized state equations

$$\begin{cases} \dot{x}_1 = \dfrac{g}{(1-k)Q} [x_3 - f(x_2)], \\[3mm] \dot{x}_2 = \dfrac{g}{kQ} x_3, \\[3mm] \dot{x}_3 = -\dfrac{1}{Q} x_3 - \dfrac{k(1-k)Q}{g}(x_1 + x_2). \end{cases} \qquad (12.40)$$

12.7.2 Hopf Bifurcation of the Equilibrium Point

First of all, we consider the following result.

---

[6]We remark that Q is the Q-factor (see Sect. 13.12) of the RLC circuit shown in Fig. 12.47, which is often called the (unloaded) tank circuit.

Every three-dimensional system $\dot{x} = f(x)$ is **dissipative** and cannot admit any repellors if

$$\nabla \cdot f(x) = \text{Tr}[J(x)] < 0 \quad \forall x,$$

where $\text{Tr}[J(x)]$ is the trace of $J(x)$.

This result holds only for three-dimensional systems and is a consequence of the divergence theorem. A proof can be found in [3].

The Jacobian matrix for the Colpitts oscillator, described by Eq. 12.40 (three-dimensional system $\dot{x} = f(x)$), is

$$J(x) = \begin{pmatrix} 0 & \dfrac{g}{(1-k)Q}e^{-x_2} & \dfrac{g}{(1-k)Q} \\[3mm] 0 & 0 & \dfrac{g}{kQ} \\[3mm] -\dfrac{k(1-k)Q}{g} & -\dfrac{k(1-k)Q}{g} & -\dfrac{1}{Q} \end{pmatrix}. \tag{12.41}$$

It is straightforward to check that $\text{Tr}[J(x)] = -\dfrac{1}{Q} < 0 \quad \forall x$; therefore, *the Colpitts oscillator cannot admit repellors*.

Computing the Jacobian matrix at the only equilibrium point, we can now analyze its stability:

$$J(0,0,0) = \begin{pmatrix} 0 & \dfrac{g}{(1-k)Q} & \dfrac{g}{(1-k)Q} \\[3mm] 0 & 0 & \dfrac{g}{kQ} \\[3mm] -\dfrac{k(1-k)Q}{g} & -\dfrac{k(1-k)Q}{g} & -\dfrac{1}{Q} \end{pmatrix}.$$

By imposing $\det[\lambda I - J(0,0,0)] = 0$, we obtain the characteristic equation[7]

$$\lambda^3 + \frac{1}{Q}\lambda^2 + \lambda + \frac{g}{Q} = 0. \tag{12.42}$$

---

[7]For a 3×3 matrix A, the characteristic polynomial has the form $p(\lambda) = \lambda^3 - Tr(A)\lambda^2 + \dfrac{1}{2}\left\{[Tr(A)]^2 - Tr(A^2)\right\}\lambda - \det(A)$, where $Tr(A)$ denotes the trace of A. More generally, the coefficients in the characteristic polynomial for an $n \times n$ matrix A can be expressed in terms of $Tr(A)$, $\det(A)$, and principal minors [2].

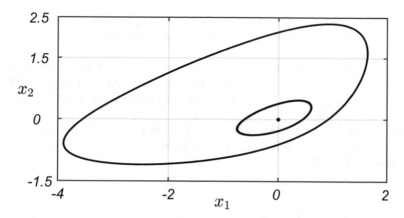

Fig. 12.48 Asymptotic trajectories projected on the (x_1, x_2)-plane for $Q = 1.6$, $k = 0.5$, and three values of g: $g = 0.8$ (stable focus), $g = 1.02$ (inner limit cycle), and $g = 1.3$ (outer limit cycle)

We want to find the parameter values, if any, such that the origin undergoes a Hopf bifurcation. Therefore, we impose that a pair of eigenvalues cross the imaginary axis and that $\lambda = j\omega$ is a solution of the characteristic equation. By substitution into Eq. 12.42, we obtain

$$-j\omega^3 - \frac{\omega^2}{Q} + j\omega + \frac{g}{Q} = 0,$$

which implies

$$\begin{cases} \omega^2 = 1, \\ g = 1. \end{cases} \tag{12.43}$$

Therefore $\lambda_{1,2} = \pm j$, and the third eigenvalue turns out to be $\lambda_3 = -\frac{1}{Q}$. You can easily check that Eq. 12.42, for $g = 1$, becomes $(\lambda^2 + 1)(\lambda + \frac{1}{Q}) = 0$.

The condition expressed by Eq. 12.43 is necessary but not sufficient to ensure the presence of a Hopf bifurcation. Two further nondegeneracy conditions must be satisfied [8]. They have been checked [5, 9], proving that the Colpitts oscillator undergoes a supercritical Hopf bifurcation for every parameter setting, due to the positiveness of Q.

Figure 12.48 shows three steady-state trajectories projected on the plane (x_1, x_2), obtained by setting $Q = 1.6$, $k = 0.5$, and

- $g = 0.8$ (stable focus);
- $g = 1.02$ (inner limit cycle);
- $g = 1.3$ (outer limit cycle).

It is apparent that below the bifurcation value, the trajectories converge to a stable focus, whereas for $g > 1$, the trajectories converge to a limit cycle, which grows with g.

Figure 12.49 shows the corresponding time plots. Also in this case, as for the Wien bridge oscillator, you can see that just after the bifurcation (Fig. 12.49b), the oscillation is nearly sinusoidal, whereas for higher values of g (Fig. 12.49c), the waveform is distorted due to the nonlinearity.

Finally, Fig. 12.50 shows the equilibrium point and limit cycle obtained for a grid of g values in the projection of the control space (g, x_1, x_3), similar to the generic plot shown in Fig. 12.33.

12.8   Small-Signal Analysis

Nonlinear components are the backbone of electronic circuits. For instance, amplifiers, oscillators, and logic gates perform their functions using the nonlinear properties of diodes and transistors, viewed as memoryless (resistive) components. Other relevant circuits, such as parametric amplifiers, make use of nonlinear capacitors and inductors.

Throughout this chapter, we have considered circuits whose nonlinearity is provided only by resistive components, whereas inductors and capacitors are always linear. For these circuits, adopting quite general assumptions, it was shown how to obtain the state equations in the form of Eq. 12.3:

$$\frac{dx}{dt} = -[D]^{-1} h(x, w(t)),$$

which, when all the input sources w are constant, can be recast in the autonomous form $\dot{x} = f(x)$. Next, it was shown how to study the stability around the equilibrium points of the circuit, obtained by solving the nonlinear algebraic equations $f(x) = 0$.

In the more general case of time-varying input sources $w(t)$, the evolution of a circuit for assigned initial conditions can be studied by solving the state equations 12.3. The solution of these nonlinear equations can almost always be obtained in a numerical form but almost never in an analytic form. This makes it difficult to infer general principles, since each circuit requires a specific treatment.

Despite the unavailability of a general analytic solution, many properties of nonlinear circuits can be studied by so-called *small-signal analysis*. In this analysis, voltages and currents within the circuit are considered with respect to a set of constant values (often called *bias*) compatible with the circuit equations. This amounts to stating that the bias values are a constant solution for the nonlinear equations of the circuit in the absence of time-varying inputs. For this *bias circuit*, the equations are reduced to an algebraic form.

Compared with the corresponding bias values, the variations, which are usually time-dependent, are assumed to be small in amplitude. This assumption settles the

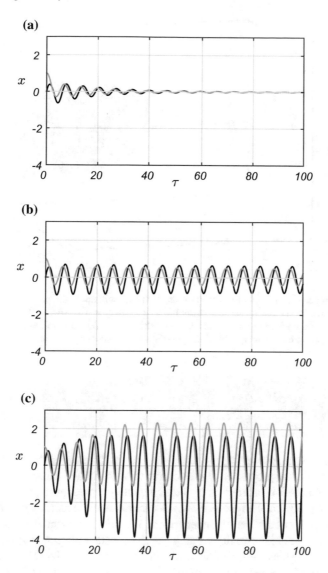

Fig. 12.49 Time plots of the normalized state variables $x_1(\tau)$ (black line) and $x_2(\tau)$ (gray line) for $Q = 1.6$, $k = 0.5$, and three values of g: **a** $g = 0.8$, **b** $g = 1.02$, and **c** $g = 1.3$. Initial conditions: $[x_1(0), x_2(0), x_3(0)] = (0, 1, 0)$

basis to approximate the original circuit equations. As a result, the small-signal voltages and currents are found simply by solving the equations of a well-defined linear equivalent circuit.

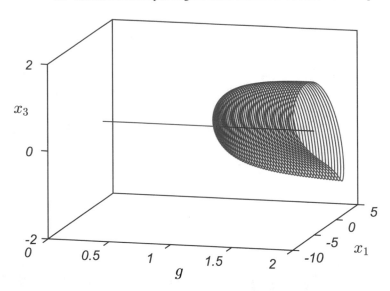

Fig. 12.50 Supercritical Hopf bifurcation for the Colpitts oscillator in the projection (g, x_1, x_3) of the control space. By increasing the control parameter g (with $Q = 1.6$ and $k = 0.5$), the equilibrium point (focus, solid line) changes its stability from stable (green) to unstable (red) and collides at $g = 1$ with a periodic solution (green limit cycles), which exists for $g > 1$ and is stable

> **Small signal**: is a time-varying signal whose amplitude is much smaller than the corresponding (constant) bias value.

This *small-signal circuit* has the same topology as the original, but its elements are *linear*; therefore, the properties of the solution can be studied using the concepts and methods available for linear circuits. This obviously represents a great advantage.

Inasmuch as the small signals represent variations with respect to the bias values, the small-signal circuit has no independent life and meaning. It must always be associated with the bias circuit, which gives the constant reference values.

In practice, the bias values within a nonlinear circuit are set by DC voltage (and/or current) sources. They can be calculated by replacing the capacitors with open circuits and the inductors with short circuits and then solving the resulting nonlinear resistive circuit. For each component of the nonlinear circuit, the bias values of the descriptive variables identify the so-called *operating point*.

Once the bias values are known, the descriptive equations of each nonlinear component are approximated by truncating at the first order a Taylor series expansion in a neighborhood of its operating point. These approximations generate the small-signal circuit and the corresponding linear equations to be solved.

For instance, consider the circuit shown in Fig. 12.51a, whose nonlinear resistor is defined by the descriptive equation $i_d = \hat{i}_d(v_d)$ depicted in Fig. 12.51b. The role

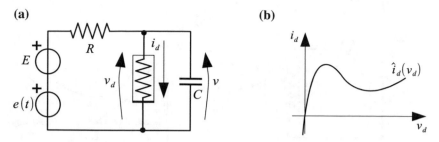

Fig. 12.51 Example: **a** a circuit with a two-terminal nonlinear resistive component; **b** resistor characteristic $i_d = \hat{i}_d(v_d)$

played by the DC voltage source $E > 0$ is to set the bias voltages and currents of the circuit, whereas $e(t)$ is a small-signal voltage source, that is, we assume that $|e(t)| \ll E$ for all t.

When $e(t) = 0$, the circuit voltages and currents are constant. In particular, no current flows through the capacitor. The values V_d, I_d for the bias circuit, shown in Fig. 12.52a, are such that

$$\begin{cases} E = RI_d + V_d, \\ I_d = \hat{i}_d(V_d), \end{cases} \tag{12.44}$$

and they can be viewed as the coordinates, in the (v_d, i_d) plane, of an intersection point Q between the straight line $i_d = \dfrac{E - v_d}{R}$ and the curve $i_d = \hat{i}_d(v_d)$. This is the *operating point* of the nonlinear resistor.

Keeping the curve $i_d = \hat{i}_d(v_d)$ fixed, for appropriate values of the circuit parameters E, R we can have one operating point Q or three possible operating points Q_1, Q_2, Q_3, as shown in Fig. 12.52b–d. In the case of a multiplicity of operating points, the small-signal analysis requires that one of them be *selected in advance* and referred to. Therefore, for a single solution or multiple solutions Q_1, Q_2, Q_3, the chosen operating point will be labeled as $Q(V_d, I_d)$.[8]

When $e(t) \neq 0$, the voltages and currents in the circuit of Fig. 12.51a become time-varying. Without loss of generality, they can be written as the sum of the constant value found in the bias circuit and a time-varying term marked by a tilde:

$$\begin{cases} i_d = I_d + \tilde{i}_d(t), \\ v_d = V_d + \tilde{v}_d(t), \\ v = V + \tilde{v}(t). \end{cases} \tag{12.45}$$

---

[8]The nature of Q as a stable or unstable equilibrium point, or its observability through circuit measurements, is a topic that does not affect the small-signal analysis. Therefore, it are not an object of discussion in this section.

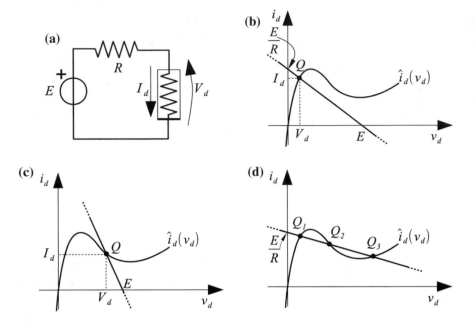

Fig. 12.52 a bias circuit; **b, c, d** the operating point(s) for three different pairs (E, R)

Notice that since $v = v_d$, we immediately have $V = V_d$ and $\tilde{v}_d(t) = \tilde{v}(t)$; moreover, $\dfrac{dv}{dt} = \dfrac{d\tilde{v}}{dt}$. With this caveat in mind, the circuit equations can be written as follows:

$$\begin{cases} E + e(t) = R\left(\left(I_d + \tilde{i}_d(t)\right) + C\dfrac{d\tilde{v}}{dt}\right) + V_d + \tilde{v}(t), \\ I_d + \tilde{i}_d(t) = \hat{i}_d\left(V_d + \tilde{v}(t)\right). \end{cases} \tag{12.46}$$

The first equation can be further simplified by removing the terms E, RI_d, V_d, which satisfy Eq. 12.44:

$$e(t) = R\left(\tilde{i}_d(t) + C\dfrac{d\tilde{v}}{dt}\right) + \tilde{v}(t).$$

The right-hand side of the second equation is nonlinear. Under the small-signal assumption, it can be approximated about the operating point Q by the first two terms of the Taylor expansion

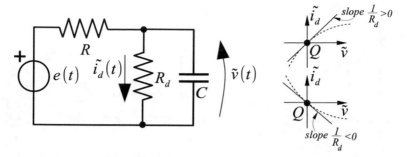

Fig. 12.53 Small-signal equivalent circuit

$$I_d + \tilde{i}_d(t) \approx \underbrace{\hat{i}_d(V_d)}_{I_d} + \underbrace{\left.\frac{d\hat{i}_d}{dv_d}\right|_{V_d}}_{\dfrac{1}{R_d}} \tilde{v}(t), \qquad (12.47)$$

where R_d denotes the *differential resistance* of the nonlinear resistor at Q. Depending on the position of Q on the curve, R_d can take positive or negative values, as shown in Fig. 12.53, right panel. The resulting small-signal equations in the neighborhood of the operating point Q are

$$\begin{cases} e(t) = R\left(\tilde{i}_d(t) + C\dfrac{d\tilde{v}}{dt}\right) + \tilde{v}(t), \\[2mm] \tilde{i}_d(t) = \dfrac{1}{R_d}\tilde{v}(t), \end{cases} \qquad (12.48)$$

which are the equations of the linear circuit shown in Fig. 12.53, left panel. This circuit has the same topology as the original one and can be obtained from it by replacing the nonlinear resistor with R_d and the bias voltage source E with a short circuit. It is valid for values of $|e(t)|$ small enough that the linearization defined in Eq. 12.47 introduces negligible errors.

Case Study
The voltage source $e(t)$ for the circuit shown in Fig. 12.51 is the rectangular pulse represented in Fig. 12.54a. The nonlinear resistor equation $i_d = \hat{i}_d(v_d)$ is

$$i_d = A_0 + \gamma\,(v_d - E_0)^3 - \alpha\,(v_d - E_0),$$

where γ and α are positive coefficients, and the terms A_0, E_0 are the co-ordinates of the resistor's operating point, as shown in Fig. 12.54b (see

(a)   **(b)**

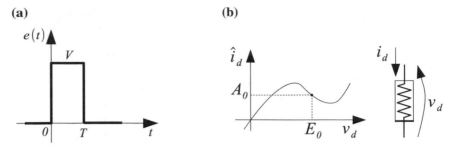

Fig. 12.54 **a** Rectangular pulse $e(t)$ for the circuit of Fig. 12.51; **b** DP characteristic of the nonlinear resistor

also Sect. 12.3). At $t = 0$, the voltage capacitor is $v(0) = E_0$, which implies $\tilde{v}(0) = 0$. Under these assumptions, find the small-signal response $\tilde{v}(t)$ and discuss its limits of validity.

Making reference to the bias circuit (Fig. 12.52a) and to the given characteristic equation, we obtain

$$\frac{E - v_d}{R} = A_0 + \gamma \, (v_d - E_0)^3 - \alpha \, (v_d - E_0),$$

which, by imposing $v_d = E_0$, implies $\dfrac{E - E_0}{R} = A_0$. At this operating point, we immediately have

$$\left. \frac{di_d}{dv_d} \right|_{E_0} = -\alpha.$$

Therefore, the small-signal circuit of Fig. 12.53 has $R_d = -\dfrac{1}{\alpha}$, and the small-signal differential equation can be recast as follows:

$$RC\frac{d\tilde{v}}{dt} + (1 - \alpha R)\tilde{v} = e(t).$$

The natural frequency

$$\lambda = -\frac{1}{RC}\,(1 - \alpha R) = -\frac{1}{R_{eq}C} \quad \text{with } R_{eq} = \frac{R}{1 - \alpha R}$$

is negative, provided that

Fig. 12.55 Circuit response under the absolute stability assumption

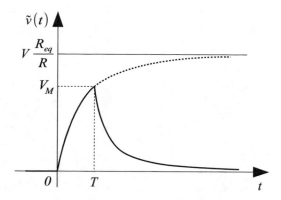

$$R < \frac{1}{\alpha}, \tag{12.49}$$

and under this assumption, the small-signal circuit is absolutely stable.

The response to $e(t)$ can now be found as follows:

- For $t < T$, $\tilde{v}(t)$ is found as the zero-state response to an input signal $V u(t)$. Therefore,

$$\tilde{v}(t) = V\frac{R_{eq}}{R}\left(1 - e^{\lambda t}\right), \qquad 0 \le t < T. \tag{12.50}$$

- For $t \ge T$, $\tilde{v}(t)$ is the zero-input response with initial condition $\tilde{v}(T) = V\frac{R_{eq}}{R}\left(1 - e^{\lambda T}\right) \doteq V_M$, that is,

$$\tilde{v}(t) = V_M e^{\lambda(t-T)}. \tag{12.51}$$

Under the absolute stability condition (12.49), expressions (12.50) and (12.51) correspond to the curve $\tilde{v}(t)$ shown in Fig. 12.55. For large values of T, the voltage V_M approaches the value $V\frac{R_{eq}}{R}$. Notice that the analytic form of these expressions is independent of whether the circuit is stable.

These results allow us to formulate a specific condition for the validity of the small-signal approximation. To obtain it, just observe that for all t, we have $0 \le \tilde{v}(t) < V\frac{R_{eq}}{R}$. Therefore, the maximum value of $\tilde{v}(t)$ must be very small compared to the voltage E_0 at the operating point:

$$V\frac{R_{eq}}{R} \ll E_0, \quad \text{that is,} \quad V \ll E_0\,(1 - \alpha R).$$

We further notice that

$$\frac{V_M}{V} < \frac{R_{eq}}{R} = \frac{1}{1 - \alpha R} = 1 + \frac{\alpha R}{1 - \alpha R}.$$

Therefore, appropriate values of α can make the above ratio quite large compared to 1.

When $R > \dfrac{1}{\alpha}$, we have $\lambda > 0$, and the small-signal circuit is unstable. Equations 12.50 and 12.51 are still valid, but it is easy to verify that $\tilde{v}(t)$ grows indefinitely over time, and the small-signal approximation becomes meaningless. In this case, the true $\tilde{v}(t)$ is obtained by considering the nonlinear terms of the DP characteristic, that is, by solving the equation

$$e(t) = RC\frac{d\tilde{v}}{dt} + R\gamma\tilde{v}^3 - \tilde{v}\,(\alpha R - 1)\,,$$

as can be checked easily.

As a further example, consider the circuit shown in Fig. 12.56. Here, the nonlinear element is a BJT. Its descriptive equations can be given in terms of the voltages v_{be}, v_{ce} and the currents i_b, i_c. By omitting unnecessary details regarding their analytic structure (see Sect. 3.2.6 in Vol. 1), these equations can be formulated as

$$\begin{cases} v_{be} = \hat{v}_{be}\,(i_b, v_{ce})\,, \\ i_c = \hat{i}_c\,(i_b, v_{ce})\,, \end{cases} \tag{12.52}$$

and are part of the following set of equations describing the circuit's behavior:

$$\begin{cases} E = R_1\left(i_b + C_1\dfrac{dv_{C_1}}{dt}\right) + v_{C_1} + e(t)\,, \\ e(t) + v_{C_1} = v_{be}\,, \\ E = R_2\left(i_c + C_2\dfrac{dv_{C_2}}{dt}\right) + v_{ce}\,, \\ v_{ce} = v_{C_2} + RC_2\dfrac{dv_{C_2}}{dt}\,, \\ v_{be} = \hat{v}_{be}\,(i_b, v_{ce})\,, \\ i_c = \hat{i}_c\,(i_b, v_{ce})\,. \end{cases} \tag{12.53}$$

As in the previous example, the voltage source E sets the bias voltages and currents for the circuit, whereas $e(t)$ is the small-signal voltage source. When $e(t) = 0$, the capacitors can be viewed as open circuits, and the voltages and currents within the circuit are constant values, denoted by capital letters. In this working condition, the

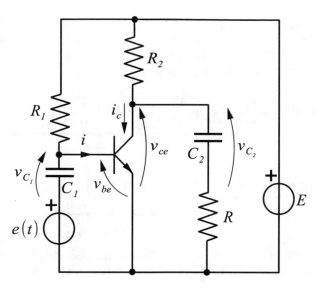

Fig. 12.56 BJT circuit

terms with time derivatives in Eq. 12.53 cancel, and we obtain the system of bias equations

$$
\begin{cases}
E = R_1 I_b + V_{C_1}, \\
V_{C_1} = V_{be}, \\
E = R_2 I_c + V_{ce}, \\
V_{ce} = V_{C_2}, \\
V_{be} = \hat{v}_{be}\left(I_b, V_{ce}\right), \\
I_c = \hat{i}_c\left(I_b, V_{ce}\right).
\end{cases}
\tag{12.54}
$$

The bias values I_b, V_{be}, I_c, V_{ce} individuate the transistor's operating point Q. The voltages V_{C_1}, V_{C_2} are the open-circuit voltages for the capacitors C_1 and C_2, respectively. Once the bias values have been obtained, we can reconsider voltages and currents in the presence of $e(t) \neq 0$ and write each of them, in full generality, as the sum of a bias value and a time-variant term marked by a tilde:

$$
\begin{aligned}
v_{be} &= V_{be} + \tilde{v}_{be}(t); \quad i_b = I_b + \tilde{i}_b(t); \quad i_c = I_c + \tilde{i}_c(t); \\
v_{ce} &= V_{ce} + \tilde{v}_{ce}(t); \quad v_{C_1} = V_{C_1} + \tilde{v}_{C_1}(t); \quad v_{C_2} = V_{C_2} + \tilde{v}_{C_2}(t).
\end{aligned}
\tag{12.55}
$$

Making use of these sums, the last two equations of Eq. 12.53, which are nonlinear, can be recast as

$$
\begin{cases}
V_{be} + \tilde{v}_{be}(t) = \hat{v}_{be}\left(I_b + \tilde{i}_b(t), V_{ce} + \tilde{v}_{ce}(t)\right), \\
I_c + \tilde{i}_c(t) = \hat{i}_c\left(I_b + \tilde{i}_b(t), V_{ce} + \tilde{v}_{ce}(t)\right),
\end{cases}
\tag{12.56}
$$

and can be approximated by truncated Taylor expansions with respect to the controlling variables i_b, v_{ce}. These expansions are written around the bias values I_b, V_{ce}, taking as valid, for all t, the small-signal assumption for all the time-varying (tilde) voltage and current terms. Therefore, we have

$$\begin{cases} V_{be} + \tilde{v}_{be}(t) \approx \underbrace{\hat{v}_{be}\,(I_b, V_{ce})}_{V_{be}} + \underbrace{\left.\frac{\partial \hat{v}_{be}}{\partial i_b}\right|_{I_b, V_{ce}} \tilde{i}_b(t) + \left.\frac{\partial \hat{v}_{be}}{\partial v_{ce}}\right|_{I_b, V_{ce}} \tilde{v}_{ce}(t)}_{\tilde{v}_{be}(t)}, \\[4mm] I_c + \tilde{i}_c(t) \approx \underbrace{\hat{i}_c\,(I_b, V_{ce})}_{I_c} + \underbrace{\left.\frac{\partial \hat{i}_c}{\partial i_b}\right|_{I_b, V_{ce}} \tilde{i}_b(t) + \left.\frac{\partial \hat{i}_c}{\partial v_{ce}}\right|_{I_b, V_{ce}} \tilde{v}_{ce}(t)}_{\tilde{i}_c(t)}. \end{cases} \tag{12.57}$$

The first-order partial derivatives calculated at the operating point (I_b, V_{ce}) are the elements h_{ij} (with $i, j = 1, 2$) of a hybrid matrix H such that

$$\begin{pmatrix} \tilde{v}_{be} \\ \tilde{i}_c \end{pmatrix} = \underbrace{\begin{pmatrix} h_{11} & h_{12} \\ h_{21} & h_{22} \end{pmatrix}}_{H} \begin{pmatrix} \tilde{i}_b \\ \tilde{v}_{ce} \end{pmatrix}, \tag{12.58}$$

which corresponds to the BJT small-signal circuit model shown in Fig. 12.57. By applying Eq. 12.55 to the first four equations of Eq. 12.53, we have

$$\begin{cases} E = R_1\left(I_b + \tilde{i}_b\right) + R_1 C_1 \dfrac{d\tilde{v}_{C_1}}{dt} + V_{C_1} + \tilde{v}_{C_1} + e(t), \\[2mm] e(t) + V_{C_1} + \tilde{v}_{C_1} = V_{be} + \tilde{v}_{be}, \\[2mm] E = R_2\left(I_c + \tilde{i}_c + C_2 \dfrac{d\tilde{v}_{C_2}}{dt}\right) + V_{ce} + \tilde{v}_{ce}, \\[2mm] V_{ce} + \tilde{v}_{ce} = V_{C_2} + \tilde{v}_{C_2} + R C_2 \dfrac{d\tilde{v}_{C_2}}{dt}. \end{cases} \tag{12.59}$$

These equations can be immediately simplified by dropping the bias terms, which balance each other based on Eq. 12.54:

$$\begin{cases} 0 = R_1 \tilde{i}_b + R_1 C_1 \dfrac{d\tilde{v}_{C_1}}{dt} + \tilde{v}_{C_1} + e(t), \\[2mm] e(t) + \tilde{v}_{C_1} = \tilde{v}_{be}, \\[2mm] 0 = R_2\left(\tilde{i}_c + C_2 \dfrac{d\tilde{v}_{C_2}}{dt}\right) + \tilde{v}_{ce}, \\[2mm] \tilde{v}_{ce} = \tilde{v}_{C_2} + R C_2 \dfrac{d\tilde{v}_{C_2}}{dt}. \end{cases} \tag{12.60}$$

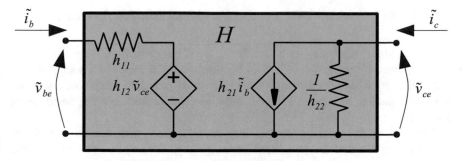

Fig. 12.57 Small-signal (linear) model for the BJT at the operating point

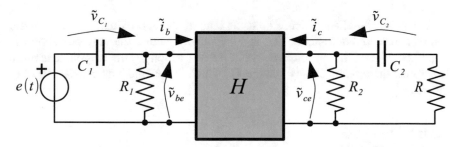

Fig. 12.58 Small-signal linear equivalent circuit corresponding to the circuit in Fig. 12.56

These equations, together with Eq. 12.58, complete the definition of the small-signal linear circuit, which is shown in Fig. 12.58. As expected, this circuit can be directly obtained from the original one by replacing the constant voltage source by a short circuit and the BJT by its small-signal circuit model shown in Fig. 12.57. As an obvious extension, every constant current source included in a circuit is replaced by an open circuit in the corresponding small-signal circuit.

Finally, notice that the bias values are always obtained by solving a nonlinear *resistive* circuit. Among the constant voltages and currents representing the bias circuit's solution, we can find the open-circuit voltages for the capacitors (e.g., $V = V_d$ in the circuit of Fig. 12.51; $V_{C_1} = V_{be}$ and $V_{C_2} = V_{ce}$ in the circuit of Fig. 12.56) and the short-circuit currents for the inductors.

All the cases considered up to now deal with *linear* capacitors and inductors, so their respective voltages and currents are (after the removal of a WSV for each linear algebraic constraint as discussed in Sect. 12.1) state variables. Therefore, their constant values found by the bias circuit correspond to equilibrium points for the state equations of the original circuit. In the presence of nonlinear capacitors and inductors, however, some kinds of nonlinearity require the choice of state variables such as charge and flow, respectively. In this case, the correspondence between the equilibrium states of an autonomous circuit and the operating points of its associated bias circuit is no longer one-to-one.

12.8.1   Relationship with the Supercritical Hopf Bifurcation

The small-signal analysis can be useful to see the supercritical Hopf bifurcation from a different perspective. When an equilibrium point undergoes a supercritical Hopf bifurcation, thus becoming unstable, a small-amplitude stable periodic solution appears around it; the onset of these oscillations can be studied using small-signal analysis.

For instance, the Wien bridge oscillator (see Sect. 12.6) is an autonomous circuit (it has no input), and the only equilibrium point is the origin. For values of the bifurcation parameter α slightly greater than 3 (supercritical Hopf bifurcation value), its small-signal equivalent circuit can be easily inferred from Fig. 12.39; it is shown in Fig. 12.59.

The small-signal equivalence holds as long as the approximation $f(v_2) \approx \alpha v_2$ can be considered acceptable, that is, qualitatively, within the gray box shown in Fig. 12.60. Notice that v_2 remains within the box as long as its amplitude is much lower than the op-amp supply voltage V_S, which is the "hidden" constant input of the circuit.

Fig. 12.59 Small-signal equivalent circuit for the Wien bridge oscillator for α slightly greater than 3

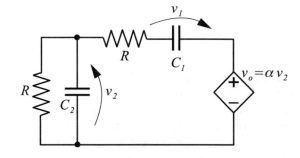

Fig. 12.60 Nonlinear function $f(v_2)$ (black solid curve) and its linear approximation αv_2 (gray dashed line). The two functions are almost overlapped within the gray box

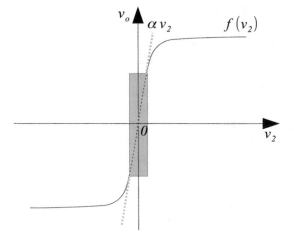

Fig. 12.61 Limits of
validity of the small-signal
approximation

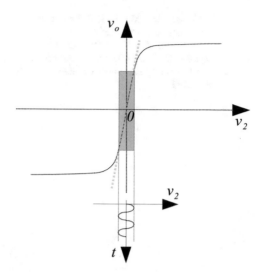

You can check that the state matrix of the small-signal circuit (for $C_1 = C_2 = C$)
is

$$A = \frac{1}{RC} \begin{pmatrix} -1 & \alpha - 1 \\ -1 & \alpha - 2 \end{pmatrix},$$

and compare it with the matrix $J(0, 0)$ found in Sect. 12.6.2 for the normalized
system.

Figure 12.61 points out that v_2 remains nearly sinusoidal as long as the small-
signal approximation is valid. The distortions noticed in Fig. 12.43c appear for α
values such that the amplitude of v_2 grows enough to prevent the use of the small-
signal approximation, that is, when the v_2 oscillations partially exit the central (almost
linear) part of the curve $f(v_2)$.

References

1. Chua LO (1980) Dynamic nonlinear networks: state of the art. IEEE Trans Circuits Syst 27:1059–
   1087
2. Meyer C (2000) Matrix analysis and applied linear algebra. SIAM, Philadelphia
3. Strogatz S (2014) Nonlinear Dynamics and Chaos with applications to physics, biology, chemistry, and engineering. Westview Press, Boulder
4. Mees AI, Chua LO (1979) The Hopf Bifurcation Theorem and Its Applications to Nonlinear
   Oscillations in Circuits and Systems. IEEE Transactions on Circuits and Systems 26:235–254
5. Maggio GM, De Feo O, Kennedy MP (1999) Nonlinear analysis of the Colpitts oscillator and
   applications to design. IEEE Trans Circuits Syst I: Fundam Theory Appl 46:1118–1130
6. De Feo O, Maggio GM, Kennedy MP (2000) The Colpitts oscillator: families of periodic solutions and their bifurcations. Int J Bifurc Chaos 10:935–958

7. De Feo O, Maggio GM (2003) Bifurcations in the Colpitts oscillator: from theory to practice. Int J Bifurc Chaos 13:2917–2934
8. Kuznetsov Yuri (2004) Elements of applied bifurcation theory. Springer, New York
9. Maggio GM, De Feo O, Kennedy MP (2004) A general method to predict the amplitude of oscillation in nearly sinusoidal oscillators. IEEE Trans Circuits Syst I: Fundam Theory Appl 51:1586–1595

Part VII
Analysis of Periodic Solutions

Chapter 13
Basic Concepts: Analysis of LTI Circuits in Sinusoidal Steady State

*Reading furnishes the mind only with materials of knowledge;
it is thinking that makes what we read ours.*
John Locke

Abstract This chapter is focused on the analysis of absolutely stable LTI circuits working in sinusoidal steady state. We show how to describe a circuit directly in the phasor domain, introduce the concepts of impedance and admittance, and describe how some properties, results, and methods introduced in Vol. 1 for memoryless circuits can be applied to memory circuits working in sinusoidal steady state. The important topic of power in sinusoidal steady state is also treated here, introducing Boucherot's theorem, the power factor correction problem, and the maximum-power-transfer theorem. Finally, the frequency response of circuits is treated, with emphasis on resonant circuits.

13.1 Sinusoidal Steady State

When an absolutely stable LTI circuit is driven by a sinusoidal input, the response becomes sinusoidal as $t \to \infty$, irrespective of the circuit's initial state.[1] All the voltages and currents associated with this *sinusoidal steady state* (or *AC steady state*) have the same frequency of the input. When we need to find only the response of this kind of circuit, we can follow different approaches, as shown in Sects. 9.4.2 and 9.5. In particular, we described a pure time-domain approach (see Case Study 2 in Sect. 9.4.2) and a mixed time-phasor-domain approach (see Case Study 1 in Sect. 9.5.2). In the latter case, the differential state equation (with real variables and coefficients) is expressed in the phasor domain in terms of algebraic equations whose unknowns are complex numbers, called phasors, which represent the input, the output, and the other circuit variables.

---

[1] For simply stable circuits, see Sect. 11.5.2.1.

© Springer Nature Switzerland AG 2020
M. Parodi and M. Storace, *Linear and Nonlinear Circuits: Basic and Advanced Concepts*,
Lecture Notes in Electrical Engineering 620,
https://doi.org/10.1007/978-3-030-35044-4_13

As a third possibility (pure phasor-domain approach), the phasor of a given variable can be obtained not by starting from its I/O relationship, as in the mixed approach, but by working from the beginning in the phasor domain, thus avoiding the use of differential equations. In this case, both the topological equations of the circuit (KVL and KCL) and the descriptive equations of the components are written directly in terms of phasors.

This kind of analysis is important for various reasons. As an example, electrical energy is distributed and made available to consumers through circuits working in sinusoidal steady state at the frequency of 50 (in Europe) or 60 Hz. As a further example, concerning a rather different point of view, the knowledge of the sinusoidal response of an LTI circuit at any frequency allows one to obtain, in principle, the response to a generic input waveform.

13.2   Circuit Equations in Terms of Phasors

In an LTI circuit working in sinusoidal steady state, all the circuit voltages $v_k(t)$ and the currents $i_k(t)$ are sinusoids at angular frequency ω. According to Sect. 9.5, the relations between v_k, i_k, and their corresponding phasors ($\dot{V}_k = V_k e^{j\varphi_k}$ and $\dot{I}_k = I_k e^{j\psi_k}$, respectively) are

$$v_k(t) = \Re\left\{\dot{V}_k e^{j\omega t}\right\} = \Re\left\{V_k e^{j(\omega t+\varphi_k)}\right\} = V_k \cos\left(\omega t + \varphi_k\right)$$
$$i_k(t) = \Re\left\{\dot{I}_k e^{j\omega t}\right\} = \Re\left\{I_k e^{j(\omega t+\psi_k)}\right\} = I_k \cos\left(\omega t + \psi_k\right).$$

13.2.1   Topological Equations

Both KVL and KCL equations can be formulated directly in terms of phasors. To show this, we can consider the KVL for a loop $\mathcal{L}$ of the circuit. For every t, assuming that the loop voltage arrows are equally oriented, we must have

$$0 = \sum_{k:v_k\in\mathcal{L}} v_k(t) = \sum_{k:v_k\in\mathcal{L}} \Re\left\{\dot{V}_k e^{j\omega t}\right\} = \Re\left\{\left[\sum_{k:v_k\in\mathcal{L}} \dot{V}_k\right] e^{j\omega t}\right\},$$

which implies that

$$\sum_{k:v_k\in\mathcal{L}} \dot{V}_k = 0.$$

A completely analogous result holds for the phasors of the currents flowing out of a cut-set $\mathcal{C}$:

$$\sum_{k:i_k\in\mathcal{C}} \dot{I}_k = 0.$$

13.2.2 Descriptive Equations

The descriptive equations of the LTI components can be easily formulated in terms of the phasors representing their descriptive variables. In so doing, the equations take the form of linear algebraic relationships among such phasors. The reference physical situation for a two-terminal component is shown in Fig. 13.1. Without loss of generality, the phasors of the descriptive variables for the two-terminal LTI element $\aleph_0$ are defined as $\dot{V} = Ve^{j\varphi}$ and $\dot{I} = Ie^{j\psi}$, which correspond to $v(t) = V\cos(\omega t + \varphi)$ and $i(t) = I\cos(\omega t + \psi)$, respectively.

Resistor: From the descriptive equation $v(t) = Ri(t)$ or $i(t) = Gv(t)$ we immediately obtain the phasor relations

$$\dot{V} = R\dot{I}; \quad \dot{I} = G\dot{V}. \tag{13.1}$$

Inasmuch as R and G are real positive numbers, we must have $\varphi = \psi$; the phasors $\dot{V}$ and $\dot{I}$ are proportional, as shown in Fig. 13.2a. As a consequence, the maximum and minimum values of sinusoids $v(t)$ and $i(t)$ occur at the same instants, as evidenced in Fig. 13.2b; $v(t)$ and $i(t)$ are said to be *in phase*.

Remark: Inasmuch as the initial condition is irrelevant, because the circuit works in (sinusoidal) steady state, the position of the time origin in Fig. 13.2b is in turn

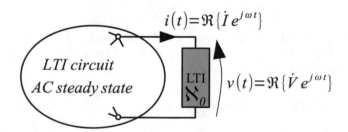

Fig. 13.1 A two-terminal LTI component $\aleph_0$ (in gray) and its descriptive variables in AC steady-state conditions

(a)   **(b)**

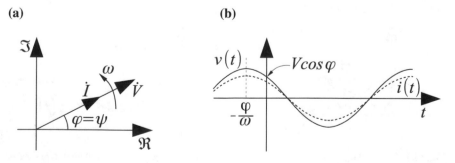

Fig. 13.2 Sinusoidal voltage and current for a resistor. **a** phasors; **b** $v(t), i(t)$

irrelevant. This means that the exact values of ψ and φ also are not important: the relevant information lies in their difference, which remains the same as time goes by. In mechanics, this is analogous to setting an arbitrary reference point to define the positions of some bodies with respect to that point, for instance on a straight line. Changing the time origin would correspond to adding some angle to the values of ψ and φ, that is, to rotate both phasors $\dot{V}$ and $\dot{I}$ in Fig. 13.2a through the same angle about the origin.

Inductor: from the component equation $v = L\dfrac{di}{dt}$, we have

$$\Re\left\{\dot{V}e^{j\omega t}\right\} = L\frac{d}{dt}\Re\left\{\dot{I}e^{j\omega t}\right\} = \Re\left\{\underbrace{j\omega L\dot{I}}_{\dot{V}}\,e^{j\omega t}\right\}.$$

By comparing the first and last terms, we obtain

$$\dot{V} = j\omega L\dot{I}. \tag{13.2}$$

Now, recalling that $j = e^{j\frac{\pi}{2}}$, the previous expression gives

$$Ve^{j\varphi} = \underbrace{\omega L I}_{V}\,e^{j\psi}e^{j\frac{\pi}{2}} \;\Rightarrow\; \psi = \varphi - \frac{\pi}{2}.$$

Therefore, the $\dot{I}$ phasor *lags* $\dot{V}$ (this term is consistent with the convention that takes as positive the counterclockwise rotation for ω) by $\frac{\pi}{2}$, and the corresponding sinusoidal terms can be written as

$$v(t) = V\cos\left(\omega t + \varphi\right); \quad i(t) = \frac{V}{\omega L}\cos\left(\omega t + \varphi - \frac{\pi}{2}\right) = \frac{V}{\omega L}\sin\left(\omega t + \varphi\right).$$

These results are summarized in Fig. 13.3.

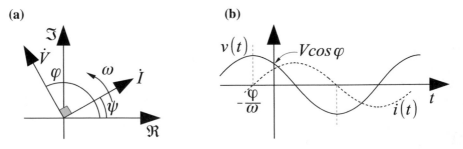

(a)   **(b)**

Fig. 13.3 Sinusoidal voltage and current for an inductor. **a** phasors; **b** $v(t)$ (solid line), $i(t)$ (dashed line)

Capacitor: from the component equation $i = C\dfrac{dv}{dt}$, we have

$$\Re\left\{\dot{I}e^{j\omega t}\right\} = C\dfrac{d}{dt}\Re\left\{\dot{V}e^{j\omega t}\right\} = \Re\left\{\underbrace{j\omega C\dot{V}}_{\dot{I}}\,e^{j\omega t}\right\}.$$

Therefore,

$$\dot{I} = j\omega C\dot{V}. \tag{13.3}$$

Furthermore, following the same line of reasoning as for the inductor, we find easily

$$Ie^{j\psi} = \underbrace{\omega C V\,e^{j\varphi}e^{j\frac{\pi}{2}}}_{I} \;\Rightarrow\; \psi = \varphi + \frac{\pi}{2}.$$

Therefore, the $\dot{I}$ phasor *leads* $\dot{V}$ by $\frac{\pi}{2}$. The corresponding sinusoidal terms can be written as

$$v(t) = V\cos{(\omega t + \varphi)}\,;\quad i(t) = \omega C V\cos\left(\omega t + \varphi + \frac{\pi}{2}\right) = -\omega C V\sin{(\omega t + \varphi)}\,.$$

These results are summarized in Fig. 13.4.

For multiterminal elements, we apply the principles described above to all their descriptive equations. For instance, a three-terminal whose descriptive equations in the time domain are

$$\begin{cases} v_1 = \alpha v_2 + L\dfrac{di_2}{dt}, \\[2mm] i_1 = \beta i_2 + C\dfrac{dv_2}{dt}, \end{cases}$$

would be described by the following equations:

(a)　　　　　　　　　　　**(b)**

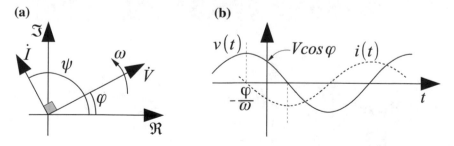

Fig. 13.4 Sinusoidal voltage and current for a capacitor. **a** phasors; **b** $v(t)$ (continuous line), $i(t)$ (dashed line)

$$\begin{cases} \dot{V}_1 = \alpha \dot{V}_2 + j\omega L \dot{I}_2, \\ \dot{I}_1 = \beta \dot{I}_2 + j\omega C \dot{V}_2, \end{cases}$$

in the phasor domain.

Two-port components are treated specifically in later sections.

13.3 Impedance and Admittance of Two-Terminal Elements

We still refer to the physical situation represented in Fig. 13.1. In Sect. 13.2.2 we found out the relationships between phasors $\dot{V}$ and $\dot{I}$ when the two-terminal $\aleph_0$ is a single (R, L, or C) component. Now we consider the more general case in which $\aleph_0$ originates from arbitrary connections between LTI components.

13.3.1 Impedance

Let us consider first the circuit shown in Fig. 13.5. The input is a sinusoidal current source $i(t)$ at angular frequency ω. The corresponding phasor is $\dot{I}_0$. The sinusoidal steady-state voltage $v(t)$ is represented by its phasor $\dot{V}$. The existence of $\dot{V}$ for any input phasor $\dot{I}_0$ means that at that given ω, the two-terminal $\aleph_0$ is current-controlled, that is, it admits the *current basis*.

> The ratio of the output phasor $\dot{V}$ and the input phasor $\dot{I}_0$ is a complex number defined as the *driving-point impedance* (or briefly, **impedance**) of $\aleph_0$ at the angular frequency ω:
>
> $$Z(j\omega) \doteq \frac{\dot{V}}{\dot{I}_0}; \quad Z(j\omega) = |Z(j\omega)| e^{j\vartheta}. \tag{13.4}$$

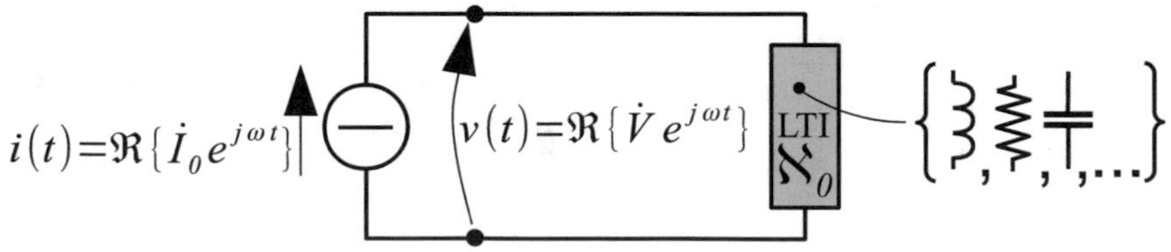

$$i(t) = \Re\{\dot{I}_0 e^{j\omega t}\} \qquad v(t) = \Re\{\dot{V} e^{j\omega t}\} \quad \boxed{\substack{\text{LTI}\\ \aleph_0}} - \left\{ \substack{\text{}} , \substack{\text{}} , \substack{\text{}} , \ldots \right\}$$

Fig. 13.5 A two-terminal LTI component $\aleph_0$ in sinusoidal steady state. Input: current source $i(t)$; output: voltage $v(t)$

For $\dot{I}_0 = I_0 e^{j\psi}$ and $\dot{V} = V e^{j\varphi}$, we have

$$|Z(j\omega)| = \left|\frac{\dot{V}}{\dot{I}_0}\right| = \frac{V}{I_0}; \quad \vartheta = \varphi - \psi. \tag{13.5}$$

Therefore, the output voltage $v(t)$ can be written as

$$v(t) = \Re\left\{Z \dot{I}_0 e^{j\omega t}\right\} = \underbrace{|Z| I_0}_{V} \Re\left\{e^{j(\vartheta+\psi)} e^{j\omega t}\right\} = |Z| I_0 \cos(\omega t + \vartheta + \psi).$$

For a given input current phasor $\dot{I}_0$ rotating at angular frequency ω, the magnitude V of the output phasor is $|Z| I_0$; the phase φ of the sinusoidal voltage is $\vartheta + \psi$.

The real part $R(\omega)$ and the imaginary part $X(\omega)$ of an impedance $Z(j\omega)$ are called **resistance** and **reactance**, respectively:

$$Z(j\omega) = R(\omega) + jX(\omega). \tag{13.6}$$

From Eq. 13.4, we directly obtain

$$R(\omega) \doteq \Re\{Z(j\omega)\} = |Z(j\omega)| \cos\vartheta;$$

$$X(\omega) \doteq \Im\{Z(j\omega)\} = |Z(j\omega)| \sin\vartheta; \tag{13.7}$$

$$\vartheta = \tan^{-1}\left\{\frac{X(\omega)}{R(\omega)}\right\}.$$

These results are summarized in Fig. 13.6.

We remark that the calculation of $V = |Z| I_0$ makes sense as long as we have $|Z| \neq \infty$ for the assigned ω value.

Taking into account Eqs. 13.1, 13.2, and 13.3, the impedances of the elementary components are immediately written as follows:

Fig. 13.6 Representation of $Z(j\omega)$ in the complex plane

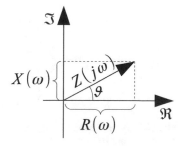

Resistor: $Z(j\omega) = R$ $(|Z| = R,\ \vartheta = 0,\ R(\omega) = R,\ X(\omega) = 0)$

Inductor: $Z(j\omega) = j\omega L$ $\left(|Z| = \omega L,\ \vartheta = \dfrac{\pi}{2},\ R(\omega) = 0,\ X(\omega) = \omega L\right)$

Capacitor: $Z(j\omega) = \dfrac{1}{j\omega C}$ $\left(|Z| = \dfrac{1}{\omega C},\ \vartheta = -\dfrac{\pi}{2},\ R(\omega) = 0,\ X(\omega) = -\dfrac{1}{\omega C}\right)$

On this basis, a generic impedance $Z(j\omega) = |Z(j\omega)|\,e^{j\vartheta} = R(\omega) + jX(\omega)$ can be classified as

resistive-inductive: for $0 < \vartheta < \dfrac{\pi}{2}$ $(R(\omega) > 0,\ X(\omega) > 0)$

resistive-capacitive: for $-\dfrac{\pi}{2} < \vartheta < 0$ $(R(\omega) > 0,\ X(\omega) < 0)$

13.3.2   Admittance

Consider now the circuit shown in Fig. 13.7, where the sinusoidal input to the LTI component $\aleph_0$ is a voltage source $v(t)$ at angular frequency ω. The voltage phasor is $\dot{V}_0 = V_0 e^{j\varphi}$. The sinusoidal steady-state current $i(t)$ is represented by its phasor $\dot{I} = I e^{j\psi}$. The existence of $\dot{I}$ for any input phasor $\dot{V}_0$ means that at that given ω, the two-terminal $\aleph_0$ is voltage-controlled, that is, it admits the *voltage basis*.

The ratio between the output phasor $\dot{I}$ and the input phasor $\dot{V}_0$ is a complex number defined as the *driving-point admittance* (or briefly, **admittance**) of $\aleph_0$ at the angular frequency ω:

$$Y(j\omega) \doteq \frac{\dot{I}}{\dot{V}_0}; \quad Y(j\omega) = |Y(j\omega)|\,e^{j\delta}. \tag{13.8}$$

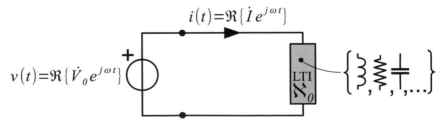

Fig. 13.7   A two-terminal LTI component $\aleph_0$ in sinusoidal steady state. Input: voltage source $v(t)$; output: current $i(t)$

For $\dot{I} = Ie^{j\psi}$ and $\dot{V}_0 = V_0 e^{j\varphi}$, we have

$$|Y(j\omega)| = \left|\frac{\dot{I}}{\dot{V}_0}\right| = \frac{I}{V_0}; \quad \delta = \psi - \varphi, \tag{13.9}$$

and the output current $i(t)$ is

$$i(t) = \Re\left\{Y\dot{V}_0 e^{j\omega t}\right\} = \underbrace{|Y|\, V_0}_{I}\, \Re\left\{e^{j(\delta+\varphi)} e^{j\omega t}\right\} = |Y|\, V_0 \cos(\omega t + \delta + \varphi).$$

Therefore, for a given input voltage phasor $\dot{V}_0$ rotating at angular frequency ω, the magnitude I of the output phasor is $|Y|\, V_0$; the phase ψ of the sinusoidal current is $\delta + \varphi$.

> The real part $G(\omega)$ and the imaginary part $B(\omega)$ of an admittance $Y(j\omega)$ are called **conductance** and **susceptance**, respectively:
>
> $$Y(j\omega) = G(\omega) + jB(\omega). \tag{13.10}$$

From Eq. 13.8, we directly obtain

$$G(\omega) \doteq \Re\{Y(j\omega)\} = |Y(j\omega)|\cos\delta;$$

$$B(\omega) \doteq \Im\{Y(j\omega)\} = |Y(j\omega)|\sin\delta; \tag{13.11}$$

$$\delta = \tan^{-1}\left\{\frac{B(\omega)}{G(\omega)}\right\}.$$

These results are summarized in Fig. 13.8.

We remark that the calculation of $I = |Y|\, V_0$ makes sense as long as for the assigned ω value, we have $|Y| \neq \infty$.

Fig. 13.8 Representation of $Y(j\omega)$ in the complex plane

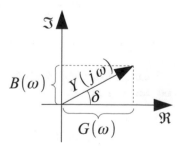

Again, considering Eqs. 13.1, 13.2, and 13.3, we immediately find the admittances of the elementary components as

Resistor: $Y(j\omega) = G$ $(|Y| = G,\ \delta = 0,\ G(\omega) = G,\ B(\omega) = 0)$

Inductor: $Y(j\omega) = \dfrac{1}{j\omega L}$ $\left(|Y| = \dfrac{1}{\omega L},\ \delta = -\dfrac{\pi}{2},\ G(\omega) = 0,\ B(\omega) = -\dfrac{1}{\omega L}\right)$

Capacitor: $Y(j\omega) = j\omega C$ $\left(|Y| = \omega C,\ \delta = \dfrac{\pi}{2},\ G(\omega) = 0,\ B(\omega) = \omega C\right)$

A generic admittance $Y(j\omega) = |Y(j\omega)|\, e^{j\delta} = G(\omega) + jB(\omega)$ can be classified as

resistive-inductive: for $-\dfrac{\pi}{2} < \delta < 0$ $(G(\omega) > 0,\ B(\omega) < 0)$

resistive-capacitive: for $0 < \delta < \dfrac{\pi}{2}$ $(G(\omega) > 0,\ B(\omega) > 0)$

Remark: The formal analogy of the two relations $V = ZI$ and $I = YV$ with those $v = Ri$ and $i = Gv$ encountered for resistors, combined with the possibility of writing KVLs and KCLs directly in terms of the voltage and current phasors, sets the basis for analyzing a circuit in sinusoidal steady state through, mutatis mutandis, the methods available for the analysis of linear memoryless circuits, described in Vol. 1.

13.3.3 Relation Between Impedance and Admittance of a Two-Terminal Component

Let us assume now that at a given angular frequency ω, the two-terminal LTI component admits both current basis and voltage basis. This means that both circuits of Figs. 13.5 and 13.7 are admitted.

Setting $\dot{I}_0 = \dot{I}$ in the first circuit, we have $\dot{V} = Z(j\omega)\dot{I}$; now setting $\dot{V}_0 = \dot{V}$ in the second circuit, we have $\dot{I} = Y(j\omega)\dot{V}$. This implies that

$$Z(j\omega) = \frac{1}{Y(j\omega)} \quad \Rightarrow \quad |Z(j\omega)| = \frac{1}{|Y(j\omega)|};\quad \vartheta = -\delta.$$

This property obviously applies for every ω value for which both bases of definition exist.

Taking into account Eqs. 13.8 and 13.11, it is easy to obtain the following relations between R, X, G, B:

$$Y = \frac{1}{Z} = \frac{1}{R + jX} = \frac{R - jX}{R^2 + X^2} \implies G = \frac{R}{R^2 + X^2}; \quad B = -\frac{X}{R^2 + X^2},$$

$$Z = \frac{1}{Y} = \frac{1}{G + jB} = \frac{G - jB}{G^2 + B^2} \implies R = \frac{G}{G^2 + B^2}; \quad X = -\frac{B}{G^2 + B^2}.$$

$$(13.12)$$

13.3.4 Series and Parallel Connections of Two-Terminal Elements

Consider the *series connection* of the two-terminal elements with impedances Z_1 and Z_2 shown in Fig. 13.9a. Inasmuch as we have

$$\dot{V} = \dot{V}_1 + \dot{V}_2; \quad \dot{V}_1 = Z_1 \dot{I}; \quad \dot{V}_2 = Z_2 \dot{I},$$

we immediately obtain $\dot{V} = (Z_1 + Z_2)\,\dot{I}$. Therefore, the impedance Z representing the series connection as in Fig. 13.9b is

$$Z = Z_1 + Z_2, \tag{13.13}$$

in complete analogy with the well-known result found for two resistors. This result can be trivially extended to the case of the series connection of any number of elements.

In terms of the admittances $Y_1 = 1/Z_1, Y_2 = 1/Z_2$, and $Y = 1/Z$, Eq. 13.13 takes the form

$$\frac{1}{Y} = \frac{1}{Y_1} + \frac{1}{Y_2} \implies Y = \frac{Y_1 Y_2}{Y_1 + Y_2}. \tag{13.14}$$

The *parallel connection* of two-terminal elements with admittances Y_1 and Y_2 is shown in Fig. 13.10a. From the relations

$$\dot{I} = \dot{I}_1 + \dot{I}_2; \quad \dot{I}_1 = Y_1 \dot{V}; \quad \dot{I}_2 = Y_2 \dot{V},$$

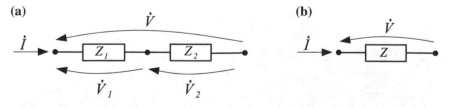

(a)

(b)

Fig. 13.9 **a** Series connection of two-terminal elements with impedances Z_1 and Z_2; **b** equivalent impedance Z

(a)   **(b)**

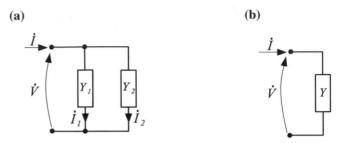

Fig. 13.10 a Parallel connection of two-terminal elements with admittances Y_1 and Y_2; **b** equivalent admittance Y

(a)   **(b)**

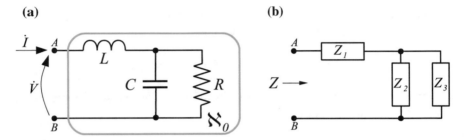

Fig. 13.11 a A composite two-terminal $\aleph_0$ with impedance Z and **b** its elementary constituents with impedances Z_1, Z_2, Z_3

we obtain $\dot{I} = (Y_1 + Y_2)\, V$ and the admittance Y representing the parallel connection as in Fig. 13.10b:

$$Y = Y_1 + Y_2. \tag{13.15}$$

This result can be directly extended to the case of the parallel connection of any number of elements.

Inasmuch as $Z_1 = 1/Y_1$, $Z_2 = 1/Y_2$, and $Z = 1/Y$, Eq. 13.15 can be recast as follows:

$$\frac{1}{Z} = \frac{1}{Z_1} + \frac{1}{Z_2} \quad \Rightarrow \quad Z = \frac{Z_1 Z_2}{Z_1 + Z_2}. \tag{13.16}$$

Case Study

Compute the impedance $Z(j\omega)$ of the two-terminal LTI component $\aleph_0$ shown in Fig. 13.11a, working at a given angular frequency ω.

The impedance $Z(j\omega)$ can be calculated by considering the elementary impedances

$$Z_1 = j\omega L; \quad Z_2 = \frac{1}{j\omega C}; \quad Z_3 = R,$$

which are the building blocks of Z, as shown in Fig. 13.11b. Therefore, we have

$$Z = Z_1 + \frac{Z_2 Z_3}{Z_2 + Z_3} = j\omega L + \frac{\frac{R}{j\omega C}}{R + \frac{1}{j\omega C}} = j\omega L + \frac{R}{1 + j\omega RC}.$$

This expression can be easily recast to evidence the real and imaginary parts:

$$Z = j\omega L + \frac{R(1 - j\omega RC)}{1 + (\omega RC)^2} = \underbrace{\frac{R}{1 + (\omega RC)^2}}_{R(\omega)} + j\,\omega \underbrace{\left(L - \frac{R^2 C}{1 + (\omega RC)^2}\right)}_{X(\omega)}.$$

The real part $R(\omega)$ (which should not be confused with the parameter R of $\aleph_0$) is the resistance of $Z(j\omega)$; the imaginary part $X(\omega)$ is its reactance. Both of them depend on the angular frequency ω and can be regarded as real functions of it. Notice that over the range $\omega \in (0, +\infty)$, we have:

- $R(\omega) > 0$;
- depending on the values of the parameters R, L, C, the signs of $X(\omega)$ and $\vartheta = \tan^{-1}\left(\frac{X(\omega)}{R(\omega)}\right)$ can change with ω. This can be shown easily by considering the term within parentheses in the expression of $X(\omega)$. This term can change sign with ω, provided that there exists a value of ω such that

$$(\omega RC)^2 L = R^2 C - L.$$

Therefore, for values of R, C, L such that $R > \sqrt{\frac{L}{C}}$, we have $X(\omega_0) = 0$ with

$$\omega_0 = \frac{1}{RC}\sqrt{\frac{R^2 C}{L} - 1},$$

which implies that

- for $\omega \in (0, \omega_0)$, $X(\omega) > 0$ and $\vartheta(\omega) \in \left(0, \frac{\pi}{2}\right)$; therefore, $\aleph_0$ is a resistive-inductive component and $\dot{V}$ leads $\dot{I}$;
- for $\omega \in (\omega_0, +\infty)$, $X(\omega) < 0$ and $\vartheta(\omega) \in \left(-\frac{\pi}{2}, 0\right)$; therefore, $\aleph_0$ is a resistive-capacitive component and $\dot{V}$ lags $\dot{I}$;
- for $\omega = \omega_0$, $X(\omega) = 0$ and $\vartheta(\omega_0) = 0$; therefore, $\aleph_0$ is a resistive component and $\dot{V}$ and $\dot{I}$ are in phase.

13.3.5   Reciprocity

The reciprocity property for two-terminal resistors (Vol. 1, Sect. 3.3) was formulated in terms of descriptive variables in the time domain. In sinusoidal steady state, every LTI two-terminal component with impedance $Z(j\omega)$ admits a reciprocity relationship between the phasors representing its descriptive variables. To show this, consider two pairs of phasors $(\dot{V}', \dot{I}')$ and $(\dot{V}'', \dot{I}'')$ such that $\dot{V}' = Z\dot{I}'$ and $\dot{V}'' = Z\dot{I}''$. We have

$$\dot{V}'\dot{I}'' = Z\dot{I}'\dot{I}'' = \dot{I}'\dot{V}'',$$

that is, a reciprocity condition in terms of phasors. A completely similar result can be obtained by considering the admittance $Y(j\omega)$.

13.4   Thévenin and Norton Equivalent Representations of Two-Terminal Elements

Consider a two-terminal $\aleph$ made up of LTI components and independent sinusoidal sources at angular frequency ω. As shown in Fig. 13.12a, in sinusoidal steady state the independent voltage and current sources inside $\aleph$ can be represented by phasors $\dot{E}_i$ and $\dot{A}_k$ ($i, k = 1, \ldots$), whereas the descriptive variables are denoted by $\dot{V}$ and $\dot{I}$. By turning off all the independent sources inside $\aleph$, we obtain the two-terminal $\aleph_0$ shown in Fig. 13.12b.

- Assuming that $\aleph$ admits the current basis, we can assign $\dot{I}$ by connecting the current source, as shown in Fig. 13.13a. The phasor $\dot{V}$ depends linearly on the whole set of independent sources, that is, the voltage sources of the set $\{\dot{E}_i\}$, the current sources $\{\dot{A}_k\}$, and the external source $\dot{I}$.
  Therefore, we can express $\dot{V}$ as

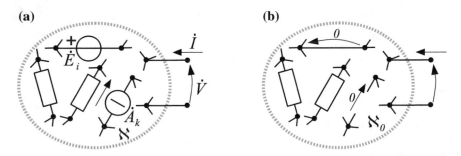

Fig. 13.12 **a** Composite two-terminal $\aleph$ and **b** its corresponding two-terminal $\aleph_0$, obtained by turning off all the independent sources

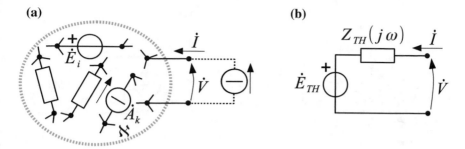

Fig. 13.13 a Assigning $\dot{I}$ to ℵ via a current source; **b** Thévenin equivalent representation for ℵ

$$\dot{V} = \underbrace{\sum_i \alpha_i \dot{E}_i + \sum_k Z_k \dot{A}_k}_{\dot{E}_{TH}} + Z_{TH}\dot{I}.$$

The voltage phasor $\dot{E}_{TH}$ and the impedance Z_{TH} are the two parameters defining the *equivalent Thévenin representation* of ℵ:

$$\dot{V} = \dot{E}_{TH} + Z_{TH}\dot{I}, \tag{13.17}$$

which corresponds to the circuit model of Fig. 13.13b.

According to Eq. 13.17, $\dot{E}_{TH}$ coincides with $\dot{V}$ when $\dot{I} = 0$. This makes it possible to adopt a simplified procedure, completely similar to the one described in Sect. 3.4.1 in Vol. 1 for memoryless two-terminals. The parameter Z_{TH} is nothing but the impedance of the two-terminal ℵ$_0$ of Fig. 13.12b.

- Assuming that ℵ admits the voltage basis, we can assign $\dot{V}$ by connecting a voltage source, as shown in Fig. 13.14a. By doing so, $\dot{I}$ can be written as a linear combination of the phasors representing the inner sources ($\{\dot{E}_i\}$ and $\{\dot{A}_k\}$), and $\dot{V}$. Analogously to the previous case, we obtain the equivalent Norton representation of ℵ:

$$\dot{I} = \dot{A}_{NR} + Y_{NR}\dot{V}, \tag{13.18}$$

which corresponds to the circuit model shown in Fig. 13.14b. The phasor $\dot{A}_{NR}$ represents the current $\dot{I}$ when the terminals of ℵ are short-circuited; Y_{NR} is the admittance of the two-terminal ℵ$_0$ shown in Fig. 13.12b. When ℵ admits both bases, we must have $Z_{TH} = \dfrac{1}{Y_{NR}}$.

(a)　　　　　　　　　　　　　　　　　**(b)**

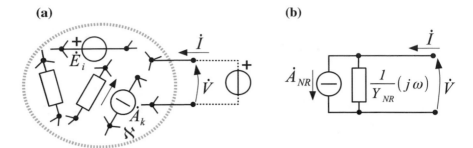

Fig. 13.14　**a** Assigning $\dot{V}$ to $\aleph$ through a voltage source; **b** Norton equivalent representation for $\aleph$

Fig. 13.15　Case Study 1

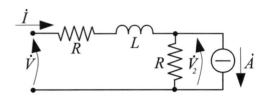

Case Study 1

Find the Thévenin and Norton equivalent representations shown in Fig. 13.16 for the two-terminal shown in Fig. 13.15, working in AC steady-state conditions.

- Making reference to the Thévenin equivalent shown in Fig. 13.16a, $\dot{E}_{TH}$ is equal to the voltage $\dot{V}_2$ when $\dot{I} = 0$, that is,

$$\dot{E}_{TH} = -R\dot{A}.$$

On the other hand, Z_{TH} is the two-terminal impedance when the independent source is turned off ($A = 0$), that is,

$$Z_{TH} = 2R + j\omega L.$$

- Setting now $\dot{V} = 0$ and making reference to the Norton equivalent shown in Fig. 13.16b, we easily obtain

$$\dot{I} = \dot{A}\frac{R}{2R + j\omega L} \quad \Rightarrow \quad \dot{A}_{NR} = -\dot{I} = -\dot{A}\frac{R}{2R + j\omega L}$$

and $Z_{NR} = Z_{TH}$.

(a)   **(b)**

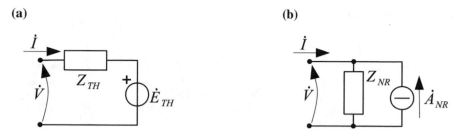

Fig. 13.16 Equivalent circuit conventions for Case Study 1: **a** Thévenin; **b** Norton

Case Study 2
Find the Thévenin equivalent shown in Fig. 13.16a for the composite two-terminal shown in Fig. 13.17, working in AC steady-state conditions.

For conciseness, we first denote by Z_1 and Z_2 the impedances

$$Z_1 = R_1 + \frac{1}{j\omega C_1} = \frac{1 + j\omega C_1 R_1}{j\omega C_1}; \quad Z_2 = \frac{R_2}{1 + j\omega C_2 R_2}$$

evidenced in Fig. 13.17.
The current $\dot{I}_1$ flows through Z_1 and Z_2. Therefore, we can write

$$\dot{I}_1 = \frac{\dot{E}}{Z_1}; \quad \dot{V} = -Z_2 \dot{I}_1,$$

and Eq. 13.17 takes the form

$$\dot{V} = -\frac{Z_2}{Z_1}\dot{E} = -\frac{j\omega C_1 R_2}{(1 + j\omega C_1 R_1)(1 + j\omega C_2 R_2)}\dot{E}.$$

This result implies that

$$\dot{E}_{TH} = -\frac{j\omega C_1 R_2}{(1 + j\omega C_1 R_1)(1 + j\omega C_2 R_2)}\dot{E}; \quad Z_{TH} = 0.$$

13.5 Two-Port Matrices

The concept of port and the definition of n-port component were introduced and discussed in Vol. 1 (Chap. 5) for memoryless elements, in particular in the linear case. In sinusoidal steady state, the descriptive variables of an LTI n-port can be represented

Fig. 13.17 Case Study 2

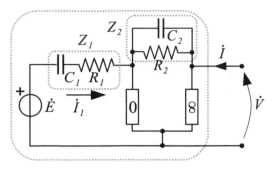

Fig. 13.18 A two-port with
its standard choice
descriptive variables
(phasors)

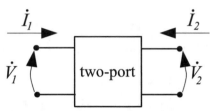

by phasors, and its n algebraic linear descriptive equations can be formulated using
matrices. The bases admitted by the n-port determine which matrices are allowed.
Each matrix entry can be found, mutatis mutandis, by following the same procedure
as for memoryless elements and can be a real or complex number.

For completeness, we briefly consider some examples of matrix-based descrip-
tions for LTI two-ports. Figure 13.18 shows a generic LTI two-port, with the phasors
representing its descriptive variables.

The resistance matrix R of a current-controlled, memoryless two-port leads to the
descriptive equation

$$\underbrace{\begin{pmatrix} \dot{V}_1 \\ \dot{V}_2 \end{pmatrix}}_{\dot{V}} = \underbrace{\begin{pmatrix} R_{11} & R_{12} \\ R_{21} & R_{22} \end{pmatrix}}_{R} \underbrace{\begin{pmatrix} \dot{I}_1 \\ \dot{I}_2 \end{pmatrix}}_{\dot{I}} \tag{13.19}$$

in terms of the phasor vectors $\dot{V}$ and $\dot{I}$ of the descriptive variables. A completely
analogous relationship holds in the case of a voltage-controlled resistive two-port
with conductance matrix G:

$$\dot{I} = G\dot{V}.$$

Matrices R and G can be viewed as particular cases of the impedance matrix Z
and admittance matrix Y, respectively. The elements of R and G are real, whereas
the entries of Z and Y are in general complex.

For instance, from the descriptive equation of the coupled inductors (Eq. 11.1) we
immediately obtain

$$\underbrace{\begin{pmatrix} \dot{V}_1 \\ \dot{V}_2 \end{pmatrix}}_{\dot{v}} = \underbrace{\begin{pmatrix} j\omega L_1 & j\omega M \\ j\omega M & j\omega L_2 \end{pmatrix}}_{Z} \underbrace{\begin{pmatrix} \dot{I}_1 \\ \dot{I}_2 \end{pmatrix}}_{\dot{I}} \quad \text{with } Z \doteq j\omega \begin{pmatrix} L_1 & M \\ M & L_2 \end{pmatrix}. \tag{13.20}$$

The impedance matrix is purely imaginary. Following the procedure described in Sect. 5.3 of Vol. 1, it is easy to show that the entries of Z and Y meet the following definitions (for $i, j = 1, 2$):

$$\begin{pmatrix} \dot{V}_1 \\ \dot{V}_2 \end{pmatrix} = \begin{pmatrix} Z_{11} & Z_{12} \\ Z_{21} & Z_{22} \end{pmatrix} \begin{pmatrix} \dot{I}_1 \\ \dot{I}_2 \end{pmatrix} \quad \Rightarrow \quad Z_{ii} = \left. \frac{\dot{V}_i}{\dot{I}_i} \right|_{\dot{I}_j = 0} ; \quad Z_{ij} = \left. \frac{\dot{V}_i}{\dot{I}_j} \right|_{\dot{I}_i = 0} \quad (i \neq j),$$

$$\tag{13.21}$$

$$\begin{pmatrix} \dot{I}_1 \\ \dot{I}_2 \end{pmatrix} = \begin{pmatrix} Y_{11} & Y_{12} \\ Y_{21} & Y_{22} \end{pmatrix} \begin{pmatrix} \dot{V}_1 \\ \dot{V}_2 \end{pmatrix} \quad \Rightarrow \quad Y_{ii} = \left. \frac{\dot{I}_i}{\dot{V}_i} \right|_{\dot{V}_j = 0} ; \quad Y_{ij} = \left. \frac{\dot{I}_i}{\dot{V}_j} \right|_{\dot{V}_i = 0} \quad (i \neq j),$$

$$\tag{13.22}$$

corresponding to obvious circuit interpretations.

Similarly, for the transmission matrix T, we have

$$\begin{pmatrix} \dot{V}_1 \\ \dot{I}_1 \end{pmatrix} = \begin{pmatrix} T_{11} & T_{12} \\ T_{21} & T_{22} \end{pmatrix} \begin{pmatrix} \dot{V}_2 \\ -\dot{I}_2 \end{pmatrix}, \tag{13.23}$$

and the matrix elements T_{ij} are immediately obtained from Eq. 5.21 by simply replacing the original descriptive variables i_i, v_j $(i, j = 1, 2)$ with their corresponding phasors.

Completely similar results hold for the hybrid matrices H and H' and for the backward transmission matrix T'.

Case Study 1
Under the hypothesis of AC steady-state conditions at angular frequency ω, find the Z matrix of the two-port shown in Fig. 13.19a and the Norton equivalent representation for the two-terminal shown in Fig. 13.19b, obtained by connecting the AC current source $\dot{A}$ to the two-port.

1. Recalling the ideal transformer equations, we can write

$$\begin{cases} \dot{I}_1 - \dfrac{\dot{V}_1}{R + j\omega L} = -n\dot{I}_2, \\ \dot{V}_2 = n\left(\dot{V}_1 - jX\left(\dot{I}_1 - \dfrac{\dot{V}_1}{R + j\omega L} \right) \right). \end{cases}$$

These equations can be recast as

$$\begin{cases} \dot{V}_1 = (R + j\omega L)\, \dot{I}_1 + n\,(R + j\omega L)\, \dot{I}_2, \\ \dot{V}_2 = n\,(R + j\omega L)\, \dot{I}_1 + n^2\,(R + j\,(\omega L + X))\, \dot{I}_2. \end{cases} \tag{13.24}$$

Therefore, the two-port matrix Z is

$$Z = \begin{pmatrix} R + j\omega L & n\,(R + j\omega L) \\ n\,(R + j\omega L) & n^2\,(R + j\,(\omega L + X)) \end{pmatrix}.$$

2. Referring now to Fig. 13.19b, we set $\dot{I}_1 = \dot{A}$ and rename the port variables $\dot{V}_2$, $\dot{I}_2$ as $\dot{V}$, $\dot{I}$. Therefore, from the second equation of Eq. 13.24, we obtain

$$\dot{V} = n\,(R + j\omega L)\,\dot{A} + n^2\,(R + j\,(\omega L + X))\,\dot{I},$$

which can be recast as

$$\dot{I} = -\frac{(R + j\omega L)}{n\,(R + j\,(\omega L + X))}\,\dot{A} + \frac{1}{n^2\,(R + j\,(\omega L + X))}\,\dot{V}.$$

Therefore, the parameters of the Norton equivalent in Fig. 13.19c are

$$\dot{A}_{NR} = \frac{(R + j\omega L)}{n\,(R + j\,(\omega L + X))}; \quad Z_{NR} = n^2\,(R + j\,(\omega L + X)).$$

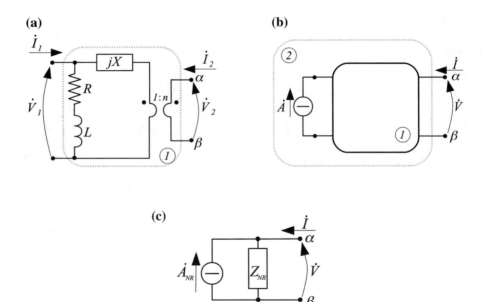

Fig. 13.19 Case Study 1: **a** structure of the two-port; **b** two-terminal element including the two-port; **c** Norton equivalent circuit to be determined

Case Study 2

Find the transmission matrix $T(j\omega)$ of the two-port shown in Fig. 13.20, working in AC steady state at angular frequency ω. Determine which bases are admitted by the two-port.

1. Inasmuch as $\dot{V}_2 = \dot{V}$, the capacitor voltage is zero, which implies that the current flowing through C also is zero. Therefore, we can easily write

$$\dot{V}_1 - j\omega L \dot{I}_1 = 0; \quad \dot{V}_2 = -R\dot{I}_1$$

or, equivalently,

$$
\begin{cases}
\dot{V}_1 = -\dfrac{j\omega L}{R}\dot{V}_2, \\[2ex]
\dot{I}_1 = -\dfrac{\dot{V}_2}{R}.
\end{cases}
$$

It is immediate to identify the transmission matrix elements by direct comparison with Eq. 13.23:

$$
T = \begin{pmatrix} -\dfrac{j\omega L}{R} & 0 \\[2ex] -\dfrac{1}{R} & 0 \end{pmatrix}.
$$

2. Inasmuch as the $\dot{I}_2$ port variable does not appear in the two-port equations, it must be present as a fixed element into any allowed basis. Therefore, the admitted bases are $(\dot{V}_1, \dot{I}_2)$, corresponding to the hybrid matrix H', and $(\dot{I}_1, \dot{I}_2)$, corresponding to the impedance matrix Z.

Fig. 13.20 Case Study 2

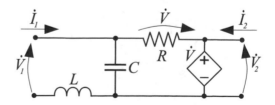

13.6   Thévenin and Norton Equivalent Representations of Two-Port Elements

The analysis of these representations faithfully follows that described in Sect. 13.4 for a two-terminal. The only difference is that in this case, $\aleph$ has two ports, whose descriptive variables are represented by the phasors $\dot{V}_1$, $\dot{I}_1$, $\dot{V}_2$, $\dot{I}_2$. This is shown in Fig. 13.21a, whereas Fig. 13.21b shows the two-port $\aleph_0$ obtained by turning off the independent sources inside $\aleph$.

- Under the assumption that $\aleph$ admits the current basis, we assign $\dot{I}_1$, $\dot{I}_2$ by connecting two current sources, as shown in Fig. 13.22a. The voltage phasors $\dot{V}_1$, $\dot{V}_2$ depend linearly on the whole set of independent sources, that is, the voltage sources of the set $\{\dot{E}_i\}$, the current sources $\{\dot{A}_k\}$, plus the external sources $\dot{I}_1$, $\dot{I}_2$. Therefore, we can write

$$\dot{V}_1 = \underbrace{\sum_i \alpha_i \dot{E}_i + \sum_k Z_{1k}\dot{A}_k}_{\dot{E}_{T1}} + Z_{11}\dot{I}_1 + Z_{12}\dot{I}_2,$$

$$\dot{V}_2 = \underbrace{\sum_i \beta_i \dot{E}_i + \sum_k Z_{2k}\dot{A}_k}_{\dot{E}_{T2}} + Z_{21}\dot{I}_1 + Z_{22}\dot{I}_2.$$

The voltage phasors $\dot{E}_{T1}$, $\dot{E}_{T2}$ and the impedance matrix

$$Z_T \doteq \begin{bmatrix} Z_{11} & Z_{12} \\ Z_{21} & Z_{22} \end{bmatrix}$$

are the parameters defining the Thévenin equivalent representation of the two-port $\aleph$:

(a)   **(b)**

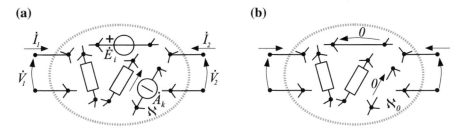

Fig. 13.21   a Composite two-port $\aleph$; **b** its corresponding two-port $\aleph_0$, obtained by turning off all the independent sources

(a)   **(b)**

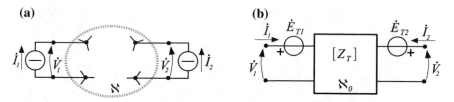

Fig. 13.22 a Assigning $\dot{I}_1$, $\dot{I}_2$; **b** Thévenin equivalent representation of ℵ

(a)   **(b)**

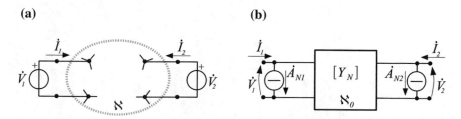

Fig. 13.23 a Assigning $\dot{V}_1$, $\dot{V}_2$; **b** Norton equivalent representation for ℵ

$$\begin{bmatrix} \dot{V}_1 \\ \dot{V}_2 \end{bmatrix} = \begin{bmatrix} \dot{E}_{T1} \\ \dot{E}_{T2} \end{bmatrix} + \begin{bmatrix} Z_{11} & Z_{12} \\ Z_{21} & Z_{22} \end{bmatrix} \begin{bmatrix} \dot{I}_1 \\ \dot{I}_2 \end{bmatrix}, \qquad (13.25)$$

which correspond to the equivalent circuit model shown in Fig. 13.22b. According to Eq. 13.25, $\dot{E}_{T1}$ and $\dot{E}_{T2}$ are the port voltages when $\dot{I}_1 = 0$ and $\dot{I}_2 = 0$, whereas Z_T is the impedance matrix of the auxiliary two-port ℵ$_0$.

• Similarly, when ℵ admits the voltage basis, we can assign $\dot{V}_1$, $\dot{V}_2$ by connecting two voltage sources, as shown in Fig. 13.23a. On so doing, $\dot{I}_1$, $\dot{I}_2$ can be written as a linear combination of $\dot{V}_1$, $\dot{V}_2$ and the phasors of the inner sources $\{\dot{E}_i\}$, $\{\dot{A}_k\}$. In complete analogy with the previous case, we obtain for ℵ the equivalent Norton representation

$$\begin{bmatrix} \dot{I}_1 \\ \dot{I}_2 \end{bmatrix} = \begin{bmatrix} \dot{A}_{N1} \\ \dot{A}_{N2} \end{bmatrix} + \underbrace{\begin{bmatrix} Y_{11} & Y_{12} \\ Y_{21} & Y_{22} \end{bmatrix}}_{Y_N} \begin{bmatrix} \dot{V}_1 \\ \dot{V}_2 \end{bmatrix}, \qquad (13.26)$$

which corresponds to the circuit model of Fig. 13.23b. From Eq. 13.26, we immediately find that $\dot{A}_{N1}$, $\dot{A}_{N2}$ are the port currents when $\dot{V}_1 = 0$ and $\dot{V}_2 = 0$, whereas Y_N is the admittance matrix of the auxiliary two-port ℵ$_0$.

Case Study 1
The two-port shown in Fig. 13.24 a is operating under sinusoidal steady-state conditions at angular frequency ω. Find the parameters of its Thévenin equivalent representation shown in Fig. 13.24b. Then find the values of α and r that for $\dot{E} = 0$ make the two-port reciprocal.

1. Referring to the currents shown in Fig. 13.25, we can write the following two equations in terms of the port variables only:

$$\begin{cases} \dot{V}_1 = R\left(\dot{I}_1 - \alpha \dot{I}_2\right) + \dot{E}, \\ \dot{V}_2 = \dot{E} + j\omega L\left(\dot{I}_2 - \left(\dfrac{\dot{V}_2 - r\dot{I}_1}{R}\right)\right). \end{cases}$$

These equations can be recast according to Eq. 13.25:

$$\begin{cases} \dot{V}_1 = R\dot{I}_1 - \alpha R\dot{I}_2 + \dot{E}, \\ \dot{V}_2 = \dfrac{j\omega Lr}{R + j\omega L}\dot{I}_1 + \dfrac{j\omega LR}{R + j\omega L}\dot{I}_2 + \dfrac{R}{R + j\omega L}\dot{E}. \end{cases}$$

Therefore, the parameters of the Thévenin representation are

$$\dot{E}_1 = \dot{E}; \qquad \dot{E}_2 = \frac{R}{R + j\omega L}\dot{E};$$

and

$$Z = \begin{pmatrix} R & -\alpha R \\ \dfrac{j\omega Lr}{R + j\omega L} & \dfrac{j\omega LR}{R + j\omega L} \end{pmatrix}.$$

2. Reciprocity requires symmetry of the Z matrix, that is, $Z_{12} = Z_{21}$:

$$-\alpha R = \frac{j\omega Lr}{R + j\omega L}.$$

This condition holds if and only if $\alpha = 0$ and $r = 0$, that is, both controlled sources are turned off. We remark that in this case, the two-port $\aleph_0$ will be composed of reciprocal components only. The reciprocity theorem (see Sect. 6.3, Vol. 1) ensures that in this case, the two-port is reciprocal.

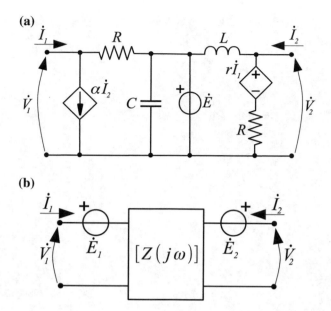

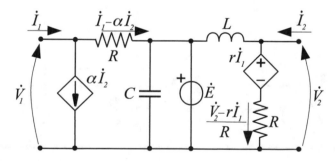

Fig. 13.24 Case Study 1: **a** composite two-port; **b** its equivalent Thévenin representation

Fig. 13.25 Solution of Case Study 1

Case Study 2

The two-port shown in Fig. 13.26a is operating under sinusoidal steady-state conditions at angular frequency ω. Obtain the parameters of its Norton equivalent representation shown in Fig. 13.26b.

1. We can find $\dot{A}_1$ and $\dot{A}_2$ by referring to Fig. 13.27a:

$$\dot{A}_1 = \dot{A} + \frac{\dot{E}}{nR}; \qquad \dot{A}_2 = -\frac{\dot{E}}{n^2 R}.$$

2. After turning off the independent sources, we obtain the two-port shown in Fig. 13.27b. Its descriptive equations are

$$\begin{cases} \dot{I}_1 = \dfrac{\dot{V}_1}{j\omega L} + \dfrac{1}{R}\left(\dot{V}_1 - \dfrac{\dot{V}_2}{n}\right) = \left(\dfrac{1}{j\omega L} + \dfrac{1}{R}\right)\dot{V}_1 - \dfrac{1}{nR}\dot{V}_2, \\[2mm] \dot{I}_2 = \dfrac{\dot{V}_2}{R} - \dfrac{1}{nR}\left(\dot{V}_1 - \dfrac{\dot{V}_2}{n}\right) = -\dfrac{\dot{V}_1}{nR} + \left(\dfrac{1}{R} + \dfrac{1}{n^2 R}\right)\dot{V}_2. \end{cases}$$

Therefore, the Y matrix (admittance matrix of $\aleph_0$) is

$$Y = \begin{pmatrix} \left(\dfrac{1}{j\omega L} + \dfrac{1}{R}\right) & -\dfrac{1}{nR} \\[4mm] -\dfrac{1}{nR} & \left(\dfrac{1}{R} + \dfrac{1}{n^2 R}\right) \end{pmatrix}.$$

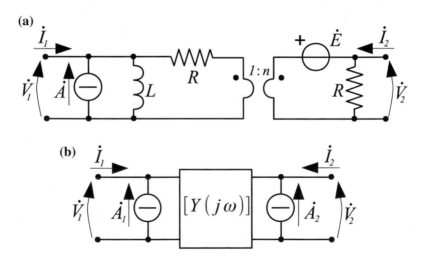

Fig. 13.26 Case Study 2: **a** composite two-port; **b** its Norton equivalent representation

(a)

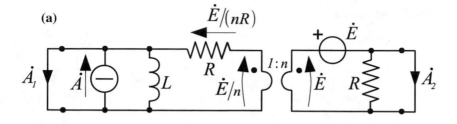

(b)

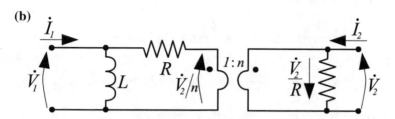

Fig. 13.27 Solution of Case Study 2. **a** Auxiliary circuit to find $\dot{A}_1$ and $\dot{A}_2$; **b** auxiliary two-port $\aleph_0$ to find the matrix Y

13.7 Sinusoidal Steady-State Power

The phasor domain analysis can also help us to understand the energetic behavior of circuits working in sinusoidal steady state. To start with, we analyze the behavior of the simplest circuit elements.

13.7.1 Two-Terminal Components

We begin by analyzing a two-terminal component working in sinusoidal steady state (see Fig. 13.28). It is characterized by descriptive variables $v(t) = V \cos(\omega t + \varphi)$ and $i(t) = I \cos(\omega t + \psi)$, corresponding to phasors $\dot{V} = V e^{j\varphi}$ and $\dot{I} = I e^{j\psi}$, respectively. The corresponding impedance $Z(j\omega)$ has magnitude $|Z(j\omega)| = \dfrac{V}{I}$ and phase $\vartheta = \varphi - \psi$.

The instantaneous power absorbed by the component is

$$p(t) = v(t)i(t) = V \cos(\omega t + \varphi) I \cos(\omega t + \psi). \qquad (13.27)$$

Inasmuch as

$$\cos\alpha \cos\beta = \frac{1}{2}\cos(\alpha - \beta) + \frac{1}{2}\cos(\alpha + \beta),$$

Fig. 13.28 Two-terminal
component working in
sinusoidal steady state

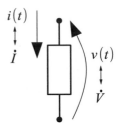

Equation 13.27 can be recast as

$$p(t) = \frac{VI}{2} \cos(\varphi - \psi) + \frac{VI}{2} \cos(2\omega t + \varphi + \psi) =$$

$$= \frac{VI}{2} \cos\vartheta + \frac{VI}{2} \cos[(2\omega t + 2\psi) + \vartheta].$$

(13.28)

Now we exploit another trigonometric equivalence:

$$\cos(\alpha + \beta) = \cos\alpha\cos\beta - \sin\alpha\sin\beta,$$

thus obtaining

$$p(t) = \frac{VI}{2} \cos\vartheta + \frac{VI}{2} \cos(2\omega t + 2\psi)\cos\vartheta - \frac{VI}{2} \sin(2\omega t + 2\psi)\sin\vartheta.$$

(13.29)

Therefore,

$$p(t) = \frac{VI}{2} \cos\vartheta\,[1 + \cos(2\omega t + 2\psi)] - \frac{VI}{2} \sin\vartheta\,\sin(2\omega t + 2\psi) \qquad (13.30)$$

is the sum of two terms, an instantaneous active power

$$p_a(t) = \underbrace{\frac{VI}{2} \cos\vartheta\,[1 + \cos(2\omega t + 2\psi)]}_{= P} \qquad (13.31)$$

and an instantaneous reactive power

$$p_r(t) = \underbrace{-\frac{VI}{2} \sin\vartheta\,\sin(2\omega t + 2\psi)}_{= Q}. \qquad (13.32)$$

The amplitudes of these oscillating powers are called **active power** (P) and **reactive power** (Q).

Remark: We already know (see Sect. 11.10.4) that the effective or RMS value of a sinusoidal waveform is its amplitude divided by $\sqrt{2}$. Therefore, we can define

$$P = V_{eff} I_{eff} \cos \vartheta,$$
$$Q = V_{eff} I_{eff} \sin \vartheta, \tag{13.33}$$

where $V_{eff} = \dfrac{V}{\sqrt{2}}$ and $I_{eff} = \dfrac{I}{\sqrt{2}}$.

Figure 13.29 shows a qualitative representation of $p_a(t)$ (panel a) and $p_r(t)$ (panel b). Both of them oscillate with angular frequency 2ω (double the one imposed by the circuit input), but with different amplitudes. Moreover, over the power period $\dfrac{T}{2} = \dfrac{\pi}{\omega}$, $p_r(t)$ has a null mean value, whereas $p_a(t)$ has a mean value P. What does this mean in practice?

(a)

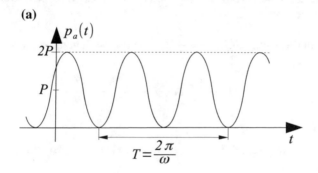

(b)

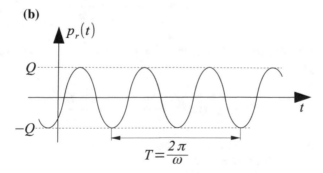

Fig. 13.29 Qualitative representations of **a** the instantaneous active power $p_a(t)$ and **b** the instantaneous reactive power $p_r(t)$ absorbed by a two-terminal element

In circuits working in AC steady state, energy storage elements such as inductors and capacitors may result in periodic reversals of the direction of energy flow, because (see Sect. 9.2.1) they are *conservative* components. The instantaneous active power is the portion of power that, averaged over a complete cycle of the AC waveform, results in net transfer of energy in one direction and can therefore be converted into other forms of energy, thus generating light, heat, or mechanical movement. The instantaneous reactive power is the portion of power that corresponds, in each cycle, to energy temporarily stored in the two-terminal element and then returned to the rest of the circuit.

Inasmuch as the frequency is fixed and the phase is not relevant in terms of energy flows, the most relevant information about the power absorbed by the two-terminal element is contained in P and Q.

Remark: In sinusoidal steady state, the mean value over the period of the instantaneous absorbed power is $\bar{p} = P$.

The active and reactive powers absorbed by the most common two-terminal passive components are:

- **Resistor**: in this case, $|Z(j\omega)| = \dfrac{V}{I} = R$ and $\vartheta = 0$, and therefore we obtain

$$P = \frac{VI}{2}\cos\vartheta = \frac{RI^2}{2} = \frac{V^2}{2R} \tag{13.34}$$

and $Q = 0$.

- **Inductor**: in this case, $|Z(j\omega)| = \dfrac{V}{I} = \omega L$ and $\vartheta = \dfrac{\pi}{2}$, and therefore we obtain $P = 0$ and

$$Q = \frac{VI}{2}\sin\vartheta = \frac{\omega L I^2}{2} = \frac{V^2}{2\omega L}. \tag{13.35}$$

- **Capacitor**: in this case, $|Z(j\omega)| = \dfrac{V}{I} = \dfrac{1}{\omega C}$ and $\vartheta = -\dfrac{\pi}{2}$, and therefore we obtain $P = 0$ and

$$Q = \frac{VI}{2}\sin\vartheta = -\frac{\omega C V^2}{2} = -\frac{I^2}{2\omega C}. \tag{13.36}$$

We defined P and Q by reasoning in the time domain. How can we compute them by reasoning in the phasor domain?

Owing to the definitions $P = \dfrac{VI}{2}\cos\vartheta$ and $Q = \dfrac{VI}{2}\sin\vartheta$, we can introduce the concept of **complex power** $P + jQ$, whose geometric interpretation is shown in Fig. 13.30.

Therefore, the complex power is a vector with amplitude $\dfrac{VI}{2}$ and phase ϑ and can be thus expressed as $P + jQ = \dfrac{VI}{2}e^{j\vartheta}$.

Fig. 13.30 Geometric interpretation of the complex power $P + jQ$

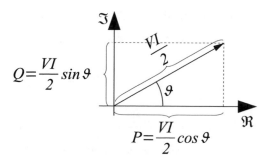

$$Q = \frac{VI}{2}\sin\vartheta$$

$$P = \frac{VI}{2}\cos\vartheta$$

The amplitude $|P + jQ| = \dfrac{VI}{2}$ is called the **apparent power**. In other words, apparent power is the product of the RMS values (see Eq. 13.31 and Sect. 11.10.4) of voltage and current.

The ratio of active power to apparent power, that is, $\cos\vartheta$, is called the **power factor**. The power factor is 1 when the voltage and current are in phase, as for the resistor. It is zero when the current lags or leads the voltage by $\dfrac{\pi}{2}$, as for inductor and capacitor, respectively. For instance, if the active power is 50 W and $\vartheta = \dfrac{\pi}{3}$ (corresponding to a resistive–inductive component), the power factor is $\cos(\dfrac{\pi}{3}) = 0.5$ and the apparent power is $\dfrac{50\,\text{W}}{0.5} = 100\,\text{W}$.

The complex power absorbed by the two-terminal can be obtained from its phasors $\dot{V}$ and $\dot{I}$ as follows:

$$P + jQ = \frac{\dot{V}\dot{I}^*}{2}; \tag{13.37}$$

therefore,

$$P = \frac{1}{2}\Re\left\{\dot{V}\dot{I}^*\right\} \tag{13.38}$$

and

$$Q = \frac{1}{2}\Im\left\{\dot{V}\dot{I}^*\right\}. \tag{13.39}$$

Equation 13.37 can be easily proved:

$$\frac{\dot{V}\dot{I}^*}{2} = \frac{1}{2}Ve^{j\varphi}\left(Ie^{j\psi}\right)^* = \frac{VI}{2}e^{j\varphi}e^{-j\psi} = \frac{VI}{2}e^{j(\varphi-\psi)} = \frac{VI}{2}e^{j\vartheta}. \tag{13.40}$$

In particular, for R, L, and C, we again achieve the following:

- **Resistor**: using the descriptive equation in the phasor domain $\dot{V} = R\dot{I}$, we obtain

$$P + jQ = \frac{\dot{V}\dot{I}^*}{2} = \frac{R|\dot{I}|^2}{2} = \frac{|\dot{V}|^2}{2R}. \tag{13.41}$$

Therefore, $P = \dfrac{R|\dot{I}|^2}{2} = \dfrac{|\dot{V}|^2}{2R}$ and $Q = 0$.

- **Inductor**: using the descriptive equation in the phasor domain $\dot{V} = j\omega L\dot{I}$, we obtain

$$P + jQ = \frac{\dot{V}\dot{I}^*}{2} = \frac{j\omega L|\dot{I}|^2}{2} = -\frac{|\dot{V}|^2}{2j\omega L}. \tag{13.42}$$

Therefore, $P = 0$ and $Q = \dfrac{\omega L|\dot{I}|^2}{2} = \dfrac{|\dot{V}|^2}{2\omega L}$.

- **Capacitor**: using the descriptive equation in the phasor domain $\dot{I} = j\omega C\dot{V}$, we obtain

$$P + jQ = \frac{\dot{V}\dot{I}^*}{2} = -\frac{j\omega C|\dot{V}|^2}{2} = \frac{|\dot{I}|^2}{2j\omega C}. \tag{13.43}$$

Therefore, $P = 0$ and $Q = -\dfrac{\omega C|\dot{V}|^2}{2} = -\dfrac{|\dot{I}|^2}{2\omega C}$.

13.7.2   Generic Components

The definition of complex power can be easily extended to generic components.

Figure 13.31 shows a generic n-terminal (panel a) and n-port (panel b) working in sinusoidal steady state and described according to the standard choice.

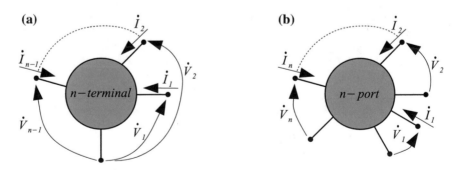

Fig. 13.31   a Generic n-terminal and **b** n-port working in sinusoidal steady state

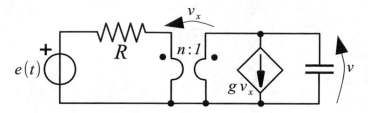

Fig. 13.32 Case Study

The complex power absorbed by the n-terminal is

$$P + jQ = \frac{1}{2} \sum_{k=1}^{n-1} \dot{V}_k \dot{I}_k^*, \tag{13.44}$$

and the complex power absorbed by the n-port is

$$P + jQ = \frac{1}{2} \sum_{k=1}^{n} \dot{V}_k \dot{I}_k^*. \tag{13.45}$$

Case Study

For the circuit shown in Fig. 13.32, working in AC steady state with $e(t) = E\cos(\omega t)$, find:

1. *the I/O relationship (state equation) for $v(t)$;*
2. *the circuit's natural frequency and absolute stability condition;*
3. *the reactive power absorbed by the capacitor;*
4. *the complex power $P_x + jQ_x$ absorbed by the VCCS.*

1. You can easily check that the I/O relationship (state equation) for $v(t)$ is

$$RC\frac{dv}{dt} + [Rg(n-1) + n^2]v = ne(t).$$

2. The circuit's natural frequency is $\lambda = -\dfrac{Rg(n-1) + n^2}{RC}$. Therefore, the absolute stability condition, by assuming $R, C > 0$, is $Rg(n-1) + n^2 > 0$. If this condition is satisfied, the circuit can reach the sinusoidal steady state.

3. The reactive power absorbed by the capacitor is (see Eqs. 13.36 and 13.43) $Q = -\dfrac{\omega C |\dot{V}|^2}{2}$. We can easily find $\dot{V}$ from the I/O relationship in the phasor domain:

$$\left\{ j\omega RC + [Rg(n-1) + n^2] \right\} \dot{V} = nE,$$

that is,

$$\dot{V} = \frac{nE}{\alpha + j\omega RC},$$

where $\alpha = Rg(n-1) + n^2$. Therefore,

$$|\dot{V}|^2 = \frac{(nE)^2}{\alpha^2 + (\omega RC)^2}$$

and

$$Q = -\frac{1}{2} \frac{\omega C (nE)^2}{\alpha^2 + (\omega RC)^2}.$$

4. The complex power absorbed by the VCCS is

$$P_x + jQ_x = \frac{\dot{V} \left(g\dot{V}_x \right)^*}{2} = \frac{\dot{V} g(n-1) \dot{V}^*}{2} = \frac{1}{2} g(n-1)|\dot{V}|^2.$$

Therefore,

$$P_x = \frac{1}{2} g(n-1) \frac{(nE)^2}{\alpha^2 + (\omega RC)^2}$$

and $Q_x = 0$.

In the above example, the VCCS absorbs a null reactive power. One might think that this is a general property of resistive components, but that is not the case.

If the component is reciprocal, the reactive power absorbed by a resistive component is null. Otherwise, the absorbed reactive power depends on the rest of the circuit.

We already know that the resistor (reciprocal two-terminal element) absorbs only active power. In contrast, the reactive power absorbed by an independent source (a nonreciprocal two-terminal element) depends on the rest of the circuit. Indeed, let us consider for instance an independent voltage source. If we connect it in parallel to a resistor (see Fig. 13.33a), it absorbs a complex power

(a)

$\dot{E}/R$

$\dot{E}$ R

(b)

$\dot{E}/(j\omega L)$

$\dot{E}$ L

Fig. 13.33 Independent voltage source working in sinusoidal steady state and connected in parallel to either **a** a resistor or **b** an inductor

Fig. 13.34 Resistive two-port working in sinusoidal steady state

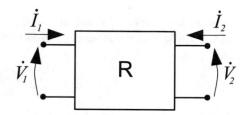

$$P + jQ = \frac{1}{2}\dot{E}\left(-\frac{\dot{E}}{R}\right)^{*} = -\frac{|\dot{E}|^2}{2R},$$

which is purely active; if we connect it in parallel to an inductor (see Fig. 13.33b), it absorbs a complex power

$$P + jQ = \frac{1}{2}\dot{E}\left(-\frac{\dot{E}}{j\omega L}\right)^{*} = -j\frac{|\dot{E}|^2}{2\omega L},$$

which is purely reactive.

As a further example, we consider a resistive two-port (shown in Fig. 13.34) described by its resistance matrix $R = \begin{pmatrix} R_{11} & R_{12} \\ R_{21} & R_{22} \end{pmatrix}$.

The complex power absorbed by the two-port is

$$P + jQ = \frac{1}{2}\left(\dot{V}_1\dot{I}_1^* + \dot{V}_2\dot{I}_2^*\right).$$

Therefore,

$$P + jQ = \frac{1}{2}\left(R_{11}\dot{I}_1 + R_{12}\dot{I}_2\right)\dot{I}_1^* + \frac{1}{2}\left(R_{21}\dot{I}_1 + R_{22}\dot{I}_2\right)\dot{I}_2^* =$$

$$= \frac{1}{2}\left(R_{11}|\dot{I}_1|^2 + R_{12}\dot{I}_1^*\dot{I}_2 + R_{21}\dot{I}_1\dot{I}_2^* + R_{22}|\dot{I}_2|^2\right).$$

Thus $Q = 0$ if $\Im\left\{R_{12}\dot{I}_1^*\dot{I}_2 + R_{21}\dot{I}_1\dot{I}_2^*\right\} = 0$. Inasmuch as $\dot{I}_1\dot{I}_2^* = \left(\dot{I}_1^*\dot{I}_2\right)^*$, the above condition is satisfied, provided that $R_{12} = R_{21}$, which is the reciprocity condition. In other words, if the resistive two-port is reciprocal, it absorbs only active power ($Q = 0$); otherwise, it can absorb nonnull reactive power, but its actual behavior depends on the rest of the circuit.

13.8   Boucherot's Theorem

Boucherot's theorem[2] is a direct consequence of Tellegen's theorem.[3]

Theorem 1 (Boucherot's theorem) *In a circuit containing n components, the kth of which absorbs an active power P_k and a reactive power Q_k, the following constraints hold:*

$$\sum_{k=1}^{n} P_k = 0 \tag{13.46}$$

and

$$\sum_{k=1}^{n} Q_k = 0. \tag{13.47}$$

Proof We assume that each component is described according to the standard choice. Therefore, the circuit graph (with L edges) can be associated with two sets of compatible[4] complex variables: the voltage phasors $\{\dot{V}_j\}$, which satisfy the KVLs, and the current phasors $\{\dot{I}_j\}$, which satisfy the KCLs. Actually, in order to compute the complex powers absorbed by all circuit elements, we should use the phasors $\{\dot{I}_j^*\}$. Are they still compatible with the graph? Owing to the compatibility of the current phasors $\{\dot{I}_j\}$, for KCL we have

$$\sum_j \dot{I}_j = 0,$$

where the sum is extended to the currents involved in a given cut-set, as shown in Sect. 13.2.1. This implies that also

$$\sum_j \dot{I}_j^* = 0.$$

---

[2]The French engineer Paul Boucherot (1869–1943) was a pioneer of AC electric power distribution. He designed induction motors and built early plants for obtaining thermal energy from the sea. He also contributed to electrical analysis, including the relationship between active and apparent power.

[3]See Sect. 2.3 in Vol. 1.

[4]Compatible means that one set of variables satisfies the KVLs and the other the KCLs; see Sect. 2.2.4 in Vol. 1.

Fig. 13.35 Case Study 1

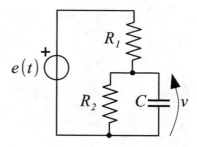

Therefore, the phasors $\{\dot{I}_j^*\}$ are also compatible with the graph, since they satisfy the KCLs.

On applying Tellegen's theorem to the compatible sets $\{\dot{V}_j\}$ and $\{\dot{I}_j^*\}$, we obtain

$$\sum_{j=1}^{L} \dot{V}_j \dot{I}_j^* = 2 \sum_{k=1}^{n} (P_k + jQ_k) = 0.$$

The last equivalence implies that

$$\sum_{k=1}^{n} P_k = 0$$

and

$$\sum_{k=1}^{n} Q_k = 0.$$

$\square$

Equation 13.46 is the most intuitive part of the theorem, since it reflects the law of conservation of energy. By contrast, Eq. 13.47 is the less trivial result of the theorem, which has remarkable consequences, as shown in the next sections.

Case Study 1
For the circuit shown in Fig. 13.35, working in AC steady state with $e(t) = E \sin(\omega t)$, find the reactive power Q delivered by the voltage source.

By virtue of Boucherot's theorem, we know that the sum of reactive powers *absorbed* by the circuit components is null:

$$Q_e + Q_{R1} + Q_{R2} + Q_C = 0.$$

Since the reactive power absorbed by each resistor is zero, the reactive power *delivered* by the voltage source coincides with $Q_C = -\dfrac{\omega C|\dot{V}|^2}{2}$.

We can find $\dot{V}$ in many ways, reasoning in either the time or the phasor domain. For instance, we can work from the beginning in the phasor domain and find the Thévenin equivalent of the composite two-terminal connected to the capacitor, as shown in Fig. 13.36. You can check that

$$\dot{E}_{TH} = \frac{\dot{E}R_2}{R_1 + R_2}$$

and

$$Z_{TH} = \frac{R_1 R_2}{R_1 + R_2}.$$

We can now apply the voltage divider rule, thus obtaining

$$\dot{V} = \dot{E}_{TH}\frac{\dfrac{1}{j\omega C}}{Z_{TH} + \dfrac{1}{j\omega C}} = \dot{E}\frac{R_2}{R_1 + R_2 + j\omega C R_1 R_2}.$$

Therefore,

$$|\dot{V}|^2 = \frac{(ER_2)^2}{(R_1 + R_2)^2 + (\omega C R_1 R_2)^2}$$

and

$$Q = -\frac{1}{2}\frac{\omega C (ER_2)^2}{(R_1 + R_2)^2 + (\omega C R_1 R_2)^2}.$$

To check your comprehension, you can try to obtain the same solution by:
- computing the impedance of the composite two-terminal connected to the voltage source $e(t)$ in Fig. 13.35;
- computing the complex power delivered by the voltage source.

Fig. 13.36 Thévenin equivalent for Case Study 1

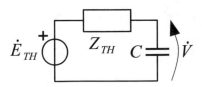

Case Study 2

For the circuit shown in Fig. 13.37, working in AC steady state with $a(t) = A \sin(\omega t)$ and $e(t) = E \cos(\omega t)$, find the active power P and the reactive power Q delivered by the independent sources.

Also in this case, we apply Boucherot's theorem. The sum of the active powers absorbed by the passive two-port (dashed rectangle in Fig. 13.37) is actually the active power absorbed by the resistor, that is,

$$P = \frac{E^2}{2R}.$$

This is also the active power delivered by the voltage source.

The reactive power absorbed by the resistor is zero, whereas $Q_C = -\dfrac{A^2}{2\omega C}$ coincides with the reactive power delivered by the current source, and $Q_L = \dfrac{E^2}{2\omega L}$ is the reactive power delivered by the voltage source.

To check your comprehension, you can try to obtain the same solution by:

- computing the impedance or admittance matrix of the two-port;
- computing the complex power delivered by the two sources.

Remark 1: If you apply the superposition principle, keep in mind that it can be applied only to find voltages and currents (or the corresponding phasors), because the power definition involves a nonlinear operation (product) on voltages and currents. To find a correct solution, we should first use the superposition principle to find the current(s) and voltage(s) involved in the power computation and then calculate the power.

Remark 2: If we imposed $Q_C + Q_L = 0$, we would obtain the relationship between the amplitudes A and E such that the two-port behaves as if it were purely resistive, which is $A = \sqrt{\dfrac{C}{L}} E.$

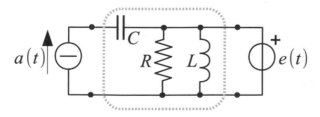

Fig. 13.37 Case Study 2

13.9 Power Factor Correction

We saw in Sect. 13.7.1 that the power absorbed by a two-terminal element (henceforth also called the **load**) is composed of two parts: the instantaneous active power (with amplitude P), which allows conversion to other forms of energy, and the instantaneous reactive power (with amplitude Q), which cannot be exploited for this practical use. Therefore, we usually modify the load in order to move its power factor toward 1: this power factor correction corresponds to making both ϑ and Q tend to a null value, or in other words, to make the load purely resistive.

We assume that the load $Z(j\omega)$ is resistive–inductive, which is quite usual for the most common electrical appliances. The connection of an appliance to a socket can be modeled as shown in Fig. 13.38a. The corresponding phasor diagram is shown in Fig. 13.38c.

How can we modify the load in order to move its power factor toward 1, keeping unaltered its voltage, as provided by the source? If we add a capacitor of proper capacitance in parallel to the load, as shown in Fig. 13.38b, we can achieve this goal. This can be easily understood by looking at the phasor diagram in Fig. 13.38d: the current flowing in the new load (labeled as $Z_R(j\omega)$ in Fig. 13.38b) now is $\dot I + \dot I_C$, and for a specific amplitude of $\dot I_C$, this sum is the projection of $\dot I$ over $\dot V$, according to the parallelogram rule for vector addition; see Appendix A.3.2.

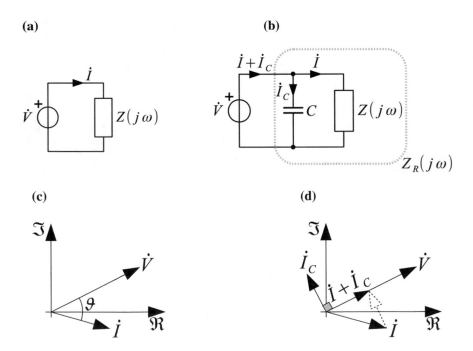

Fig. 13.38 Connection between a voltage source and a load (**a**) before and (**b**) after power factor correction. Corresponding phasor diagrams (**c**) before and (**d**) after power factor correction

The network angular frequency ω and voltage amplitude $|\dot{V}|$ are fixed, because their values depend on the country you live in; see Sect. 3.2.2, Vol. 1. Therefore, we act on C to properly set the amplitude of $\dot{i}_C$.

We can follow two possible approaches to find the correct C value. The first approach requires that we know the value of ϑ. If that is the case, we can impose that $Z_R(j\omega)$ does not absorb any reactive power:

$$Q_C + Q_Z = 0. \tag{13.48}$$

Owing to Boucherot's theorem, stating that the sum of the reactive powers absorbed by voltage source, capacitor, and load is null ($Q_V + Q_C + Q_Z = 0$) means that the voltage source does not deliver any reactive power to $Z_R(j\omega)$.

But

$$Q_C = -\frac{\omega C |\dot{V}|^2}{2} \tag{13.49}$$

and

$$Q_Z = \frac{1}{2}\Im\left\{\dot{V}\left(\frac{\dot{V}}{Z(j\omega)}\right)^*\right\} = \frac{1}{2}\Im\left\{\dot{V}\frac{\dot{V}^*}{|Z|e^{-j\vartheta}}\right\} = \frac{1}{2}\frac{|\dot{V}|^2 \sin\vartheta}{|Z|}. \tag{13.50}$$

Therefore, imposing the constraint of Eq. 13.48 means imposing

$$C = \frac{\sin\vartheta}{\omega|Z|}, \tag{13.51}$$

and this is the capacitance value ensuring a perfect alignment between current and voltage phasors in $Z_R(j\omega)$, which in this way is purely resistive, as desired.

We remark that the value of C provided by Eq. 13.51 is independent of the amplitude of $|\dot{V}|$, but it depends on the network frequency. Therefore, in general it depends on the country you live in.

The second approach does not focus on ϑ, but on the internal structure of $Z(j\omega)$ or its analytic expression. In this case, we impose that $Z_R(j\omega)$ have no imaginary part, thus obtaining again a purely resistive load, as desired. The method is illustrated in the following case studies.

Case Study 1
The circuit shown in Fig. 13.39 works in AC steady state with $e(t) = E\cos(\omega t)$. The original load $Z(j\omega)$ has the internal structure shown within the dashed box. Find the capacitance C such that by adding the capacitor between nodes A and B, the power factor of the modified load is 1.

First of all, we find an analytical expression for $Z(j\omega)$ (see Fig. 13.40):

$$Z(j\omega) = \frac{\dot{V}}{\dot{I}} = R + j\omega L(\alpha + 1).$$

We now compute an expression for the parallel connection of $Z(j\omega)$ and the capacitor

$$Z_R(j\omega) = \frac{\dfrac{Z}{j\omega C}}{Z + \dfrac{1}{j\omega C}} = \frac{Z}{1 + j\omega C Z} = \frac{R + j\omega L(\alpha + 1)}{1 - \omega^2 LC(\alpha + 1) + j\omega RC}$$

and separate its real and imaginary parts:

$$Z_R(j\omega) = \frac{[R + j\omega L(\alpha + 1)][1 - \omega^2 LC(\alpha + 1) - j\omega RC]}{[1 - \omega^2 LC(\alpha + 1)]^2 + (\omega RC)^2}.$$

Finally, we impose that the imaginary part of $Z_R(j\omega)$ be null,

$$-\omega R^2 C + \omega L(\alpha + 1)[1 - \omega^2 LC(\alpha + 1)] = 0,$$

thus obtaining

$$C = \frac{(\alpha + 1)L}{R^2 + [\omega L(\alpha + 1)]^2}.$$

Fig. 13.39 Case Study 1

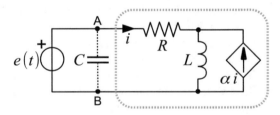

Fig. 13.40 Solution of Case Study 1

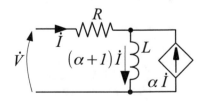

Case Study 2

The circuit shown in Fig. 13.41 a works in AC steady state with $e(t) = E\sin(\omega t)$. *Find the admittance* $Y(j\omega)$ *of the two-terminal load and the admittance to be connected in parallel to it in such a way that the modified load has a power factor equal to* 1.

Making reference to Fig. 13.41b, we have

$$\begin{cases} \dot{V} = (R_1 + R_2)\,\dot{I}_2, \\ \dot{I} = j\omega C_1 \dot{V} + j\omega C_2 R_2 \dot{I}_2 + \dot{I}_2. \end{cases}$$

Therefore, the resulting descriptive equation for the two-terminal is

$$\dot{I} = \left(j\omega \left(C_1 + C_2 \frac{R_2}{R_1 + R_2} \right) + \frac{1}{R_1 + R_2} \right) \dot{V},$$

and its admittance is

$$Y(j\omega) = \frac{1}{R_1 + R_2} + j\omega \underbrace{\left(C_1 + C_2 \frac{R_2}{R_1 + R_2} \right)}_{C_{eq}}.$$

Inasmuch as the reactive power absorbed by $Y(j\omega)$ is negative,

$$Q = \frac{1}{2}\Im\left\{\dot{V}\dot{I}^*\right\} = -\frac{1}{2}\left|\dot{V}\right|^2 \omega C_{eq} < 0,$$

the admittance to be connected in parallel to the original load must be an inductor L such that

$$\Im\left\{\frac{1}{j\omega L} + Y(j\omega)\right\} = 0 \quad \Rightarrow \quad -\frac{1}{\omega L} + \omega C_{eq} = 0,$$

that is,

$$L = \frac{1}{\omega^2 C_{eq}} = \frac{R_1 + R_2}{\omega^2\left((R_1 + R_2)\,C_1 + R_2 C_2\right)}.$$

You can easily verify that for this value of L, the corresponding reactive power is equal in absolute value to that absorbed by $Y(j\omega)$, but it has opposite sign, and therefore the resulting power factor is equal to 1.

Fig. 13.41 Case Study 2: **a** circuit; **b** the two-terminal load and its voltage and current phasors

(a)

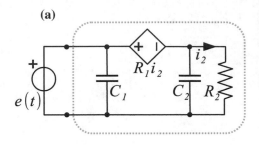

(b)

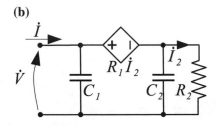

Case Study 3

Under the assumption of AC steady state at angular frequency ω, consider the two-terminal of Fig. 13.17 connected to the load with impedance $Z = R + jX$ shown in Fig. 13.42. Find the value of C_2 such that the nullor delivers only active power.

The complex power delivered by the nullor in the general case is

$$P + jQ = \frac{1}{2}\dot{V}\dot{i}_\infty^*,$$

where

$$\dot{i}_\infty = \frac{\dot{V}}{Z} + \frac{\dot{V}}{Z_2}$$

and

$$Z_2 = \frac{R_2}{1 + j\omega C_2 R_2}.$$

Therefore, the complex power expression can be written as

$$P + jQ = \frac{1}{2} |\dot{V}|^2 \left(\frac{1}{Z^*} + \frac{1}{Z_2^*} \right) = \frac{1}{2} |\dot{V}|^2 \left(\frac{R + jX}{R^2 + X^2} + \frac{1 - j\omega R_2 C_2}{R_2} \right).$$

The term Q is null, provided that

$$\frac{X}{R^2 + X^2} - \frac{\omega R_2 C_2}{R_2} = 0 \;\; \Rightarrow \;\; C_2 = \frac{X}{\omega \left(R^2 + X^2 \right)}.$$

Remark 1: In all cases considered up to now, we wanted to obtain a power factor correction keeping unaltered the voltage amplitude provided to a resistive–inductive [resistive–capacitive] load by the source, as shown in Fig. 13.43a. Therefore, we added a capacitor [an inductor] in parallel to the original load, thus keeping fixed the voltage phasor $\dot{V}$ and projecting the current phasor over it, as shown in Fig. 13.38d. In other cases, the power source is represented by a current source, as shown in Fig. 13.43b, and we want to obtain a power factor correction keeping unaltered the amplitude of the current flowing through the load: if so, the component must be added in series to the original load.

Remark 2: If the load is resistive-capacitive, in both cases shown in Fig. 13.43 the additional component must be an inductor.

These concepts are exemplified in the next case study.

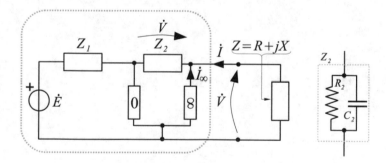

Fig. 13.42 Case Study 3: the two-terminal of Fig. 13.17 connected to Z

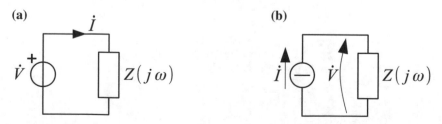

Fig. 13.43 Connection between power sources and a load: **a** voltage source and **b** current source

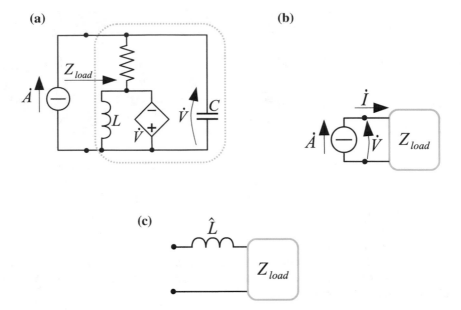

Fig. 13.44 Case Study 4: **a** circuit; **b** compact representation; **c** power factor correction

Case Study 4
The circuit shown in Fig. 13.44 a works in sinusoidal steady state at angular frequency ω. Find an expression for the complex power absorbed by Z_{load} and calculate the inductance $\hat{L}$ to be connected in series to Z_{load} in such a way to obtain a power factor equal to 1.

From the circuit equations we easily get the relationship

$$\dot{A} = \dot{V}\left(\frac{2}{R} + j\omega C\right),$$

which implies that

$$Z_{load} = \frac{\dot{V}}{\dot{A}} = \frac{R}{2 + j\omega RC} = \frac{R(2 - j\omega RC)}{4 + (\omega RC)^2}.$$

Therefore, Z_{load} is resistive–capacitive, which justifies the use of the series inductance $\hat{L}$ to correct the power factor.
We can calculate $\hat{L}$ by imposing

$$\Im\left\{j\omega \hat{L} + Z_{load}\right\} = 0 \quad \Rightarrow \hat{L} = \frac{R^2 C}{4 + (\omega RC)^2}.$$

Fig. 13.45 Model of a circuit composed of voltage source, electrical wires/cables, and load (appliance)

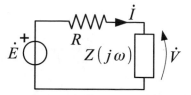

13.9.1 Advantages for the Consumer

Figure 13.45 shows a more realistic model of a circuit composed of voltage source, electrical wires/cables, and load (appliance). The wires are represented by a resistor R, which takes into account their inherent resistivity.

The load is $Z = |Z|e^{j\vartheta}$.

The current phasor $\dot{I}$ can be expressed as

$$\dot{I} = \frac{\dot{E}}{R+Z}. \tag{13.52}$$

Therefore, the active power absorbed by the wires is

$$P_R = \frac{R|\dot{I}|^2}{2} = \frac{1}{2}\frac{R|\dot{E}|^2}{|R+Z|^2}. \tag{13.53}$$

On the other hand, the active power absorbed by the load is

$$P_Z = \frac{1}{2}\Re\left\{\dot{V}\dot{I}^*\right\} = \frac{1}{2}\Re\left\{\frac{\dot{E}Z}{R+Z}\left(\frac{\dot{E}}{R+Z}\right)^*\right\} = \frac{1}{2}\frac{|\dot{E}|^2}{|R+Z|^2}\Re\{Z\} \tag{13.54}$$

with $\Re\{Z\} = |Z|\cos\vartheta$.

Finally, we compute the ratio

$$\frac{P_Z}{P_R} = \frac{|Z|\cos\vartheta}{R}. \tag{13.55}$$

The most advantageous situation for the consumer (minimal losses) obtains when this ratio is maximum, that is, when $\vartheta = 0$, by assuming that R and $|Z|$ are fixed.

13.9.2 Advantages for the Utility Company and Motivation for High-Voltage Transmission Lines

Figure 13.46 sketches a power grid, with a complete path from energy production in an industrial facility for the generation of electric power to energy consumption in industrial or domestic plants.

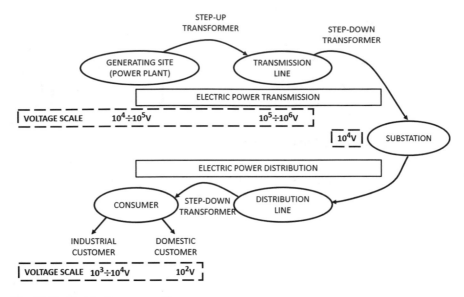

Fig. 13.46 Sketch of a power grid

Electric power transmission is the movement of electrical energy from a generating site, such as a power station (also commonly referred to as a power plant or powerhouse or generating plant), to an electrical substation.

Most power stations contain one or more *generators*, which are rotating machines that convert mechanical power into electrical power. Sources of mechanical energy include steam turbines, gas turbines, water turbines, and internal combustion engines. The first electromagnetic generator, the Faraday disk, was invented in 1831 by the British scientist Michael Faraday. Most of today's generators are still based on Faraday's law of electromagnetic induction. The energy source harnessed to turn the generator varies from fossil fuels (coal, oil, natural gas) to nuclear power to renewable sources such as solar, wind, wave, and hydroelectric.

At the power stations, the sinusoidal voltage produced by a generator has an amplitude ranging from a few kilovolts to 30 kV, depending on the size of the unit. The generator terminal voltage is then stepped up by the power station transformer to a higher voltage (from about 100 kV to about 800 kV AC, varying by the transmission system and by the country) for transmission over long distances. The interconnected lines that facilitate this movement are known as an electric *power transmission system*. This is distinct from the local wiring between substations and customers, which is typically referred to as an *electric power distribution system*.

Most transmission lines are overhead and transmit high-voltage three-phase sinusoidal currents. Why is electricity transmitted at high voltages, higher than 100 kV, thus requiring a significant cost for transformers? The main reason is to reduce the energy loss that occurs in long-distance transmission.

Fig. 13.47 Circuit modeling
electric power transmission

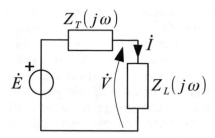

To prove this statement, we refer to the circuit shown in Fig. 13.47, where the voltage source models the sinusoidal high voltage to be supplied, the impedance $Z_T(j\omega) = R_T + jX_T$ models the transmission line, and the impedance $Z_L(j\omega)$ (with phase ϑ) represents the substation with the connected load.

First of all, we show that this physical system can be treated as a lumped circuit within circuit theory, without having to resort to distributed circuits. In Sect. 1.1, Vol. 1, we stated that a circuit is lumped if its physical dimensions are negligible compared to the smallest wavelength of interest in the electric variables involved. In the case under study, we deal with sinusoids at a frequency $f = 50\,\text{Hz}$ or $60\,\text{Hz}$ (depending on the country), traveling at a speed close to the speed of light $c \approx 3 \cdot 10^8$ m/s. This means that their wavelength $(= \dfrac{c}{f})$ is either 6000 Km, for $f = 50\,\text{Hz}$, or 5000 Km, for $f = 60\,\text{Hz}$. In both cases, transmission lines with lengths of up to hundreds of kilometers can be modeled as shown in Fig. 13.47.

The substation must be able to deliver at any time a given amount of active power

$$P = \frac{1}{2}|\dot{V}||\dot{I}| \cos \vartheta \tag{13.56}$$

in order to fulfill the contractual customer demand.

The power loss due to transmission can be estimated as

$$P_T = \frac{1}{2}R_T|\dot{I}|^2. \tag{13.57}$$

By obtaining $|\dot{I}|$ from Eq. 13.56, one can recast Eq. 13.57 as

$$P_T = \frac{1}{2}R_T \left(\frac{2P}{|\dot{V}| \cos \vartheta}\right)^2 = \frac{2R_T P^2}{|\dot{V}|^2 \cos^2 \vartheta}. \tag{13.58}$$

Therefore, inasmuch as R_T (depending on the material the transmission line wires are made of) and P are fixed, the power loss is minimal for both power factor and voltage amplitude $|\dot{V}|$ as high as possible.

The first condition is satisfied if each load connected to the power distribution network absorbs only active power. Utilities usually do not charge consumers for reactive power losses, since they do no real work for the consumer. However, if there are inefficiencies of the customer's load that cause the power factor to fall below a certain level, utilities may charge customers in order to cover an increase in their power plant fuel use and their degraded line and plant capacity.

The second condition is satisfied if we transmit high-voltage signals, since the higher the value of $|\dot{E}|$, the greater that of $|\dot{V}|$ for fixed impedances. Of course, the adopted high-voltage values are a trade-off between the above need, the cost of the transformers, and safety considerations.

13.10   Theorem on the Maximum Power Transfer

Figure 13.48 shows a model of another physically important case, in which we have to select or design a passive load impedance $Z_L(j\omega)$ so that it absorbs the maximum active power. The rest of the circuit provides energy to the load and is represented by its Thévenin or Norton equivalent, as shown in Fig. 13.48. The parameters of the assigned equivalent (source phasor $|\dot{E}|$ or $|\dot{A}|$ and passive impedance $Z(j\omega)$ or admittance $Y(j\omega)$) are assigned and fixed.

As an example, a microphone is a device that can be represented by its Thévenin equivalent. An amplifier connected to the microphone should have an input impedance $Z_L(j\omega)$ such that it absorbs the maximum fraction of the signal power, as shown in Fig. 13.49.

Theorem 2 (Maximum-power-transfer theorem) *The optimum load impedance* $Z_L(j\omega)$ *that absorbs the maximum active power is equal to the complex conjugate of* $Z(j\omega)$, *that is,* $Z_L = Z^*$.

The proof is given by making reference to the Thévenin equivalent, but it holds, mutatis mutandis, also for the Norton equivalent.

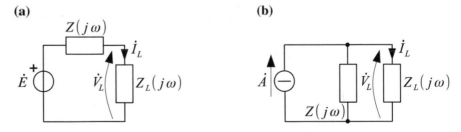

(a)   **(b)**

Fig. 13.48 Circuit modeling power transfer from a source (represented by its **a** Thévenin or **b** Norton equivalent) to a load

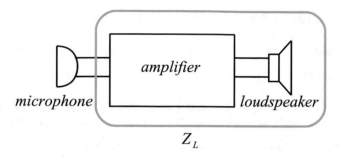

Fig. 13.49 Example

Proof We have $\dot{V}_L = \dot{E} \dfrac{Z_L}{Z_L + Z}$ (voltage divider) and $\dot{I}_L = \dfrac{\dot{E}}{Z_L + Z}$; therefore,

$$P = \frac{1}{2} \Re \left\{ \dot{V}_L \dot{I}_L^* \right\} = \frac{1}{2} \frac{|\dot{E}|^2}{|Z_L + Z|^2} \Re \left\{ Z_L \right\}. \tag{13.59}$$

By assuming $Z = R + jX$ and $Z_L = R_L + jX_L$, Eq. 13.59 can be recast as

$$P = \frac{|\dot{E}|^2}{2} \frac{R_L}{(R + R_L)^2 + (X + X_L)^2}. \tag{13.60}$$

By assumption, R, X, and $|\dot{E}|$ are assigned, whereas R_L and X_L have to be selected in order to maximize P. Therefore, we can consider P as a function of two variables, $P(R_L, X_L)$.

Maximizing $P(R_L, X_L)$ with respect to X_L is the easiest task. Indeed, we have to minimize $(X + X_L)^2$, which is minimal (null) for $X_L = -X$. We remark that if we assume that we are working with passive impedances, this means that the load has a nature complementary to that of the Thévenin impedance: if the Thévenin impedance is resistive–inductive, the load must be resistive–capacitive and conversely.

With this choice of X_L, the absorbed power becomes

$$P(R_L) = \frac{|\dot{E}|^2}{2} \frac{R_L}{(R + R_L)^2}. \tag{13.61}$$

To maximize $P(R_L)$ with respect to R_L (≥ 0 due to the Z_L passivity assumption), we impose the optimum condition $\dfrac{dP}{dR_L} = 0$:

$$\frac{dP}{dR_L} = \frac{|\dot{E}|^2}{2} \frac{(R + R_L)^2 - 2(R + R_L)R_L}{(R + R_L)^4} = \frac{|\dot{E}|^2}{2} \frac{R - R_L}{(R + R_L)^3} = 0, \tag{13.62}$$

which provides $R_L = R \geq 0$. You can easily check that this condition corresponds to a maximum of $P(R_L)$.

In summary, the maximum power transfer condition is $Z_L = R - jX = Z^*$, which corresponds to $P = \dfrac{|\dot{E}|^2}{8R}$.   □

When the condition $Z_L = Z^*$ holds, we say that Z_L is *conjugately matched* with the source impedance, or more compactly, that the load is matched to the source.

Remark: Although we did not impose any condition about the power factor, the overall impedance connected to the voltage source is $Z + Z_L = 2R$, purely real. This means that the voltage source "sees" a resistive equivalent load, with a power factor equal to 1.

Case Study

The two-terminal shown in Fig. 13.50a works in sinusoidal steady state at angular frequency ω. Find the value of the impedance Z that absorbs the maximum active power when connected to the two-terminal.

We choose to make reference to the Norton equivalent representation of the two-terminal. The impedance Z_{NR} of Fig. 13.50b is easily found by setting $\dot{A} = 0$ in the two-terminal and finding the descriptive equation $\dot{V} = Z_{NR}\dot{I}$:

$$Z_{NR} = \frac{j\omega C R^2}{2\,(1 + j\omega C R)} = \underbrace{\frac{\omega C R^2}{2\,(1 + \omega^2 C^2 R^2)}}_{H(\omega)} (\omega C R + j).$$

Therefore, we must have

$$Z = Z_{NR}^* = H\,(\omega)\,(\omega C R - j).$$

Notice that knowing the value of the equivalent Norton current source

$$\dot{A}_{NR} = \frac{2\dot{A}}{j\omega C R}$$

is completely inessential to determining Z.

(a)

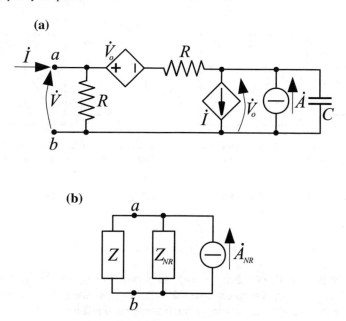

(b)

Fig. 13.50 Case Study: **a** two-terminal element; **b** its Norton equivalent connected to Z

13.11 Frequency Response

The so-called *frequency response* $H(j\omega)$ of an LTI absolutely stable circuit working in sinusoidal steady state is a function that allows one to obtain the behavior of the corresponding output variable for any value of the sinusoidal input angular frequency ω. This function lays the foundation for a vast number of applications, some of which are highlighted in this section. We add that in principle, the knowledge of $H(j\omega)$ for all values of ω allows the determination of the response to any input signal for which the Fourier transform exists.

13.11.1 Network Functions of a Circuit

The impedance and admittance discussed in Sect. 13.3 refer to a network $\aleph_0$ connected to a sinusoidal input source. When the internal structure of $\aleph_0$ consists of several components, it is possible to consider as output phasor one of the descriptive variables $\dot{V}_k$ and $\dot{I}_k$ for the kth element, whereas the input phasor represents the sinusoid impressed by the source connected to $\aleph_0$. In this way, for every k, you can define the complex numbers

$$\frac{\dot{V}_k}{\dot{I}_0}; \quad \frac{\dot{I}_k}{\dot{I}_0},$$

(a)   **(b)**

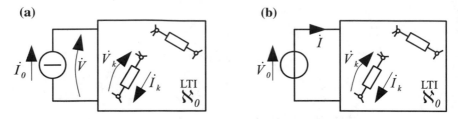

Fig. 13.51 LTI network $\aleph_0$ in sinusoidal steady state with **a** current input source and **b** voltage input source. Here $\dot{V}_k$ and $\dot{I}_k$ are the possible output phasors for the kth element in both circuits

for an input phasor $\dot{I}_0$ (current source), as shown in Fig. 13.51a, and

$$\frac{\dot{I}_k}{\dot{V}_0}; \quad \frac{\dot{V}_k}{\dot{V}_0},$$

for an input phasor $\dot{V}_0$ (voltage source), as shown in Fig. 13.51b. In both cases, the second term is a (complex) dimensionless number, whereas the first term has the physical dimension of an impedance in the first case and an admittance in the second. We can consider, as particular cases, the circuit of Fig. 13.5 with $\dot{V}_k \equiv \dot{V}$, and the circuit of Fig. 13.7 with $\dot{I}_k \equiv \dot{I}$; in these cases, the ratios $\dot{V}_k/\dot{I}_0$ and $\dot{I}_k/\dot{V}_0$ identify the driving-point impedance and admittance, respectively, of $\aleph_0$.

For a given circuit, the ratios $\dot{V}_k/\dot{I}_0$, $\dot{I}_k/\dot{I}_0$ and $\dot{I}_k/\dot{V}_0$, $\dot{V}_k/\dot{V}_0$ depend on the angular frequency of the input source. Therefore, it is usual to think of them as functions of $j\omega$ and refer to them with the common name of *network* or *transfer functions*, often denoted by $H(j\omega)$.

13.11.2   Some Properties of the Network Functions

We first consider the case in which the output variable is a state variable $x_i(t)$, the ith component of the state vector x of the circuit. Let $u(t)$ be the input source. Therefore, the I/O relationship between u and x_i can be written, according to Eq. 11.42, as

$$a_n \frac{d^n x_i}{dt^n} + \cdots + a_1 \frac{dx_i}{dt} + a_0 x_i = b_m \frac{d^m u}{dt^m} + \cdots + b_1 \frac{du}{dt} + b_0 u,$$

where all coefficients $\{a_n, \ldots, a_0\}$; $\{b_m, \ldots, b_0\}$ are *real* and with suitable physical dimensions. Under sinusoidal steady state at angular frequency ω, the corresponding algebraic relationship between phasors $\dot{U}$ and $\dot{X}_i$ is

$$\underbrace{\left(a_n (j\omega)^n + \cdots + a_1 (j\omega) + a_0\right)}_{D_i(j\omega)} \dot{X}_i = \underbrace{\left(b_m (j\omega)^m + \cdots + b_1 (j\omega) + b_0\right)}_{N_i(j\omega)} \dot{U}.$$

$$\text{(13.63)}$$

Therefore, the corresponding network function is

$$H_i(j\omega) = \frac{\dot{X}_i}{\dot{U}} = \frac{N_i(j\omega)}{D_i(j\omega)},$$

and both polynomials $N_i(j\omega)$ and $D_i(j\omega)$ have real coefficients.

Now we consider the case of a nonstate output variable. In the absence of algebraic constraints ($M = 0$), Eq. 11.76 written for a single input $u(t)$ and a single output $y(t)$ can be recast as

$$y(t) = \sum_{i=1}^{n} C_i x_i(t) + d u(t),$$

where n is the size of the state vector x. The coefficients C_i and d are real. In sinusoidal steady state, the corresponding expression for the output phasor $\dot{Y}$ is

$$\dot{Y} = \sum_{i=1}^{n} C_i \dot{X}_i + d\dot{U} = \underbrace{\left(\sum_{i=1}^{n} C_i \frac{N_i(j\omega)}{D_i(j\omega)} + d \right)}_{H(j\omega)} \dot{U}.$$

Inasmuch as $N_i(j\omega)$ and $D_i(j\omega)$ are polynomials whose coefficients are real, as well as C_i and d, the resulting transfer function $H(j\omega)$ can also be easily recast as the ratio of two polynomials $N(j\omega)$ and $D(j\omega)$ with real coefficients:

$$H(j\omega) = \frac{\dot{Y}}{\dot{U}} = \frac{N(j\omega)}{D(j\omega)}.$$

> For an LTI circuit under sinusoidal steady state, every network function $H(j\omega)$ can be written as the ratio of two polynomials $N(j\omega)$ and $D(j\omega)$. Both polynomials have *real* coefficients.

This result remains valid even when the circuit has M algebraic constraints. To verify this, we can simply apply the above approach, mutatis mutandis, to Eq. 11.77. Therefore, the polynomials N and D of a network function H depend on the network $\aleph_0$ and on the *kind* of its input (current or voltage), but *not* on the input phasor $\dot{I}_0$ or $\dot{V}_0$, respectively. In practice, the easiest way to obtain the network functions follows from the circuit equations in terms of phasors, which lead to the solution through a purely algebraic process.

Fig. 13.52 Case Study 1:
circuit analyzed in
Sect. 11.5.2

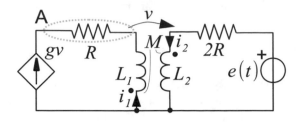

Case Study 1
The circuit shown in Fig. 13.52 works in sinusoidal steady state at angular
frequency ω. Find the following network functions:

$$H_1(j\omega) = \frac{\dot{I}_1}{\dot{E}}; \quad H_2(j\omega) = \frac{\dot{I}_2}{\dot{E}}; \quad H_3(j\omega) = \frac{\dot{V}}{\dot{E}}.$$

From the analysis developed in Sect. 11.5.2, we have the following circuit
equations in terms of phasors:

$$\begin{cases} j\omega M \dot{I}_1 + (2R + j\omega L_2)\,\dot{I}_2 = \dot{E}; \\[2mm] \left(j\omega\,(L_1 + M) + \dfrac{1}{g} \right)\dot{I}_1 + j\omega\,(L_2 + M)\,\dot{I}_2 = 0. \end{cases}$$

The unknown phasors $\dot{I}_1$ and $\dot{I}_2$ can be found, for instance, by applying
Cramer's rule. To this end, we first calculate some determinants:

$$\Delta_0 = \det \left(\begin{matrix} j\omega M & (2R + j\omega L_2) \\[2mm] j\omega\,(L_1 + M) + \dfrac{1}{g} & j\omega\,(L_2 + M) \end{matrix} \right) =$$

$$= (j\omega)^2\,(M^2 - L_1 L_2) - j\omega \left(\frac{L_2}{g} + 2R\,(L_1 + M) \right) - \frac{2R}{g};$$

$$\Delta_1 = \det \left(\begin{matrix} \dot{E} & (2R + j\omega L_2) \\ 0 & j\omega\,(L_2 + M) \end{matrix} \right) = j\omega\,(L_2 + M)\,\dot{E};$$

$$\Delta_2 = \det \left(\begin{matrix} j\omega M & \dot{E} \\[2mm] j\omega\,(L_1 + M) + \dfrac{1}{g} & 0 \end{matrix} \right) = -\dot{E} \left(j\omega\,(L_1 + M) + \frac{1}{g} \right).$$

Therefore, since $\dot{I}_1 = \dfrac{\Delta_1}{\Delta_0}$ and $\dot{I}_2 = \dfrac{\Delta_2}{\Delta_0}$, we have

$$H_1(j\omega) = \frac{\dot{I}_1}{\dot{E}} = \frac{j\omega\,(L_2 + M)}{\Delta_0},$$

$$H_2(j\omega) = \frac{\dot{I}_2}{\dot{E}} = \frac{-\left(j\omega\,(L_1 + M) + \dfrac{1}{g}\right)}{\Delta_0}.$$

Finally, the voltage phasor $\dot{V}$ can be found as

$$\dot{V} = -\frac{1}{g}\dot{I}_1 = -\frac{1}{g}\frac{\Delta_1}{\Delta_0},$$

from which we obtain

$$H_3(j\omega) = \frac{\dot{V}}{\dot{E}} = -\frac{1}{g}\frac{j\omega\,(L_2 + M)}{\Delta_0}.$$

Notice that the denominators of the three network functions H_1, H_2, and H_3 contain the same polynomial term Δ_0.

Two important properties apply to the function $H(j\omega) = |H(j\omega)|\,e^{j\vartheta(j\omega)}$. They derive from N and D being polynomials with real coefficients:

- the magnitude $|H(j\omega)|$ is an *even* function of ω:

$$|H(j\omega)| = |H(-j\omega)|\,; \tag{13.64}$$

- the phase $\vartheta(j\omega)$ is an *odd* function of ω:

$$\vartheta(j\omega) = -\vartheta(-j\omega). \tag{13.65}$$

To demonstrate this, let us consider, for example, the polynomial $N(j\omega)$, which can be written, in full generality, as

$$
\begin{aligned}
N(j\omega) &= p_0 \cdot (j\omega)^m + p_1 \cdot (j\omega)^{m-1} + \cdots + p_{m-1} \cdot j\omega + p_m = \\
&= \underbrace{\left(p_m + p_{m-2} \cdot (j\omega)^2 + p_{m-4} \cdot (j\omega)^4 + \cdots\right)}_{\alpha(\omega^2)} + j\omega \cdot \underbrace{\left(p_{m-1} + p_{m-3} \cdot (j\omega)^2 + \cdots\right)}_{\beta(\omega^2)} = \\
&= \alpha(\omega^2) + j\omega \cdot \beta(\omega^2).
\end{aligned}
$$

Similarly, we obtain $D(j\omega) = \gamma(\omega^2) + j\omega \cdot \delta(\omega^2)$. Therefore,

$$H(j\omega) = \frac{\alpha(\omega^2) + j\omega \cdot \beta(\omega^2)}{\gamma(\omega^2) + j\omega \cdot \delta(\omega^2)} = \frac{\alpha\gamma + \omega^2\beta\delta}{\gamma^2 + \omega^2\delta^2} + j\omega\frac{\beta\gamma - \alpha\delta}{\gamma^2 + \omega^2\delta^2}.$$

Inasmuch as both fractions are *even* functions of ω, we have

$$\Re\{H(j\omega)\} = \Re\{H(-j\omega)\} \quad (= \Re\{H^*(j\omega)\}),$$

$$\Im\{H(j\omega)\} = -\Im\{H(-j\omega)\} \quad (= -\Im\{H^*(j\omega)\}),$$

which imply that

$$|H(j\omega)| = \sqrt{(\Re\{H(j\omega)\})^2 + (\Im\{H(j\omega)\})^2} =$$
$$= \sqrt{(\Re\{H(-j\omega)\})^2 + (\Im\{H(-j\omega)\})^2} = |H(-j\omega)|;$$

$$\vartheta(j\omega) = \tan^{-1}\left(\frac{\Im\{H(j\omega)\}}{\Re\{H(j\omega)\}}\right) = \tan^{-1}\left(\frac{-\Im\{H(-j\omega)\}}{\Re\{H(-j\omega)\}}\right) = -\vartheta(-j\omega),$$

thus proving Eqs. 13.64 and 13.65.

Thanks to these properties, the functions $|H(j\omega)|$ and $\vartheta(j\omega)$ are completely known once their behavior is characterized for $\omega \in [0, +\infty)$.

Case Study 2

The circuit shown in Fig. 13.53 works in sinusoidal steady state at angular frequency ω. Find an expression for the network function $H(j\omega) = \dot{V}/\dot{E}$ and study the corresponding magnitude and phase functions.

The network function is obtained as

$$H(j\omega) = \frac{\dot{V}}{\dot{E}} = \frac{Z_2}{Z_2 + Z_1} \quad \text{with} \quad Z_2 = \frac{R_2\frac{1}{j\omega C}}{R_2 + \frac{1}{j\omega C}}; \quad Z_1 = R_1.$$

After few manipulations, $H(j\omega)$ can be recast as the ratio of polynomials with real coefficients

$$H(j\omega) = \frac{R_2}{(R_1 + R_2) + j\omega C R_1 R_2}.$$

From this expression we obtain

$$|H(j\omega)| = \frac{R_2}{\sqrt{(R_1 + R_2)^2 + (\omega C R_1 R_2)^2}} = \underbrace{\frac{R_2}{R_1 + R_2}}_{H_0} \cdot \frac{1}{\sqrt{1 + \left(\omega C \dfrac{R_1 R_2}{R_1 + R_2}\right)^2}}.$$
(13.66)

We notice that

$$|H(j0)| = \frac{R_2}{R_1 + R_2} \doteq H_0; \qquad \frac{d}{d\omega}|H(j\omega)|\Big|_{\omega=0} = 0;$$

$$\frac{d}{d\omega}|H(j\omega)|\Big|_{\omega>0} < 0; \qquad \lim_{\omega \to \infty}|H(j\omega)| = 0.$$

Moreover, at the angular frequency ω_0 such that $\omega_0 C \dfrac{R_1 R_2}{R_1 + R_2} = 1$, we have

$$|H(j\omega_0)| = \frac{H_0}{\sqrt{2}}; \qquad \omega_0 = \frac{1}{C\dfrac{R_1 R_2}{R_1 + R_2}}.$$

This set of general results allows us to plot the function $|H(j\omega)|$ shown in Fig. 13.54.

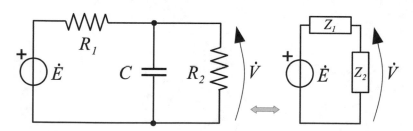

Fig. 13.53 Case Study 2: low-pass filter

Fig. 13.54 Case Study 2: plot of $|H(j\omega)|$

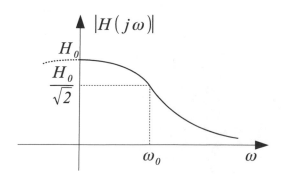

The shape of this function justifies the name *low-pass filter* for this circuit: at low frequencies ($0 \leq \omega \leq \omega_0$), the ratio between the amplitudes of the output $v(t)$ and the input $e(t)$ (both sinusoidal) ranges between H_0 and $\dfrac{H_0}{\sqrt{2}}$; at higher frequencies, this ratio drops monotonically to zero. In other words, for $\omega > \omega_0$, the higher the frequency, the lower the amplitude of $v(t)$, which becomes more and more negligible with respect to the amplitude of $e(t)$.

The term ω_0 is called the *cutoff frequency*, and the frequency range $[0, \omega_0]$ is the filter *bandwidth*.

From the expression for $H(j\omega)$, the phase function $\vartheta(j\omega)$ is seen to be

$$\vartheta(j\omega) = -\tan^{-1}\left(\frac{\omega R_1 C}{1 + \dfrac{R_1}{R_2}}\right).$$

We notice that

$$\vartheta(j0) = 0; \quad \vartheta(j\omega_0) = -\arctan(1) = -\frac{\pi}{4}; \quad \frac{d}{d\omega}\vartheta(j\omega) < 0; \quad \lim_{\omega \to \infty} \vartheta(j\omega) = -\frac{\pi}{2}.$$

With this in mind, we can represent $\vartheta(j\omega)$ as shown in Fig. 13.55.

Fig. 13.55 Case Study 2: plot of $\vartheta(j\omega)$

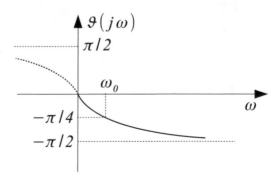

Case Study 3

In the circuit of Fig. 13.56, (just as in Case Study 2), let the input be $e(t) = \underbrace{E\cos(\omega_e t)}_{e_1(t)} + \underbrace{0.1 E\cos(10\omega_e t)}_{e_2(t)}$ *and the cutoff frequency of the low-pass filter be* ω_0.

1. *Find an expression for the steady-state voltage* $v(t)$;
2. *for* $\omega_0 = \omega_e$, *compare the normalized waveforms* $\dfrac{e(t)}{E}$ *and* $\dfrac{v(t)}{H_0 E}$.

1. We write the network function $H(j\omega)$ computed in Case Study 2 in the compact form

$$H(j\omega) = \frac{H_0}{\sqrt{1 + \left(\dfrac{\omega}{\omega_0}\right)^2}} e^{j\vartheta(j\omega)}; \qquad \vartheta(j\omega) = \tan^{-1}\left(-\frac{\omega}{\omega_0}\right).$$

The expressions for H_0 and ω_0 in terms of the circuit parameters were obtained in the previous case study.

The steady-state expression for $v(t)$ can be written as

$$v(t) = \underbrace{V_1 \cos(\omega_e t + \varphi_1)}_{v_1(t)} + \underbrace{V_2 \cos(10\omega_e t + \varphi_2)}_{v_2(t)}.$$

Inasmuch as the circuit is linear, $v_1(t)$ and $v_2(t)$ can be found as the sinusoidal steady-state voltages corresponding to $e_1(t)$ and $e_2(t)$, respectively. Therefore, we have

$$e_1(t) = \Re\left\{E e^{j\omega_e t}\right\}; \qquad \dot{V}_1 = E\, H(j\omega_e),$$

$$v_1(t) = \Re\left\{\dot{V}_1 e^{j\omega_e t}\right\} = \underbrace{\frac{E H_0}{\sqrt{1 + \left(\dfrac{\omega_e}{\omega_0}\right)^2}}}_{V_1} \cos(\omega_e t + \varphi_1); \qquad \varphi_1 = \tan^{-1}\left(-\frac{\omega_e}{\omega_0}\right),$$

and

$$e_2(t) = 0.1 E \cos(10\omega_e t); \qquad \dot{V}_2 = 0.1 E\, H(j 10\omega_e),$$

$$v_2(t) = \underbrace{\frac{0.1 E H_0}{\sqrt{1 + \left(\dfrac{10\omega_e}{\omega_0}\right)^2}}}_{V_2} \cos(10\omega_e t + \varphi_2); \qquad \varphi_2 = \tan^{-1}\left(-\frac{10\omega_e}{\omega_0}\right).$$

2. By setting $\omega_0 = \omega_e = \dfrac{2\pi}{T}$, the angular frequency $10\omega_e$ is well beyond the cutoff frequency ω_0. Therefore, in terms of amplitude (see Fig. 13.56), the term $v_2(t)$ affects $v(t)$ much less than $e_1(t)$ affects $e(t)$. This is shown in Fig. 13.57, where the contributions of $e_1(t)$ and $e_2(t)$ to the waveform $\dfrac{e(t)}{E}$ are clearly evident, whereas $\dfrac{v(t)}{H_0 E}$ is almost coincident with a sinusoid at angular frequency ω_e.

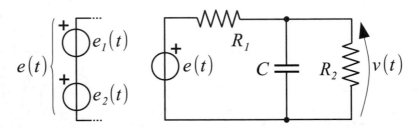

Fig. 13.56 Case Study 3: low-pass filtering example

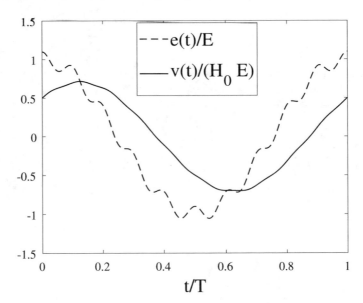

Fig. 13.57 Plot of the normalized voltages $\dfrac{e(t)}{E}$ (dashed line) and $\dfrac{v(t)}{H_0 E}$ (continuous line) in the circuit of Fig. 13.56

As far as the phase is concerned, the phase angles values for $v_1(t)$ and $v_2(t)$ are

$$\varphi_1 = \tan^{-1}(-1) = -\frac{\pi}{4} \text{ rad} \equiv -45°;$$

$$\varphi_2 = \tan^{-1}(-10) = -1.4711 \text{ rad} \equiv -84.29°,$$

respectively.

As a consequence, the sinusoidal terms at ω_e and $10\omega_e$ in the expression for $v(t)$ are shifted in time with respect to the corresponding terms in $e(t)$. The time shifts can be obtained as follows:

$$\cos(\omega_e t + \varphi_1) = \cos\left(\omega_e\left(t + \frac{\varphi_1}{\omega_e}\right)\right) = \cos\left(\frac{2\pi}{T}(t - T_1)\right);$$

$$\cos(10\omega_e t + \varphi_2) = \cos\left(10\omega_e\left(t + \frac{\varphi_2}{10\omega_e}\right)\right) = \cos\left(\frac{20\pi}{T}(t - T_2)\right),$$

where $T_1 = \dfrac{|\varphi_1| T}{2\pi} = \dfrac{T}{8} = 0.125T$ and $T_2 = \dfrac{|\varphi_2| T}{20\pi} = 0.0234T$. The resulting sinusoids are plotted in Fig. 13.58.

We remark that since $\vartheta(j\omega)$ is a nonlinear function of ω, the time shift changes with ω; therefore, we have $T_1 \neq T_2$.

13.12 Resonant Circuits and the Q Factor

The general properties (state equations, natural frequencies, stability) of the RLC series oscillator shown in Fig. 13.59 were already studied in Sect. 11.6. Here, we focus on the circuit's behavior under steady-state conditions to highlight its properties as a *resonant circuit* and as a *filter*. The phasor $\dot{E}$ represents the voltage of the source, operating at angular frequency ω. The impedance $Z(j\omega)$ connected to the source is

$$Z(j\omega) = R + j\omega L + \frac{1}{j\omega C} = R + j\left(\omega L - \frac{1}{\omega C}\right).$$

Fig. 13.58 The time-shifted sinusoidal terms of **a** $v_1(t)$, $e_1(t)$, and **b** $v_2(t)$, $e_2(t)$

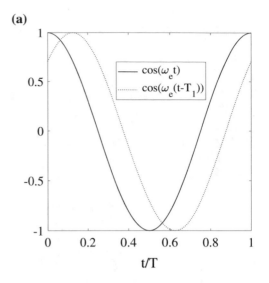

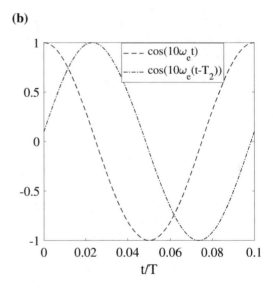

Fig. 13.59 Series resonant circuit

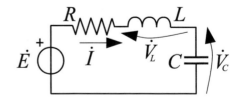

Fig. 13.60 Representation
of the voltage phasors
$\dot{E}, \dot{V}_L, \dot{V}_C$ at $\omega = \omega_0$

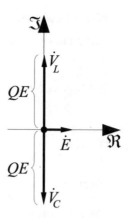

The imaginary part of $Z(j\omega)$ becomes zero at the angular frequency $\omega_0 = \dfrac{1}{\sqrt{LC}}$.
With this in mind, we can recast $Z(j\omega)$ as

$$Z(j\omega) = R + j\left(\frac{\omega}{\omega_0}\omega_0 L - \frac{1}{\frac{\omega}{\omega_0}\omega_0 C}\right) = R\left(1 + j\underbrace{\frac{1}{R}\sqrt{\frac{L}{C}}}_{Q}\left(\frac{\omega}{\omega_0} - \frac{\omega_0}{\omega}\right)\right) =$$

$$= R\left(1 + jQ\left(\frac{\omega}{\omega_0} - \frac{\omega_0}{\omega}\right)\right).$$

The dimensionless term $Q = \dfrac{1}{R}\sqrt{\dfrac{L}{C}} = \dfrac{\omega_0 L}{R} = \dfrac{1}{R\omega_0 C}$ is called the *quality factor* of the resonant circuit. Unfortunately, this coefficient has commonly the same notation as the reactive power, but in this section Q denotes only the quality factor.

We begin by analyzing the circuit's behavior at $\omega = \omega_0$, that is, at *resonance*. We obviously have $Z(j\omega_0) = R$ and

$$\dot{I} = \frac{\dot{E}}{R}; \quad \dot{V}_L = j\omega_0 L\dot{I} = jQ\dot{E}; \quad \dot{V}_C = \frac{1}{j\omega_0 C}\dot{I} = -jQ\dot{E}.$$

Therefore, at resonance we have $\dot{V}_L = -\dot{V}_C$. The amplitude of both these phasors is $V_L = V_C = QE$. Assuming (without loss of generality) $\dot{E} = E$, these voltage phasors are symbolically represented in the complex plane as shown in Fig. 13.60. The usual values for Q in a resonant circuit, which is designed to oscillate, range from a few tens to around 100 units. This implies that at resonance, $V_L, V_C \gg E$; in spite of this, since $\dot{V}_L + \dot{V}_C = 0$, the series connection of L and C behaves like a short circuit.

We can now study the circuit's behavior for $\omega \neq \omega_0$. Denoting by ω_n the normalized frequency $\dfrac{\omega}{\omega_0}$ and by $Z_n(j\omega_n)$ the normalized impedance $\dfrac{Z}{R}$, we have

$$Z_n(j\omega_n) = 1 + jQ\left(\omega_n - \frac{1}{\omega_n}\right) = |Z_n|\,e^{j\vartheta}; \quad Z = R\,|Z_n|\,e^{j\vartheta},$$

$$|Z_n(j\omega_n)| = \sqrt{1 + Q^2\left(\omega_n - \frac{1}{\omega_n}\right)^2}; \quad \vartheta(j\omega_n) = \tan^{-1}\left(Q\left(\omega_n - \frac{1}{\omega_n}\right)\right).$$

Therefore:

- When $0 < \omega_n < 1$, we have $\tan(\vartheta) = \left(Q\dfrac{\omega_n^2 - 1}{\omega_n}\right) < 0$. Therefore, ϑ ranges between $-\dfrac{\pi}{2}$ and 0, and Z is of resistive–capacitive type.
- When $\omega_n > 1$, we have $\tan(\vartheta) > 0$, ϑ ranges between 0 and $\dfrac{\pi}{2}$, and Z is of resistive–inductive type.

The current $\dot{I}$ can be studied by introducing the network function

$$H(j\omega) = \frac{\dot{I}}{\dot{E}} = \frac{1}{Z(j\omega)},$$

for which, taking into account the previously introduced terms, we can define the normalized form

$$H_n(j\omega_n) = \frac{1}{Z_n(j\omega_n)} = \frac{1}{1 + jQ\left(\omega_n - \dfrac{1}{\omega_n}\right)} = \frac{e^{-j\vartheta(j\omega_n)}}{\sqrt{1 + Q^2\left(\omega_n - \dfrac{1}{\omega_n}\right)^2}},$$

obtained by computing the dimensionless product $RH(j\omega)$. We easily obtain

$$\lim_{\omega_n \to 0} |H_n(j\omega_n)| = 0; \quad \lim_{\omega_n \to \infty} |H_n(j\omega_n)| = 0; \quad |H_n(j1)| = 1; \quad \frac{d}{d\omega_n}|H_n(j\omega_n)|\bigg|_{\omega_n=1} = 0.$$

Moreover, when ω_n is such that $Q\left(\omega_n - \dfrac{1}{\omega_n}\right) = \pm 1$, we have $|H_n(j\omega_n)| = \dfrac{1}{\sqrt{2}}$. These ω_n values satisfy the equations

$$Q\omega_n^2 + \omega_n - Q = 0 \quad \text{and} \quad Q\omega_n^2 - \omega_n - Q = 0.$$

Both of these equations have a positive and a negative solution. Due to the symmetry of $|H(j\omega)|$, we are interested in the positive solutions only:

$$\omega_{n1} = \frac{1}{2Q}\left(-1 + \sqrt{1 + 4Q^2}\right) \quad \text{for the first equation;}$$

$$\omega_{n2} = \frac{1}{2Q}\left(1 + \sqrt{1 + 4Q^2}\right) \quad \text{for the second equation.}$$

On the basis of these results, we can trace the curve $|H_n(j\omega_n)|$ as shown in Fig. 13.61. Within the normalized frequency domain $[\omega_{n1}, \omega_{n2}]$, the magnitude of the network function ranges between $\frac{1}{\sqrt{2}}$ and 1. This means that for all ω_n within this frequency domain, the amplitude of the sinusoidal current ranges between a maximum of $I_M = \frac{E}{R}$ (at the resonance frequency) and the value $\frac{I_M}{\sqrt{2}}$ at both ω_{n1} and ω_{n2}. Out of this frequency domain, that is, for $\omega_n < \omega_{n1}$ or for $\omega_n > \omega_{n2}$, the current amplitude rapidly drops to very small values, as shown in Fig. 13.61. For these reasons, the resonant circuit can be viewed as a *band-pass filter*. The width of the normalized frequency domain $[\omega_{n1}, \omega_{n2}]$ is

$$\omega_{n2} - \omega_{n1} = \frac{1}{2Q}(1 - (-1)) = \frac{1}{Q}.$$

By multiplying this result by ω_0, we obtain the so-called *bandwidth B* of the resonant circuit:

$$B = \omega_2 - \omega_1 = \frac{\omega_0}{Q}.$$

Therefore, increasing values of Q correspond to ever smaller bandwidth values. The curves $H_n(j\omega_n)$ plotted in Fig. 13.62 highlight the influence of Q on the shape of the resonance curve and on its bandwidth.

When the voltage source contains a sum of sinusoidal terms working at different frequencies, the resonant circuit selects the frequency terms within the range $[\omega_1, \omega_2]$, and the sinusoidal terms that contribute the most to the circuit current are only those working at such frequencies.

Fig. 13.61 General characteristics of the normalized network function $|H_n(j\omega_n)|$ for $\omega_n \in [0, +\infty)$

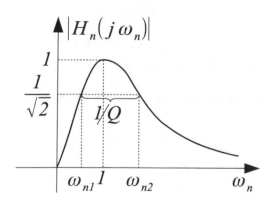

Fig. 13.62 Plots of the normalized network function $|H_n\,(j\omega_n)|$ for different values of Q

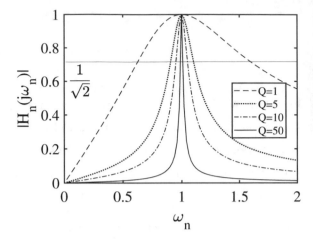

The quality factor Q plays a role also in the energy balance of the circuit at resonance. We assume that $\omega = \omega_0$ and $\dot{I} = Ie^{j\varphi}$. As a consequence, $\dot{V}_C = \dfrac{1}{j\omega_0 C}\dot{i}$ and

$$i(t) = \Re\left\{\dot{I}e^{j\omega_0 t}\right\} = I\cos\left(\omega_0 t + \varphi\right);$$
$$v_c(t) = \frac{I}{\omega_0 C}\Re\left\{-je^{j(\omega_0 t+\varphi)}\right\} = \frac{I}{\omega_0 C}\sin\left(\omega_0 t + \varphi\right).$$

Therefore, the energy stored in the circuit is

$$w_s(t) = \frac{1}{2}Cv_C^2(t) + \frac{1}{2}Li^2(t) = \frac{1}{2}C\left(\frac{I}{\omega_0 C}\right)^2\sin^2\left(\omega_0 t + \varphi\right) + \frac{1}{2}LI^2\cos^2\left(\omega_0 t + \varphi\right) =$$

$$= \frac{1}{2}LI^2\left(\underbrace{\frac{1}{\omega_0^2 LC}}_{1}\sin^2\left(\omega_0 t + \varphi\right) + \cos^2\left(\omega_0 t + \varphi\right)\right) = \frac{1}{2}LI^2,$$

that is, a constant value. This energy can now be compared with the energy w_d dissipated by the resistor during the period $T = \dfrac{2\pi}{\omega_0}$. Inasmuch as the average power dissipated over T is $P_d = \dfrac{1}{2}RI^2$, we have

$$w_d = T\frac{1}{2}RI^2.$$

Therefore,

$$\frac{w_s}{w_d} = \frac{L}{RT} = \frac{1}{2\pi}\frac{\omega_0 L}{R} = \frac{1}{2\pi}\frac{1}{\omega_0 C R} = \frac{1}{2\pi}Q.$$

This result shows that Q is proportional to the ratio between w_s and w_d. At resonance, the higher the value of Q, the smaller the energy loss in a period T with respect to the energy stored in the conservative elements. When the resistance R of the series resonant circuit approaches zero, the quality factor Q becomes infinite.

The interpretation of Q in an energetic perspective allows us to show a link with the damping factor ζ encountered in Sects. 11.6 and 11.7. By comparing the two expressions, we find that $\zeta = \frac{R}{2}\sqrt{\frac{C}{L}} = \frac{1}{2Q}$.

These results are valid, mutatis mutandis, also for oscillators of a nonelectrical nature, such as the mechanical ones discussed in Sect. 11.7. In other words, the Q factor is often used to characterize the response of a harmonic oscillator, whatever its nature is. A higher Q implies a lower damping factor, and the oscillations die out more slowly. A system with $Q < \frac{1}{2}$ is *overdamped* and does not oscillate but decays exponentially (see Sect. 11.7), whereas $Q > \frac{1}{2}$ characterizes an *underdamped* oscillator: a high-quality (high-Q) bell rings for a very long time after being struck. A pendulum, a bell, or a mass–spring oscillator immersed in the air have a high Q; when they are immersed in water or in oil, their Q drops to very small values. A purely oscillatory system, such as a mechanical oscillator in the absence of friction or a pure LC oscillator, has an infinite Q factor.

Remark: for a nonforced (autonomous) underdamped harmonic oscillator of second order, the Q factor is strictly related to the position of the eigenvalues (natural frequencies) in the complex plane. For instance, for the series resonant circuit, the eigenvalues are (see Eq. 11.66)

$$\lambda_{1,2} = \frac{-RC \pm \sqrt{(RC)^2 - 4LC}}{2LC} = \omega_0 \frac{-1 \pm \sqrt{1 - 4Q^2}}{2Q} = \omega_0\left(-\zeta \pm \sqrt{\zeta^2 - 1}\right).$$

For fixed L and C values, it is evident that the lower the value of R, the closer $\lambda_{1,2}$ is to the imaginary axis and the higher the value of Q. In the limit case for $R = 0$, we obtain a pure LC oscillator and Q is infinite.

The parallel resonant circuit shown in Fig. 13.63 can be studied following the same guidelines as those adopted in the series resonant case. Here we can make reference to the network function $H(j\omega) = \frac{\dot{V}}{\dot{A}} = Z(j\omega)$. As before, the resonant frequency is $\omega_0 = \frac{1}{\sqrt{LC}}$, and we define the normalized frequency $\omega_n = \frac{\omega}{\omega_0}$. Dividing Z by R, we obtain the normalized network function

Fig. 13.63 Parallel resonant circuit

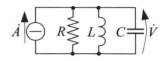

Fig. 13.64 Problem 13.1

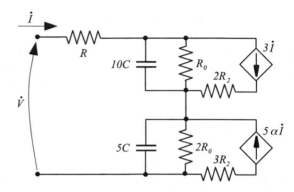

$$H_n(j\omega_n) \doteq \frac{Z(j\omega_n)}{R} = \frac{1}{1 + jQ\left(\omega_n - \dfrac{1}{\omega_n}\right)},$$

where the quality factor is $Q = \dfrac{R}{\omega_0 L} = \omega_0 C R$, which becomes infinite as $R \to \infty$ and the resistor behaves as an open circuit. It is easy to verify that $|H_n(j\omega_n)|$ can still be represented as shown in Fig. 13.61. Finally, Q maintains the relations with the bandwidth B, the damping factor ζ, and the energy ratio $\dfrac{w_s}{w_d}$ found in the series case. You are invited to check these results.

13.13 Problems

13.1 For the composite two-terminal shown in Fig. 13.64, working in sinusoidal steady state at angular frequency ω:

- find the impedance $Z(j\omega)$;
- assume $R_0 = R/10$ and $\alpha > -1$, and discuss the behavior of the resistance and reactance in terms of α.

13.2 Consider the circuit shown in Fig. 13.65, working in sinusoidal steady state.

1. Find the admittance $Y(j\omega)$ of the composite two-terminal connected to $\dot{E}$;
2. find the active and reactive power absorbed by the controlled source;
3. find the impedance to be connected in series to the composite two-terminal such that $\dot{E}$ delivers only active power.

Fig. 13.65 Problem 13.2

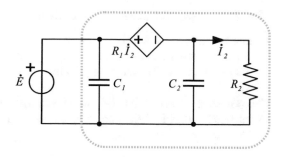

Fig. 13.66 Problem 13.3

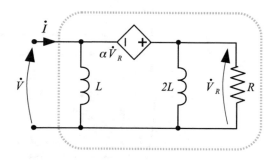

Fig. 13.67 Problem 13.4

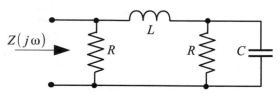

13.3 Consider the composite two-terminal shown in Fig. 13.66. Assume that the controlled source has gain $\alpha \neq 1$ and find:

1. the impedance $Z(j\omega)$ of the composite two-terminal;
2. the conditions that α must satisfy to let $Z(j\omega)$ be of resistive–inductive type;
3. the capacitance C of the capacitor to be connected in parallel to $Z(j\omega)$ with $\alpha = 0$ in order to obtain for the resulting element a purely resistive behavior.

13.4 For the composite two-terminal shown in Fig. 13.67, find:

1. the impedance $Z(j\omega)$;
2. the inductance L such that the two-terminal behaves like a resistor.

13.5 For the two-port shown in Fig. 13.68a:

1. find the impedance matrix $[Z(j\omega)]$;
2. connect a voltage source $\dot{E}$ to the second port, as shown in Fig. 13.68b, and find, if such exists, the Thévenin's equivalent circuit shown in Fig. 13.68c.

13.6 Given the composite two-terminal shown in Fig. 13.69a, working under sinusoidal steady state, find the parameters $\dot{E}_{TH}$ and Z_{TH} of the Thévenin equivalent

(a)

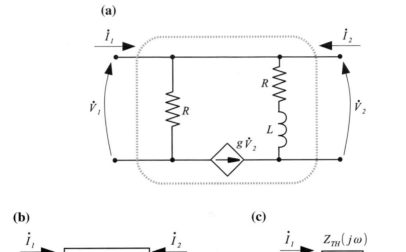

(b)   **(c)**

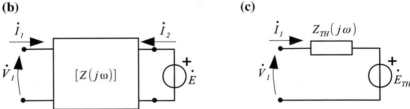

Fig. 13.68 Problem 13.5

(a)   **(b)**

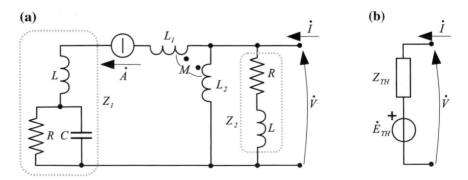

Fig. 13.69 Problem 13.6

circuit shown in Fig. 13.69b. Hint: preliminarily compute the impedances Z_1 and Z_2 shown in Fig. 13.69a.

13.7 Given the composite two-terminal shown in Fig. 13.70a, working under sinusoidal steady state, find the parameters $\dot{A}_{NR}$ and Z_{NR} of the Norton equivalent circuit shown in Fig. 13.70b.

13.8 Given the two-terminal shown in Fig. 13.71a, working under sinusoidal steady state, find the parameters $\dot{A}_{NR}$ and Z_{NR} of the Norton equivalent circuit shown in Fig. 13.71b.

13.9 The two-terminal shown in Fig. 13.72 operates in sinusoidal steady state. Find the parameters $\dot{A}_{NR}$ and Z_{NR} of the Norton equivalent circuit shown in Fig. 13.71b.

(a)   **(b)**

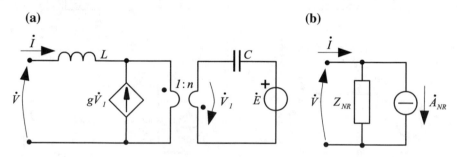

Fig. 13.70 Problem 13.7

(a)   **(b)**

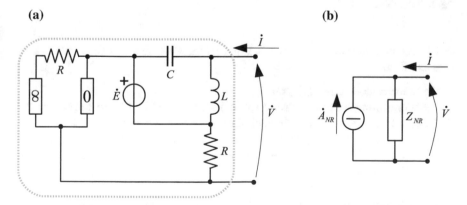

Fig. 13.71 Problem 13.8

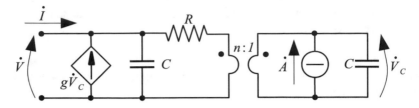

Fig. 13.72 Problem 13.9

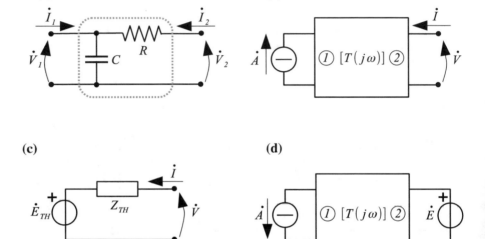

Fig. 13.73 Problem 13.10

13.10 Consider the two-port shown in Fig. 13.73a, working under sinusoidal steady state and find:

1. the transmission matrix $[T(j\omega)]$;
2. after connecting a current source $\dot{A}$ to the first port as in Fig. 13.73b: the parameters $\dot{E}_{TH}$ and Z_{TH} of the Thévenin equivalent circuit shown in Fig. 13.73c;
3. after connecting the two-port to a current and a voltage source as in Fig. 13.73d: the active and the reactive power absorbed by the two-port, by setting $\dot{A} = A$ and $\dot{E} = E$ for simplicity.

13.11 Consider the two-port shown in Fig. 13.74a, working under sinusoidal steady state and find:

1. the transmission matrix $[T(j\omega)]$;
2. after connecting a current source $\dot{A} = A$ to the second port as in Fig. 13.74b: the parameters $\dot{E}_{TH}$, Z_{TH} for the Thévenin equivalent circuit and $\dot{A}_{NR}$, Z_{NR} for the Norton equivalent circuit, as shown in Fig. 13.74c;

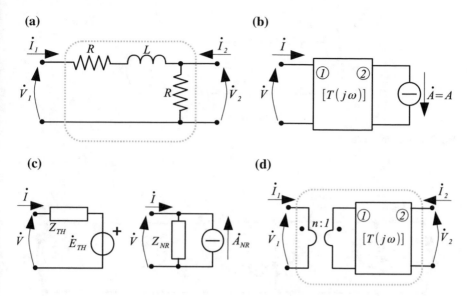

Fig. 13.74 Problem 13.11

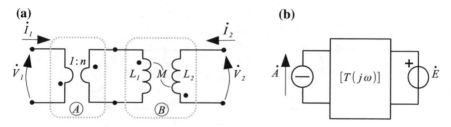

Fig. 13.75 Problem 13.12

3. after connecting the two-port in cascade to an ideal transformer as shown in Fig. 13.74d: the impedance matrix of the resulting two-port.

13.12 Consider the cascade connection of two-port A with two-port B, shown in Fig. 13.75a. In sinusoidal steady-state conditions, find:

1. the transmission matrix $T_A(j\omega)$ of two-port A;
2. the transmission matrix $T_B(j\omega)$ of two-port B in the case of closely coupled inductors;
3. the transmission matrix $[T(j\omega)]$ of the overall two-port;
4. the active and reactive power entering the overall two-port when a current source $\dot{A} = -jA$ is connected to the first port and a voltage source $\dot{E} = E$ to the second port, as shown in Fig. 13.75b.

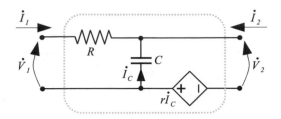

Fig. 13.76 Problem 13.13

(a)

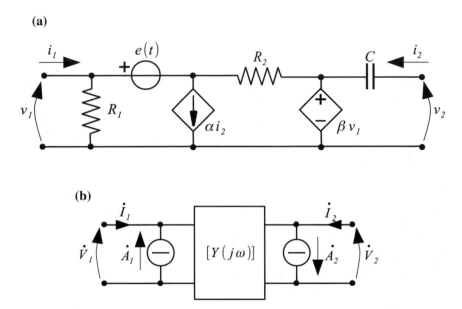

(b)

Fig. 13.77 Problem 13.14

13.13 Find the transmission matrix $[T(j\omega)]$ of the two-port shown in Fig. 13.76.

13.14 Assume that the network shown in Fig. 13.77a operates as a two-port in sinusoidal steady-state conditions. Find:

1. the parameters of its Norton equivalent model, following the conventions shown in Fig. 13.77b;
2. the reciprocity conditions for $[Y(j\omega)]$.

13.15 Assume that the two-port shown in Fig. 13.78a operates in sinusoidal steady-state conditions. Find the parameters of its Thévenin equivalent model, following the conventions shown in Fig. 13.78b.

13.16 Assume that the network shown in Fig. 13.79 operates as a two-port in sinusoidal steady-state conditions. Find the parameters of its Thévenin equivalent model, following the conventions shown in Fig. 13.78b

(a)

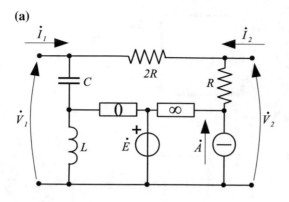

(b)

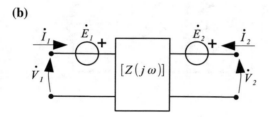

Fig. 13.78 Problem 13.15

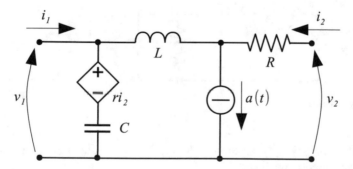

Fig. 13.79 Problem 13.16

13.17 The two-port shown in Fig. 13.80 operates in sinusoidal steady-state conditions. Setting $\dot{E} = E$ and $\dot{A} = -jA$, find the active and the reactive power delivered by the two-port.

13.18 The two-port shown in Fig. 13.81 operates in sinusoidal steady-state conditions. Find the active and the reactive power absorbed by the nullor, assuming $\dot{E} = E$ and $\dot{A} = -jA$.

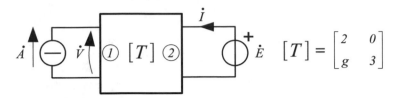

Fig. 13.80 Problem 13.17

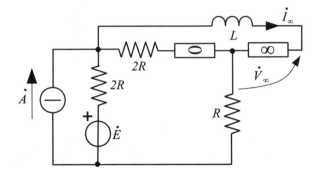

Fig. 13.81 Problem 13.18

(a)   **(b)**

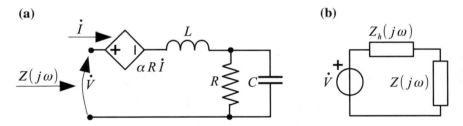

Fig. 13.82 Problem 13.19

13.19 The composite two-terminal shown in Fig. 13.82a operates in sinusoidal steady-state conditions. Find:

1. the impedance $Z(j\omega)$;
2. the value of the inductance that brings the power factor of $Z(j\omega)$ to 1;
3. the impedance Z_h to be connected in series to Z so that Z absorbs the maximum active power from the voltage source in Fig. 13.82b.

13.20 The circuit shown in Fig. 13.83 is working in sinusoidal steady-state conditions. Find the impedance $Z_1(j\omega)$ such that the active power absorbed by $Z_2(j\omega)$ is maximum.

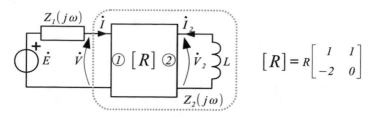

Fig. 13.83 Problem 13.20

(a)   **(b)**

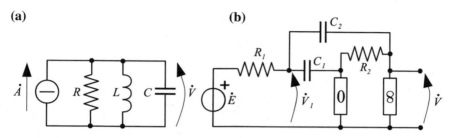

Fig. 13.84 Problem 13.21

13.21 The circuits shown in Fig. 13.84 operate in sinusoidal steady-state conditions. As shown in Sect. 13.11.2, the network function $H_1(j\omega) = \dot{V}/\dot{A}$ for the circuit in Fig. 13.84a can be written as

$$H_1(j\omega) = \frac{R}{1 + jQ\left(\dfrac{\omega}{\omega_0} - \dfrac{\omega_0}{\omega}\right)}$$

with $\omega_0 = 1/\sqrt{LC}$ and $Q = \omega_0 C R = R/(\omega_0 L)$.

Analyze the circuit of Fig. 13.84b and find an expression for the network function $H_2(j\omega) = \dot{V}/\dot{E}$. Show that by properly defining ω_0 and Q, $H_2(j\omega)$ can be written as $H_1(j\omega)$, apart from a constant coefficient. Find expressions for ω_0, Q, and the constant coefficient in terms of the physical parameters of the circuit.

Chapter 14
Advanced Concepts: Analysis of Nonlinear Oscillators

We shall not cease from exploration
And the end of all our exploring
Will be to arrive where we started
And know the place for the first time.
T. S. Eliot, Little Gidding

Abstract This chapter is focused on nonlinear oscillators. In particular, we provide a set of tools (Poincaré section, Floquet multipliers, Lyapunov exponents) for the analysis of nonlinear oscillators, with emphasis on the study of limit cycles. Moreover, the most common bifurcations that involve limit cycles are described. Examples of analysis of nonlinear oscillators—even forced, coupled, and networked—are provided.

Note to the Reader. Some of the concepts illustrated in this chapter, mainly in its last sections, show just the tip of the iceberg. Whenever a topic is only briefly touched upon, we refer the reader to other books or scientific papers where he/she can find either a deeper treatment or a more complete survey. This is the reason why the list of references at the end of this chapter is the longest of the whole book.

14.1 Periodic Solutions and (Limit) Cycles

An oscillator produces a periodic fluctuation of fixed amplitude, waveform, and period. Clocks, computers, radios, synthesizers, watches all contain mechanical or electronic oscillators. Also, nature offers a large variety of periodic phenomena, ranging from the motion of celestial bodies to the intermittent discharge of water ejected by geysers, the beating of a heart, the periodic firing of a neuron, and so forth.

The amplitude of periodic solutions in *linear oscillators* is set entirely by the initial conditions, and any perturbation brings the solution to follow a different closed curve.

© Springer Nature Switzerland AG 2020
M. Parodi and M. Storace, *Linear and Nonlinear Circuits: Basic and Advanced Concepts*,
Lecture Notes in Electrical Engineering 620,
https://doi.org/10.1007/978-3-030-35044-4_14

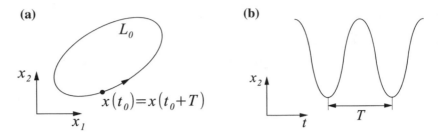

Fig. 14.1 a Example of 2D periodic solution in the state plane; **b** corresponding time evolution of one of the state variables

For instance, as evidenced in Chap. 11 and in Sect. 12.2.1, linear circuits of second or higher order can admit oscillations and therefore periodic solutions.

> A **cycle** is a periodic solution, namely a nonequilibrium closed trajectory L_0 such that by choosing on the cycle an initial condition $x(t_0) = x_0 \in L_0$, we have $x(t_0 + T) = x_0$ (periodicity condition) for all t_0. The minimal T with this property is called the **period** of the cycle L_0.

Figure 14.1 shows an example of a two-dimensional cycle L_0 of period T in the state space (panel a) and corresponding time evolution of the T-periodic state variable x_2 (panel b).

By contrast, in stable *nonlinear oscillators*, the asymptotic fluctuation is determined by the structure of the oscillator itself, and any (sufficiently small) perturbation decays in time and the fluctuation is pushed back to the standard amplitude and period.

> A **limit cycle** is an isolated cycle, namely a closed trajectory in a neighborhood of which there are no other cycles.

For instance, the trajectories shown in Fig. 12.21 are cycles, but not limit cycles, because they are not isolated. In contrast, the asymptotic trajectory shown in Fig. 12.28 is a limit cycle, since neighboring trajectories are not closed: they converge toward the limit cycle.

If all neighboring trajectories approach the limit cycle L_0, we say that L_0 is *stable* or *attracting*.

A circuit (or more generally, a system) that admits a stable limit cycle can exhibit *self-sustained oscillations*, that is, it can oscillate even in the absence of forcing periodic inputs, producing a standard oscillation with fixed period, waveform, and amplitude.

Only nonlinear circuits can generate limit cycles, whereas linear circuits can generate cycles, but not limit cycles.

14.2   Poincaré Section

The first tool we provide is the so-called Poincaré section.

Given an n-dimensional system $\dot{x} = f(x)$, a **Poincaré section**[1] Σ is a smooth manifold (usually of dimension $n - 1$) of the state space transverse to the flow in a specified direction, that is, all trajectories starting on Σ evolve intersecting Σ at a nonzero angle.

Figure 14.2 shows two examples of Poincaré sections. Figure 14.2a is concerned with a second-order system and shows a trajectory (black line) approaching a stable limit cycle L_0 (light gray line); the dark gray thick line Σ is a Poincaré section that is intersected (filled dots) by the trajectory and by L_0 in only one direction.

Figure 14.2b is concerned with a third-order system and shows a trajectory (black line) approaching a stable limit cycle L_0 (light gray line); the gray surface Σ is a Poincaré section that is intersected (black circles) by the trajectory and by L_0 in only one direction.

The word "transverse" excludes the possibility that a trajectory either misses the Poincaré section or is tangent to it: there *must* be intersection at a nonzero angle.

(a)   **(b)**

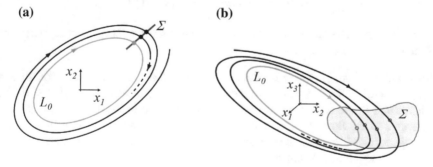

Fig. 14.2 Examples of Poincaré sections **a** in a 2D system (thick gray line Σ) and **b** in a 3D system (gray surface Σ)

---

[1] It is named after Jules Henri Poincaré (1854–1912) a French mathematician, theoretical physicist, engineer, and philosopher of science. In his research on the three-body problem, he became the first person to discover a chaotic deterministic system, which laid the foundations of modern chaos theory. He is also considered one of the founders of the field of topology.

Fig. 14.3 Poincaré section
Σ transverse to the flow
induced by the vector field
$f(x)$

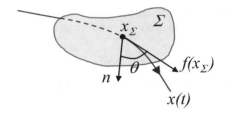

Therefore, the normal vector n of Σ at a given intersection point x_Σ cannot be orthogonal to the vector $\dot{x}(= f(x))$, which is tangent to the trajectory, as shown in Fig. 14.3. In other words, the angle θ between n and $f(x)$ must be different from $\dfrac{\pi}{2}$.

The Poincaré section is an extremely useful tool from at least two points of view. On the one hand, it allows us to introduce the so-called **Poincaré map**, widely used in the analysis of nonlinear dynamical systems [1, 2] and obtained by following trajectories from one intersection with Σ to the next. On the other hand, it allows us to classify stable limit cycles on the basis of the number of their distinct intersections with Σ, as will be explained later in greater detail.

For a system $\dot{x} = f(x)$ with fixed parameters, the simplest choice of Σ is a hyperplane orthogonal to the limit cycle L_0 at a given point $x_0 \in L_0$. In general, Σ is not a plane, but a surface described by the generic equation $g(x) = 0$.

For a system $\dot{x} = f(x; p)$ depending on parameters, the Poincaré section can be parameter-independent if the vector field obeys specific symmetries, but usually it is parameter-dependent.

Theorem 14.1 *Given a dynamical system $\dot{x} = f(x; p)$ and a function $h(x)$,[2] the surface $g(x) = 0$ with $g(x) = \nabla h \cdot f$ is transverse to the flow in a specified direction, provided that $\dfrac{d^2 h[x(t)]}{dt^2} < 0$ (or > 0) for all $x \in \{x : g(x) = 0\}$.*

✂ **Shortcut.** The proof can be skipped without compromising the comprehension of the next sections.

Proof The surface $g(x) = 0$ is transverse to the flow if $\nabla g \cdot f \neq 0$ for all $x \in \{x : g(x) = 0\}$. Indeed, ∇g is orthogonal to the surface described by $g(x) = 0$, and f is tangent to the trajectory. Therefore, this condition ensures that the surface is transverse to the flow. This geometric property is exemplified in Fig. 14.4, where the trajectory $x(t)$ crosses the surface $g(x) = 0$ (labeled as Σ) at a point x_Σ, with $\nabla g(x_\Sigma)$ orthogonal to Σ but not to $f(x_\Sigma)$.

It is known that $\nabla g \cdot f = \nabla(\nabla h \cdot f) \cdot f$ and $\nabla h \cdot f = \displaystyle\sum_{i=1}^{n} \dfrac{\partial h}{\partial x_i} f_i$.

Therefore,

$$\nabla(\nabla h \cdot f) \cdot f = f^T H_h f + (\nabla h)^T J_f f, \tag{14.1}$$

---

[2]We assume the existence and continuity of all derivatives under consideration for h.

Fig. 14.4 Geometric interpretation of the transversality condition

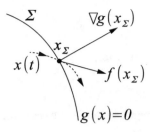

where

$$H_h(x) = \begin{pmatrix} \dfrac{\partial^2 h}{\partial x_1^2} & \dfrac{\partial^2 h}{\partial x_1 \partial x_2} & \cdots & \dfrac{\partial^2 h}{\partial x_1 \partial x_n} \\ \dfrac{\partial^2 h}{\partial x_2 \partial x_1} & \dfrac{\partial^2 h}{\partial x_2^2} & \cdots & \dfrac{\partial^2 h}{\partial x_2 \partial x_n} \\ \vdots & \vdots & \ddots & \vdots \\ \dfrac{\partial^2 h}{\partial x_n \partial x_1} & \dfrac{\partial^2 h}{\partial x_n \partial x_2} & \cdots & \dfrac{\partial^2 h}{\partial x_n^2} \end{pmatrix} = H_h^T(x)$$

is the symmetric Hessian matrix of $h(x)$ [3] and

$$J_f(x) = \begin{pmatrix} \dfrac{\partial f_1}{\partial x_1} & \dfrac{\partial f_1}{\partial x_2} & \cdots & \dfrac{\partial f_1}{\partial x_n} \\ \dfrac{\partial f_2}{\partial x_1} & \dfrac{\partial f_2}{\partial x_2} & \cdots & \dfrac{\partial f_2}{\partial x_n} \\ \vdots & \vdots & \ddots & \vdots \\ \dfrac{\partial f_n}{\partial x_1} & \dfrac{\partial f_n}{\partial x_2} & \cdots & \dfrac{\partial f_n}{\partial x_n} \end{pmatrix}$$

is the Jacobian matrix of f.

Inasmuch as

$$\frac{dh}{dt} = \sum_{i=1}^{n} \frac{\partial h}{\partial x_i} \frac{dx_i}{dt}$$

and

$$\begin{aligned} \frac{d^2 h}{dt^2} &= \sum_{i=1}^{n} \left[\frac{d}{dt}\left(\frac{\partial h}{\partial x_i}\right) \frac{dx_i}{dt} + \frac{\partial h}{\partial x_i} \frac{d^2 x_i}{dt^2} \right] = \\ &= \left(\frac{dx}{dt}\right)^T H_h \frac{dx}{dt} + (\nabla h)^T \frac{d^2 x}{dt^2} = \\ &= f^T H_h f + (\nabla h)^T \frac{d^2 x}{dt^2}, \end{aligned}$$

we obtain

$$f^T H_h f = \frac{d^2h}{dt^2} - (\nabla h)^T \frac{d^2x}{dt^2}. \tag{14.2}$$

Moreover,

$$J_f f = J_f \dot{x} = \frac{df}{dt} = \frac{d^2x}{dt^2}, \tag{14.3}$$

and therefore we have (from Eqs. 14.1–14.3)

$$\nabla(\nabla h \cdot f) \cdot f = f^T H_h f + (\nabla h)^T J_f f = \frac{d^2h}{dt^2} - (\nabla h)^T \frac{d^2x}{dt^2} + (\nabla h)^T \frac{d^2x}{dt^2},$$

that is,

$$\nabla(\nabla h \cdot f) \cdot f = \frac{d^2h}{dt^2}.$$

In conclusion, the surface $\nabla h \cdot f$ is transverse to the flow, provided that $\dfrac{d^2h}{dt^2} < 0$
(or > 0) for all x belonging to the surface.   □

In particular, we can choose $h(x) = x_j$, that is, the jth state variable. With this
choice, ∇h is a null vector, with the exception of its (unitary) jth component. Therefore, the Poincaré section is a manifold transverse to the flow defined by the equation

$$\nabla h \cdot f = f_j = \dot{x}_j = 0.$$

This is the locus of maxima, minima, and inflection points of $x_j(t)$. Among these
points, the condition $\dfrac{d^2h}{dt^2} < 0 \, (> 0)$ specifies a direction of the flow and is satisfied
only by the points corresponding to the maxima (minima) of $x_j(t)$.

For a dynamical system $\dot{x} = f(x; p)$, a **possible choice for the Poincaré
section** that is transverse to the flow in a specified direction for every value of
p is the manifold defined by the conditions

$$\begin{cases} f_j = \dot{x}_j = 0, \\[2mm] \dfrac{df_j}{dt} = \ddot{x}_j < 0 \ \ (\text{or} > 0), \end{cases} \qquad j \in \{1, 2, \ldots, n\}, \tag{14.4}$$

that is, the locus of points corresponding to the maxima (or minima) of $x_j(t)$.

For instance, Fig. 14.5 shows an example for the van der Pol oscillator, using
Eq. 12.24 with $\varepsilon = 0.1$. In this example, with respect to the general framework
defined above, the state vector is $(x_1, x_2) = (x, y)$. The curve $g(x) = 0$ is the null-

Fig. 14.5 Van der Pol
oscillator (see Eq. 12.24)
with $\varepsilon = 0.1$: **a** intersections
of the trajectory (black line)
starting from the initial
condition $(1, 2)$ with the
Poincaré section (portion of
the nullcline $\dot{x} = 0$,
nonlinear dashed gray line);
b corresponding time
evolution of $x(\tau)$

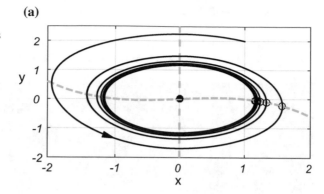

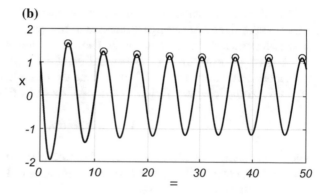

cline $\dot{x} = 0$ (nonlinear dashed gray curve in panel a). The trajectory (black solid
line) shown in panel a, starting from the initial condition $(1, 2)$, crosses this curve
many times in the chosen direction, that is, the portion of the nullcline such that
$\ddot{x} < 0$, which is the Poincaré section. The intersections are marked by black circles,
corresponding to the maxima of $x(\tau)$ marked in panel b.

14.3 Floquet Multipliers

Let $x^0(t)$ be a periodic (with period T) solution of the nth-order system $\dot{x} = f(x)$,
corresponding to a cycle L_0.

A generic solution of the system can be represented as $x(t) = x^0(t) + w(t)$, where
$w(t)$ is a vector of length n representing a deviation from the periodic solution.

By analyzing the evolution of a small perturbation $w(t)$ away from $x^0(t)$, we can
check the stability of the periodic solution. We have

$$\dot{w}(t) = \dot{x}(t) - \dot{x}^0(t) = f(x^0(t) + w(t)) - f(x^0(t)) = J(t)w(t) + \mathcal{O}(\|w\|^2),$$
$$\tag{14.5}$$

where $J(t)$ is the Jacobian matrix of the system, computed on $x^0(t)$: $J(t) = J(x)|_{x=x^0(t)}$. This matrix is in turn periodic with period T: $J(t + T) = J(t)$.

The time-varying linearized system obtained by neglecting the $\mathcal{O}(\|w\|^2)$ terms in Eq. 14.5,

$$\dot{w} = J(t)w(t), \tag{14.6}$$

is called the **variational equation** about the cycle L_0.

The variational equation is the main (linear) part of the system governing the evolution of perturbations in a neighborhood of the cycle L_0. The stability of L_0 depends on the properties of this equation, whose solutions are not necessarily periodic [4].

We take n linearly independent initial conditions $w^{(1)}(t_0), \ldots, w^{(n)}(t_0)$, which form a basis of the state space $\mathbb{R}^n$. If we take the n solutions $w^{(1)}(t), \ldots, w^{(n)}(t)$ of Eq. 14.6 starting from these initial conditions, we can write Eq. 14.6 in matrix form,

$$\dot{W} = J(t)W(t), \tag{14.7}$$

where the columns of $W(t)$ are the n solutions: $W(t) = \left(w^{(1)}(t) \cdots w^{(n)}(t) \right)$. By construction, $\det[W(t)] \neq 0$ and $W(t)$ is a **fundamental matrix** [4, 5]. Moreover, Liouville's theorem[3] [5] ensures that

$$\det[W(t)] = \det[W(t_0)]e^{\int_{t_0}^{t} \text{Tr}[J(\tau)]d\tau}. \tag{14.8}$$

As a particular case, if we choose as the n initial conditions the orthonormal basis of $\mathbb{R}^n$ with standard ordering, we have $W(t_0) = I_n$ (identity matrix of size n), and $W(t)$ is the so-called **principal fundamental matrix**.

By virtue of the superposition principle, for any given constant vector c of length n, $w(t) = W(t)c$ is a solution of the linear Eq. 14.6, because it is a linear combination of solutions. If we evaluate this solution at $t = t_0$ and choose the principal fundamental matrix, we obtain $c = w(t_0)$ and therefore

$$w(t) = W(t)w(t_0). \tag{14.9}$$

We can choose $t_0 = 0$ without loss of generality. Moreover, we know that $J(t)$ is periodic with period T, whereas the perturbation $w(t)$, in general, is not periodic.

---

[3] Joseph Liouville (1809–1882) was a French mathematician. He worked in a number of different fields in mathematics, including number theory, complex analysis, differential geometry, and topology, but also mathematical physics and even astronomy: the crater Liouville on the Moon is named after him.

Equation 14.9 computed at $t = T$ with $t_0 = 0$ provides

$$w(T) = W(T)w(0). \tag{14.10}$$

This yields a propagator matrix such that

$$w(mT) = [W(T)]^m w(0) \quad m = 1, 2, \ldots . \tag{14.11}$$

Hence, to determine the stability of the periodic solution $x^0(t)$, we need only to determine the eigenvalues of the matrix $W(T)$. For instance, by assuming that $[W(T)]^m$ is diagonalizable, we have

$$[W(T)]^m = P \begin{pmatrix} \mu_1^m & \cdots & 0 \\ \vdots & \ddots & \vdots \\ 0 & \cdots & \mu_n^m \end{pmatrix} P^{-1},$$

and the stability condition is $|\mu_j| \le 1$ for all $j = 1, \ldots, n$.

The matrix $W(T)$ is called a **monodromy matrix**, and its eigenvalues govern the behavior of a perturbation in a neighborhood of the periodic solution $x^0(t)$: if any of the eigenvalues are greater than one in magnitude, the periodic solution is unstable.

Now we show that one of the eigenvalues of $W(T)$ is unitary. We know that $x^0(t)$ is a solution of the dynamical system $\dot{x} = f(x)$, and therefore it satisfies[4]

$$\frac{dx^0(t)}{dt} = f(x^0(t)), \tag{14.12}$$

so

$$\frac{d^2 x^0(t)}{dt^2} = J(t) \frac{dx^0(t)}{dt}. \tag{14.13}$$

Hence $\dfrac{dx^0(t)}{dt}$ (vector tangent to the cycle L_0) is a solution to Eq. 14.6. Therefore, we can choose $w(t) = \dfrac{dx^0(t)}{dt}$ and substitute it into Eq. 14.10:

$$\frac{dx^0(t)}{dt}\bigg|_{t=T} = W(T) \frac{dx^0(t)}{dt}\bigg|_{t=0}.$$

Moreover, $\dfrac{dx^0(t)}{dt}$ is periodic with period T, and thus $\dfrac{dx^0(t)}{dt}\bigg|_{t=T} = \dfrac{dx^0(t)}{dt}\bigg|_{t=0}$, which implies that

---

[4]Note for the reader: here we use Leibniz's notation $\dfrac{dx}{dt}$ instead of Newton's notation $\dot{x}$ to make the formulas more legible.

$$\left.\frac{dx^0(t)}{dt}\right|_{t=0} = W(T)\left.\frac{dx^0(t)}{dt}\right|_{t=0}.$$

This means that $\dfrac{dx^0(t)}{dt}$ is an eigenvector corresponding to a unitary eigenvalue, as anticipated. Moreover, $\dfrac{dx^0(t)}{dt}$ spans a one-dimensional invariant subspace, which is quite intuitive, since the cycle L_0 corresponding to the solution $x^0(t)$ is an invariant set.

The **Floquet multipliers**[5] (or characteristic numbers) of the periodic solution $x^0(t)$ are the eigenvalues $\mu_1, \ldots, \mu_{n-1}, 1$ of the monodromy matrix.

A cycle L_0 is a **stable limit cycle** if and only if $|\mu_j| < 1$ for $i = 1, \ldots, n-1$.

Figure 14.6 shows a bunch of trajectories (gray lines) starting from initial conditions (black dots) distributed over a circle surrounding a stable limit cycle (black line) in a 3D system. The trajectories describe a sort of funnel as they evolve around the cycle.

Another important property of the Floquet multipliers is that **the product of all multipliers of any cycle is positive**.

This can be shown using Eq. 14.8. This equation, applied to the monodromy matrix, allows us to express the determinant of the monodromy matrix in terms of the matrix $J(t)$, through the so-called **Liouville formula**:

$$\det[W(T)] = e^{\int_0^T \operatorname{Tr}(J(t))dt} > 0. \tag{14.14}$$

We remark that

$$\det[W(T)] = 1 \cdot \mu_1 \cdot \mu_2 \cdots \mu_{n-1}$$

and

$$\operatorname{Tr}[J(t)] = \nabla \cdot f(x^0(t)) = \left.\frac{\partial f_1}{\partial x_1}\right|_{x=x^0(t)} + \left.\frac{\partial f_2}{\partial x_2}\right|_{x=x^0(t)} + \cdots + \left.\frac{\partial f_n}{\partial x_n}\right|_{x=x^0(t)}.$$

---

[5] They are named after Achille Marie Gaston Floquet (1847–1920), a French mathematician, mainly known for his work in theory of differential equations.

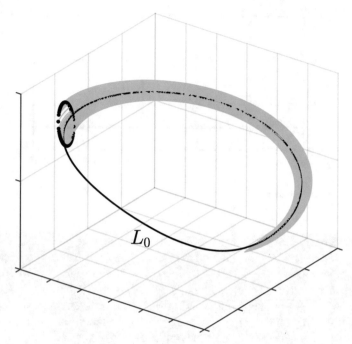

Fig. 14.6 Example of stable limit cycle: the trajectories (gray lines) starting from the black dots shrink around the cycle L_0 (black line)

As a consequence, the Liouville formula can be recast as

$$\mu_1 \mu_2 \cdots \mu_{n-1} = e^{\int_0^T \nabla \cdot f(x^0(t))dt} > 0, \tag{14.15}$$

and this proves that the product of all multipliers of any cycle is positive.

We remark that the Liouville formula holds even for nondistinct Floquet multipliers: a repeated Floquet multiplier is counted according to its own multiplicity.

For $n = 2$, the Liouville formula can also provide the value of the multiplier μ_1.

Case Study: Van der Pol Oscillator

Analyze the stability of the periodic solution of the van der Pol oscillator through the Floquet multipliers.

We already know that the Jacobian matrix is

$$J(x) = \begin{pmatrix} \varepsilon(1 - 3x^2) & -1 \\ 1 & 0 \end{pmatrix}.$$

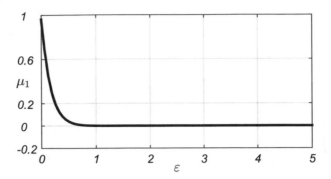

Fig. 14.7 Plot of μ_1 versus ε

Therefore, in the limit case $\varepsilon = 0$ (recall that $\varepsilon > 0$), we have $\mathrm{Tr}[J(t)] = 0$ for all t and $\mu_1 = 1$. This is the only ε value for which we can compute the Floquet multipliers analytically. If we increase ε, we can compute the Floquet multipliers only using a numerical tool based on the integration of a trajectory. Figure 14.7 shows the nonunitary multiplier μ_1 for different values of ε. It is apparent that μ_1 tends to 0 by increasing ε. In this condition, the limit cycle not only is stable, but also has the maximum degree of stability, because μ_1 is as far as possible from the edge of stability, that is, the unit circle.

14.4   Poincaré Map

Another remarkable point is concerned with the relationship between Floquet multipliers and the Poincaré map. Figure 14.8a shows a stable limit cycle L_0 (gray line), a trajectory approaching it (black line), and a Poincaré section Σ (gray surface) in a 3D system. The intersections of the trajectory with the Poincaré section Σ define a sequence of points $(\ldots, \bar{x}_k, \bar{x}_{k+1}, \ldots)$ converging to the intersection x^* between L_0 and Σ. Of course, we could represent the intersection points in a lower-dimensional system of $n - 1$ coordinates (say, z) defined on the domain Σ, as shown in Fig. 14.8b. Each point of this sequence can be mapped to the next one by a function $P : \Sigma \to \Sigma$, which is called a Poincaré map:

$$\bar{z}_{k+1} = P(\bar{z}_k),$$

where $\bar{z}_k$ corresponds (in the Σ coordinate system) to $\bar{x}_k$. The above equation describes the evolution of a *discrete-time system*. As such, the stability of an equilibrium point of P depends on the position of the eigenvalues of the Jacobian matrix of P computed at the equilibrium point *with respect to the unit circle*: for instance,

(a) **(b)**

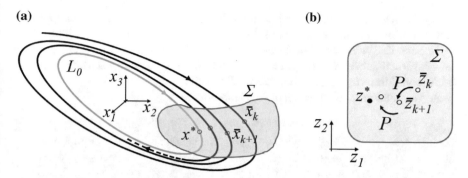

Fig. 14.8 Examples of intersections between a trajectory approaching a stable limit cycle L_0 and a Poincaré section (gray surface Σ) in a 3D system: **a** original n-dimensional coordinate system and **b** local $(n-1)$-dimensional coordinate system

the equilibrium point is (locally) attractive if and only if all eigenvalues lie strictly within the unit circle.

If L_0 is a stable limit cycle, the sequence defined by the Poincaré map starting from a given initial condition converges to the equilibrium point z^*, corresponding (in the Σ coordinate system) to x^*. In other words, x^* and z^* are the same point in two different reference systems. In general, **a cycle L_0 whose intersection with a Poincaré section is x^* shares the same stability properties with the equilibrium point z^* of the corresponding Poincaré map.**

This means that the stability of both L_0 and z^* depends on the position with respect to the unit circle of the $n-1$ eigenvalues of the Jacobian matrix of the map P computed at z^*. It is possible to show [2] that the Floquet multipliers $\mu_1, \ldots, \mu_{n-1}$ (all but the unitary multiplier) coincide with these eigenvalues. Moreover, these multipliers are independent of the position of x^* on L_0, the Poincaré section Σ, and the choice of local coordinates z on it [2].

Since an analytic expression of P is usually unknown, the Poincaré map represents a conceptual tool of great importance, but from a practical standpoint the information about the stability of a limit cycle is typically obtained numerically, from Poincaré sections, Floquet multipliers, or other tools, such as the Lyapunov exponents treated in the next section.

14.5 Stability of Generic Invariant Sets: Lyapunov Exponents

According to the Poincaré–Bendixson theorem [1, 6], the dynamical possibilities in a second-order ($n = 2$) circuit/system described by a state equation $\dot{x} = f(x)$, with f Lipschitz-continuous, are limited, because every trajectory must eventually diverge, converge to a stable equilibrium point (stationary steady state), or approach a closed orbit (periodic steady state). Nothing more complicated is possible.

In Sect. 12.3 we defined an invariant set of a dynamical system $\dot{x} = f(x)$ as a flow-invariant subset S of the state space such that the trajectory starting from any initial condition $x(t_0) \in S$ evolves within S for all $t > t_0$.

A particular case of an invariant set is the attractor.

> An **attractor** is a closed invariant set $A \subset \mathbb{R}^n$ that has an open neighborhood $\mathcal{B}(A)$, called the **basin of attraction** of A, such that all trajectories starting within $\mathcal{B}(A)$ converge to A as $t \to \infty$.

Up to now, we have met many examples of invariant sets: equilibrium point, limit cycle, stable/unstable/center manifold, and (by definition) trajectories. Among them, stable and attractive equilibrium points and limit cycles are also attractors.

Further examples of invariant sets and attractors require state spaces with $n \geq 3$, that is, the presence of at least three state variables. In this case, trajectories may indeed eventually wander around in a bounded region without settling down to an equilibrium point or a closed orbit. This absence of periodicity can produce either *quasiperiodic* solutions converging in the state space to a *stable torus* or *chaotic solutions* approaching in the state space a *strange attractor*.

> In geometry, a **torus** (plural *tori*) is a surface generated by revolving a circle in a three-dimensional space about an axis coplanar with the circle. If the axis of revolution does not touch the circle, the surface has a ring shape and is called a torus of revolution.

In practice, a torus usually has a donut shape, as shown in Fig. 14.9, and is characterized by two frequencies (one, say ω_A, for revolution along the circle, and one, say

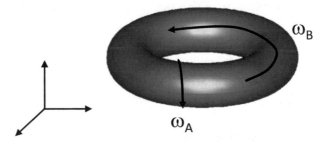

Fig. 14.9 Example of torus (of revolution), with angular frequencies ω_A along the circle and ω_B along the ring; the ratio $\dfrac{\omega_A}{\omega_B}$ is an irrational number

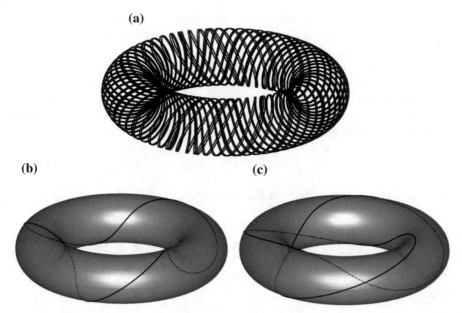

Fig. 14.10 **a** Quasiperiodic and **b, c** periodic trajectories, obtained by setting $\dfrac{\omega_A}{\omega_B}$ to **a** an irrational value, **b** $1/3$, **c** $2/3$

ω_B, for the rotation along the ring) whose ratio is an irrational number, which leads to a quasiperiodic behavior (see Sect. 9.5.3). Indeed, in this case every trajectory starting on the torus will densely fill the torus's surface, in the limit for $t \to \infty$. An example is shown in Fig. 14.10a, where the trajectory is drawn for a finite time. In contrast, if the ratio $\dfrac{\omega_A}{\omega_B}$ is a rational number $\dfrac{p}{q}$, the trajectory return back to its initial condition after q turns around the circle and p turns around the ring, thus describing a *knot cycle*. Figure 14.10 shows two examples, for $\dfrac{\omega_A}{\omega_B} = \dfrac{1}{3}$ (panel b) and $\dfrac{\omega_A}{\omega_B} = \dfrac{2}{3}$ (panel c).

So far, we have introduced and discussed methods for the analysis of the stability of stationary (equilibrium point) or periodic (limit cycle) solutions. How can we evaluate the stability of a generic invariant set?

Up to now, we have resorted mainly to linearization. For evaluating the stability of an equilibrium point x^*, we analyze the eigenvalues of the Jacobian matrix of the system computed at $x = x^*$; see Sect. 12.4; for a limit cycle $x^0(t)$, we analyze the Floquet multipliers (see Sect. 14.3), which are obtained by linearizing the system about $x^0(t)$.

For generic invariant sets, we can follow the same line of reasoning and analyze the evolution of a small perturbation $w(t)$ away from the trajectory, which leads us to define the so-called *Lyapunov exponents*.[6]

We consider a generic solution of the system $\dot{x} = f(x)$, with initial condition $x(t_0) = x_0$, that is, a trajectory $\tilde{x}(t)$ starting from x_0.

To determine whether trajectories converge toward (or diverge away from) $\tilde{x}(t)$, we once more linearize the system about $\tilde{x}(t)$ and study the time evolution of a perturbation $w(t)$. To this end, we represent a generic trajectory as

$$x(t) = \tilde{x}(t) + w(t) \tag{14.16}$$

and study the evolution of the perturbation $w(t)$:

$$\dot{w}(t) = \dot{x}(t) - \dot{\tilde{x}}(t) = f(\tilde{x}(t) + w(t)) - f(\tilde{x}(t)) =$$
$$= f(\tilde{x}(t)) + J(t)w(t) + \mathcal{O}(\|w\|^2) - f(\tilde{x}(t)),$$

where $J(t)$ is the (time-varying) Jacobian matrix of the system. Therefore,

$$\dot{w}(t) = J(t)w(t) + \mathcal{O}(\|w\|^2),$$

and in the limit of infinitesimally small perturbation, this provides the *variational equation* about the solution $\tilde{x}(t)$:

$$\dot{w}(t) = J(t)w(t). \tag{14.17}$$

Using a fundamental matrix $W(t)$ as in Sect. 14.3, for any given constant vector c of length n, $w(t) = W(t)c$ is a solution of the linear Eq. 14.17, owing to the superposition principle. We already know that W is invertible, and therefore we can obtain

$$c = [W(t_0)]^{-1}w(t_0).$$

Finally, given an initial condition $w(t_0) = w_0$, we have

$$w(t) = W(t)[W(t_0)]^{-1}w_0 \doteq \tilde{W}(t)w_0. \tag{14.18}$$

Of course, if $W(t)$ is the principal fundamental matrix, we have $W(t_0) = I_n$ and $\tilde{W}(t) = W(t)$.

The properties of the matrix $\tilde{W}(t)$—which usually cannot be computed analytically and has to be calculated numerically—can be used to determine the asymptotic behavior of $w(t)$ and hence the stability of $\tilde{x}(t)$.

---

[6]They are named after Aleksandr Mikhailovich Lyapunov (1857–1918), a Russian mathematician and physicist, mainly known for his development of the stability theory of dynamical systems, as well as for his many contributions to mathematical physics and probability theory.

The ith **Lyapunov exponent** (or characteristic exponent) in the direction $p_i = \dfrac{w_0}{\|w_0\|}$ $(i = 1, \ldots, n)$ of an infinitesimally small perturbation w_0 about the trajectory $\tilde{x}(t)$ started from x_0 is given as (see Eq. 14.18)

$$\lambda_i = \lim_{w_0 \to 0} \lim_{t \to \infty} \frac{1}{t} \ln \frac{\|w(t)\|}{\|w_0\|} = \lim_{t \to \infty} \frac{1}{t} \ln \|\tilde{W}(t) p_i\|,$$

provided that the set $\{p_i\}$ forms an orthonormal basis and $\sum_{i=1}^{n} \lambda_i$ is minimized.

The set $\{\lambda_i\}$ of all Lyapunov exponents is also called the **Lyapunov spectrum**.

Remarks

- The Lyapunov exponents are, by definition, real numbers.
- The Lyapunov spectrum is naturally ordered from the largest Lyapunov exponent to the smallest one: $\lambda_1 > \lambda_2 > \cdots > \lambda_n$. If so, λ_1 is called the *maximal Lyapunov exponent*.
- The Lyapunov exponents are independent of the initial condition x_0, with the exception of some rare cases [7].
- The ith Lyapunov exponent λ_i can be interpreted geometrically as a measure of the exponential rate of growth (if $\lambda_i > 0$) or decay (if $\lambda_i < 0$) of the infinitesimal perturbation along the p_i direction as a time average with $t \to \infty$. In other words, we monitor the long-term evolution of an infinitesimal n-sphere of initial conditions; the sphere becomes an n-ellipsoid due to the action of the flow. Since the orientation of the ellipsoid changes continuously as it evolves, the directions associated with a given exponent vary with time. One cannot, therefore, speak of a well-defined direction associated with a given exponent.
- Altogether, the computation of the Lyapunov spectrum requires evolving n linearly independent perturbations, with the problem that all trajectories, under the action of the flow, tend to align to the same direction as t grows. However, this numerical drawback can be counterbalanced by orthonormalizing the evolutions of vectors p_i along the trajectory, with the help of the Gram–Schmidt procedure [8, 9].

In summary, the Lyapunov exponents measure the growth rates of small perturbations about a generic trajectory. A (bounded) trajectory that does not converge towards a fixed point is characterized by at least one zero Lyapunov exponent: it corresponds to the invariance along its own trajectory.

For *nonstrange attractors*, the number of zero Lyapunov exponents coincides with the geometric dimension of the attractor. For instance:

- for a stable equilibrium point (with geometric dimension zero), all Lyapunov exponents are negative;

- for a stable limit cycle (a curve with geometric dimension one), there is one zero Lyapunov exponent (due to the invariance along the cycle), and all the others are negative;
- for a stable torus (a surface with geometric dimension two), there are two zero Lyapunov exponents (due to the invariance on the torus), and all the others are negative.

For *strange attractors* in three-dimensional systems, we have:

- $\lambda_1 > 0$, accounting for the *sensitivity to initial conditions*,[7] which is a fingerprint of chaos;
- $\lambda_2 = 0$, accounting for the invariance along the trajectory;
- $\lambda_3 < 0$, with $|\lambda_3| > \lambda_1$, accounting for the overall attractive behavior.

14.5.1 Particular Cases

Equilibrium Point

Let x^* be a hyperbolic equilibrium point of the n-dimensional system $\dot{x} = f(x)$, set $A = J(x)|_{x=x^*}$, and set $t_0 = 0$ without loss of generality. According to Sect. 12.4, the evolution of the small perturbation $w(t)$ away from x^* is governed by the following linear system:

$$\dot{w} = Aw,$$

whose general solution is

$$w(t) = e^{At}c.$$

The corresponding solution to the initial value problem with $w(0) = w_0$ is

$$w(t) = e^{At}w_0.$$

For the sake of simplicity, we assume that the eigenvalues of A are simple and call them $\hat{\lambda}_1, \hat{\lambda}_2, \ldots, \hat{\lambda}_n$. Consequently, the eigenvalues of e^{At} are $e^{\hat{\lambda}_1 t}, e^{\hat{\lambda}_2 t}, \ldots, e^{\hat{\lambda}_n t}$. Therefore, by choosing w_0 in the direction of the ith eigenvector, the ith Lyapunov exponent can be computed as follows:

$$
\begin{aligned}
\lambda_i &= \lim_{w_0 \to 0} \lim_{t \to \infty} \frac{1}{t} \ln \frac{\|e^{\hat{\lambda}_i t} w_0\|}{\|w_0\|} = \lim_{t \to \infty} \frac{1}{t} \ln |e^{\hat{\lambda}_i t}| = \\
&= \lim_{t \to \infty} \frac{1}{t} \ln |e^{\Re\{\hat{\lambda}_i\}t} e^{j\Im\{\hat{\lambda}_i\}t}| = \lim_{t \to \infty} \frac{1}{t} \ln e^{\Re\{\hat{\lambda}_i\}t} = \Re\{\hat{\lambda}_i\}.
\end{aligned}
$$

---

[7] Sensitivity to initial conditions means that in a chaotic system, two arbitrarily close initial conditions generate significantly different trajectories, after a possibly similar initial transient. Thus, an arbitrarily small change, or perturbation, of the current trajectory may lead to significantly different future behavior.

Inasmuch as the stability of x^* is determined by the real parts of $\hat{\lambda}_1, \hat{\lambda}_2, \ldots, \hat{\lambda}_n$, this means that the Lyapunov exponent λ_i embeds the same information about stability as $\hat{\lambda}_i$.

Limit Cycle

If we consider a limit cycle $x^0(t)$, we know (see Eq. 14.11) that

$$w(mT) = [W(T)]^m w(0) \qquad m = 1, 2, \ldots,$$

with principal fundamental matrix $W(t)$ and monodromy matrix $W(T)$, whose eigenvalues μ_i are the Floquet multipliers.

Therefore, by choosing w_0 in the direction of the ith element of the orthonormal basis of $\mathbb{R}^n$, the ith Lyapunov exponent can be computed as follows:

$$\lambda_i = \lim_{w_0 \to 0} \lim_{m \to \infty} \frac{1}{m} \ln \frac{\|[W(T)]^m w_0\|}{\|w_0\|} = \lim_{m \to \infty} \frac{1}{m} \ln(|\mu_i|^m) = \ln |\mu_i|.$$

It is apparent that if μ_i lies within the unit circle, λ_i is negative. Therefore, from a stability standpoint, the Lyapunov exponent λ_i embeds the same information as the Floquet multiplier μ_i.

Case Study: Lyapunov Exponents for the van der Pol Oscillator

Analyze the stability of the periodic solution of the van der Pol oscillator through the numerical computation of the Lyapunov exponents.

We used the algorithm described in [10]. The Lyapunov exponents computed for a grid of values of the parameter ε are shown in Fig. 14.11: λ_1 (solid curve) and λ_2 (dashed curve). It is apparent that $\lambda_1 = 0$ for every value of ε; this Lyapunov exponent corresponds to the unitary Floquet multiplier. The second Lyapunov exponent, namely λ_2, tends to 0 for $\varepsilon \to 0$; this corresponds to the fact that in the limit case $\varepsilon = 0$, we have two unitary Floquet multipliers. If we increase ε, λ_2 decreases; this corresponds to the increasing stability of the limit cycle, also witnessed by the nonunitary Floquet multiplier $\mu_1 \to 0$; see Fig. 14.7.

14.6 Bifurcations

We already treated the topic of bifurcations in Sects. 10.4 and 14.6, focusing on bifurcations of equilibrium points. In general, bifurcations can also involve invariant sets of different kinds, such as limit cycles, tori, or strange attractors.

The qualitative theory of dynamical systems originated in Poincaré's work, developed toward the end of the nineteenth century. Most of the results on bifurcations of

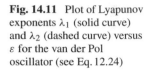

Fig. 14.11 Plot of Lyapunov exponents λ_1 (solid curve) and λ_2 (dashed curve) versus ε for the van der Pol oscillator (see Eq. 12.24)

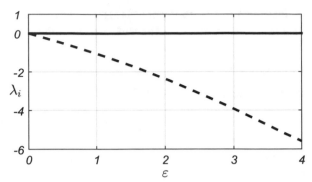

continuous-time systems, like those considered in this book, are due to the group of Russian mathematicians working on dynamics at Gorky (now Nizhny Novgorod), in particular Andronov and Leontovich [11] and at Moscow, in particular Kolmogorov [12] and Arnold [13].

Recent and complete treatments of this topic can be found in the already cited references [1, 2, 6]. Numerical aspects are well described in [2, 14].

A bifurcation of an equilibrium (or limit cycle) is called **local** if it occurs when a continuous change in a parameter causes one or more eigenvalues of the Jacobian matrix associated with the equilibrium (or limit cycle) to cross the edge of stability.

All the bifurcations considered up to now have been local: in the fold bifurcation of equilibria (see Sect. 12.5.1), one real eigenvalue of the Jacobian matrix computed at one of the colliding equilibrium points becomes null, thus originating a one-dimensional center manifold in the system linearized about each equilibrium; in the Hopf bifurcation (see Sect. 12.5.2), a pair of complex conjugate eigenvalues of the Jacobian matrix computed at the bifurcating equilibrium point cross the imaginary axis, thus originating a two-dimensional center manifold in the system linearized about the equilibrium.

Similarly, for circuits of order higher than one, the simplest way to have a bifurcation for a limit cycle L_0 is that one or more Floquet multipliers—or correspondingly, one or more eigenvalues of the Jacobian matrix $J(x)|_{x=x^*}$ of the Poincaré map whose equilibrium point x^* corresponds to L_0—cross the edge of stability, that is, the unit circle.

In the simplest cases, in the complex plane we have one Floquet multiplier equal to 1 (see Fig. 14.12a) or equal to -1 (see Fig. 14.12b) or a pair of complex conjugate multipliers that cross the unit circle, as sketched in Fig. 14.12c.

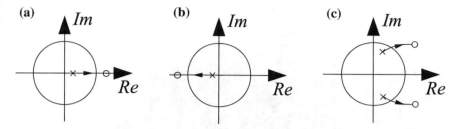

Fig. 14.12 Changes of Floquet multipliers that determine change of stability of limit cycles: **a** fold bifurcation; **b** flip bifurcation; **c** Neimark–Sacker bifurcation. Only the multipliers responsible for the bifurcation are shown. The crosses (dots) mark the initial (final) positions of the multipliers. The transition from initial to final position is due to a continuous change in a bifurcation parameter

A **global** bifurcation causes changes in the state portrait that cannot be confined to a small neighborhood of an equilibrium (or limit cycle), as is the case with local bifurcations. A global bifurcation cannot be revealed by eigenvalue degeneracies and is the result of the global behavior of the system.

Usually, in the control space $(x; p)$ a global bifurcation corresponds to a collision between stable/unstable manifolds of one or more saddle points (or saddle limit cycles).

In the following subsections, we illustrate the most common bifurcations (both local and global) involving limit cycles.

14.6.1 Fold Bifurcation of Limit Cycles

The first case of local bifurcation is the so-called *fold bifurcation of limit cycles* (or tangent bifurcation of limit cycles or saddle-node bifurcation of limit cycles), which corresponds to the collision of two limit cycles in the control space, which in turn corresponds to the collision of the two corresponding equilibria of the Poincaré map defined on a Poincaré section transversal to both cycles.

For instance, for a second-order system with a parameter p, Figs. 14.13 and 14.14 show, for three values of p, a qualitative example pointing out:

- the limit cycles L_1 (stable) and L_2 (unstable) and two sections (Σ_1 corresponding to $p = p_1$ and Σ_2 corresponding to $p = p_2 > p_1$) in the control space $(x; p)$ (Fig. 14.13); the two cycles are distinct for $p < p_2$ (green and red thick cycles on Σ_1), collide at $p = p_2$ (blue thick line on Σ_2), and disappear for $p > p_2$;
- the position of the Floquet multipliers of L_1 and L_2 in the complex plane for $p = p_1$ and $p = p_2$ (Fig. 14.14).

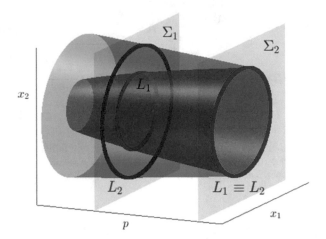

Fig. 14.13 Fold bifurcation of limit cycles for a second-order circuit/system: by changing (increasing in the figure) the control parameter p, two limit cycles L_1 (stable, green) and L_2 (unstable, red) collide in the control space (p, x_1, x_2) at $p = p_2$. The Poincaré section Σ_1 [Σ_2] is the plane $p = p_1$ [$p = p_2$]

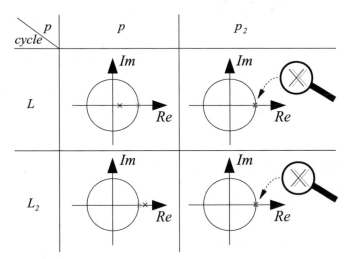

Fig. 14.14 Fold bifurcation of limit cycles for a second-order circuit/system: Floquet multipliers of L_1 and L_2 for $p = p_1$ and $p = p_2$. The unitary multiplier is marked with a gray cross

In terms of eigenvalue degeneracy, the nonunitary Floquet multiplier of L_1 [L_2] is smaller [larger] than 1, so that when the two cycles collide the same multiplier must be equal to 1.

Varying the parameter in the opposite direction, this bifurcation explains the sudden birth of a pair of cycles with *nonnull amplitude*, in contrast to the Hopf bifurcation case.

14.6.2 Flip Bifurcation

The *flip bifurcation* or period-doubling bifurcation corresponds to the collision of two particular limit cycles in the control space, one (say, L_2) tracing twice the other (say, L_1) with double period. In other words, at the bifurcation, L_1 has period T and L_2 has period $2T$.

It is a local bifurcation, and in terms of eigenvalue degeneracy, the nonunitary Floquet multiplier of L_1 is negative, so that at the bifurcation it must be equal to -1. Therefore, since the product of all multipliers must be positive according to the Liouville formula (see Sect. 14.3), this bifurcation requires at least a three-dimensional state space in order to have a second negative Floquet multiplier.

In the supercritical case for three-dimensional systems, L_1 changes its stability across the bifurcation and L_2 appears (stable) when L_1 becomes unstable. In the subcritical case, L_1 again changes its stability across the bifurcation, but L_2 appears (unstable) when L_1 becomes stable.

Notice that in systems of dimension higher than 3 we can have flip bifurcations involving only unstable cycles.

We illustrate this bifurcation using as a case study the Colpitts oscillator introduced in Sect. 12.7. We set $k = 0.5$ and $Q = 1.6$ and analyze the oscillator dynamics by changing the control parameter g. Figure 14.15 shows, in the projections (g, x_1, x_3) (panel a) and (g, x_2, x_3) (panel b) of the control space, the stable cycle L_1 (thin light green lines), which becomes unstable (thin red lines) by increasing g. After the flip bifurcation (at about $g = 2.1$), the stable cycle L_2 (thick dark green lines) appears, which is characterized by the presence of two loops.

Figure 14.16 shows the simulation results corresponding to the (stable) 1-cycle L_1, obtained by setting $g = 1.8$. The two nonunitary Floquet multipliers of this cycle can be computed numerically, and they turn out to be real and, according to the Liouville formula, both negative, with values between -1 and 0. The asymptotic trajectory in the state space is shown in panel a, together with the Poincaré section $\dot{x}_2 = 0$ (i.e., $x_3 = 0$, according to Eq. 12.39), whereas panel b displays the time evolution of $x_2(\tau)$ and panel c the one-sided amplitude spectrum $|X_2(f)|$ of $x_2(\tau)$, obtained with a sampling frequency $f_s = 10\,\text{kHz}$. We remark that f is the normalized frequency corresponding to the normalized time τ.

On the Poincaré section there is only one intersection of the cycle L_1 in a specified direction, corresponding to the maxima (or minima) of $x_2(\tau)$. Notice that x_2 oscillates (nearly sinusoidally) with period T around a constant bias value. Therefore, its Fourier series has two main components, one constant and one sinusoidal at frequency $f_1 = \dfrac{1}{T}$. Correspondingly, its spectrum has a peak at zero frequency (mean value) and a second peak at f_1.

Figure 14.17 shows the simulation results corresponding to the (stable) 2-cycle L_2, obtained by setting $g = 2.2$. The asymptotic trajectory in the state space is shown in panel a (together with the Poincaré section $\dot{x}_2 = 0$), whereas panel b displays the time evolution of $x_2(\tau)$, pointing out the presence of a second oscillation (there are two distinct maxima/minima within one period) and panel c the single-sided amplitude

Fig. 14.15 Supercritical flip bifurcation for the Colpitts oscillator in the projections **a** (g, x_1, x_3) and **b** (g, x_2, x_3) of the control space. By increasing the control parameter g (with $k = 0.5$ and $Q = 1.6$), the stable cycle L_1 (thin light green lines) becomes unstable (thin red lines). The stable cycle L_2 (thick dark green lines) is characterized by the presence of two loops

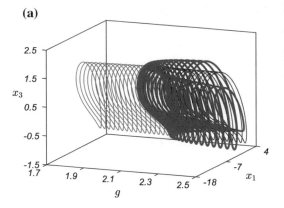

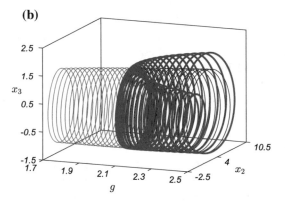

spectrum $|X_2(f)|$ of $x_2(\tau)$, obtained with a sampling frequency $f_s = 10\,\text{kHz}$. It is apparent that a subharmonic frequency component appears at the left of the main peak at nonzero frequency: this is due to the fact that now $x_2(\tau)$ has doubled its period and is nearly the sum of a constant term and two sinusoids, one at frequency f_2 close to f_1 (the main frequency can change by varying g) and one at frequency $\frac{f_2}{2}$.

On the Poincaré section there are two distinct intersections in a specified direction, corresponding to the cycle L_2 in the state space and to the two distinct maxima (or minima) of $x_2(\tau)$ over one period. The flip bifurcation corresponds to the collision of these two intersections, as evidenced in Fig. 14.18, which shows the x_2 coordinate of the asymptotic trajectory intersections (maxima of $x_2(\tau)$) for a grid of values of g, that is, a one-dimensional *bifurcation diagram* with respect to g. For $g < 1$ there is a stable equilibrium at the origin, at $g = 1$ this equilibrium undergoes a supercritical Hopf bifurcation, and for increasing values of g we have an infinite sequence of flip bifurcations at values $\{g_i\}$: $g_1 \approx 2.1$, $2.4 < g_2 < 2.5$, and so forth. These bifurcation

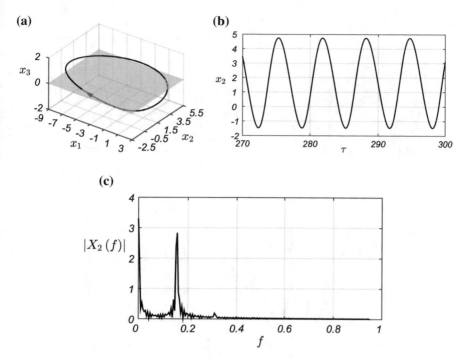

Fig. 14.16 **a** 1-cycle (obtained with $g = 1.8$) in the state space; **b** corresponding time evolution of $x_2(\tau)$; **c** single-sided amplitude spectrum of $x_2(\tau)$

values accumulate at a critical value p_∞, after which the attractor is no longer a limit cycle and becomes a *strange attractor*.

It is apparent that the Colpitts oscillator undergoes a *period-doubling cascade*, or Feigenbaum cascade, which is a sequence of flip bifurcations governed by a rate that tends to a constant limit, independent of the specific system under examination:

$$\lim_{i \to \infty} \frac{g_i - g_{i-1}}{g_{i+1} - g_i} = \delta,$$

where $\delta (= 4.6692...)$ is a universal constant called the *Feigenbaum constant*.

At each flip bifurcation, a stable cycle doubles its period and becomes unstable, actually a saddle cycle, since only one out of the three Floquet multipliers leaves the unit circle. In this way, saddle cycles of longer and longer periods accumulate in the state space. This constitutes one of the most common *routes to chaos*: a strange attractor is basically an aperiodic trajectory visiting a bounded region of the state space densely filled of saddle cycles, repelling along some directions (*stretching*) and attracting along others (*folding*).

The vertical dashed lines mark g values corresponding to a 1-cycle (Fig. 14.16), 2-cycle (Fig. 14.17), and 4-cycle (Fig. 14.19).

(a)   **(b)**

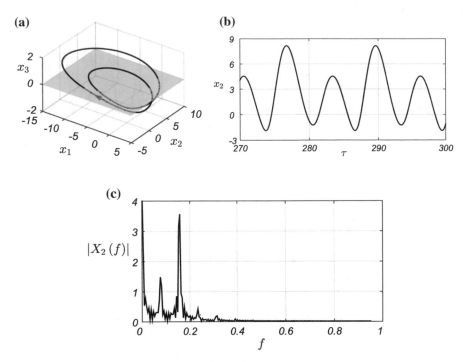

Fig. 14.17   a 2-cycle (obtained with $g = 2.2$) in the state space; **b** corresponding time evolution of $x_2(\tau)$; **c** single-sided amplitude spectrum of $x_2(\tau)$

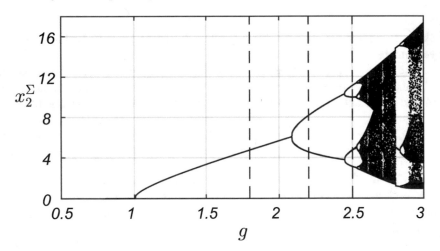

Fig. 14.18   One-dimensional bifurcation diagram with respect to g

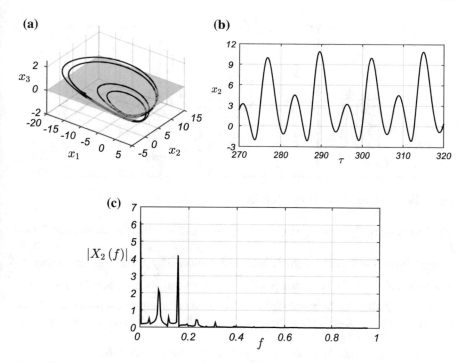

Fig. 14.19 **a** a 4-cycle (obtained with $g = 2.5$) in the state space; **b** corresponding time evolution of $x_2(\tau)$; **c** single-sided amplitude spectrum of $x_2(\tau)$

The spectrum of the 4-cycle contains other evident peaks, corresponding to sub-harmonic sinusoidal components, whereas the spectrum of the strange attractor tends to become continuous.

14.6.3 Neimark–Sacker Bifurcation

As shown in Fig. 14.12, in addition to the fold of cycles (real Floquet multiplier that crosses 1) and the flip (real Floquet multiplier that crosses -1), there is a third way for having Floquet multipliers that leave the unit circle; it is the so-called Neimark–Sacker bifurcation, whereby a pair of complex conjugate Floquet multipliers cross the unit circle in the complex plane, thus causing a local bifurcation.

This bifurcation corresponds to the collision of a limit cycle L (corresponding to a periodic solution) and a torus T (corresponding to a quasiperiodic solution) in the control space. When supercritical, the Neimark–Sacker bifurcation explains how a stable limit cycle can become unstable and generate a stable torus, by varying a parameter. On a Poincaré section Σ transverse to the flow, a stable equilibrium (intersection of L with Σ) bifurcates into an unstable equilibrium and a small regular

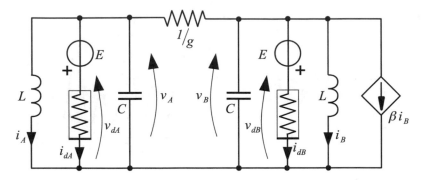

Fig. 14.20 Two coupled van der Pol oscillators

set of points densely covering (for $t \to \infty$) a closed curve (the intersection of $\mathcal{T}$ with Σ).

We remark that the existence of a torus is determined by the position of the Floquet multipliers on the unit circle, with phase $\pm\theta$: if the ratio $\dfrac{\theta}{2\pi}$ is a rational number p/q, the trajectory after the Neimark–Sacker bifurcation converges to a periodic solution corresponding to a *knot limit cycle* (see Fig. 14.10b, c); otherwise, it converges to a torus (see Fig. 14.10a).

As a consequence, in a two-parameter space, the Neimark–Sacker bifurcation curve separates a region $\mathcal{R}_1$ in which the system has periodic regimes (corresponding to stable limit cycles) from another region $\mathcal{R}_2$ in which the asymptotic regime is either quasiperiodic (corresponding to a stable torus) or periodic (corresponding to a stable knot limit cycle). In $\mathcal{R}_2$, the subregions where the attractor is a knot cycle are delimited by two fold bifurcations of cycles merging on the Neimark–Sacker curve and are called *Arnold tongues*. The Arnold tongues are infinite in number; generically, there is a tongue for each possible (p, q) pair, where p is the number of complete oscillations over one period of the slowest state component and q is the number of complete oscillations over one period of the fastest state component. However, only a few of them can be easily detected numerically or experimentally, since the others are very thin.

The Arnold tongues provide a key to explaining the phenomenon known as *phase locking*, which we illustrate using as a case study two coupled van der Pol oscillators. In periodically forced systems this phenomenon is known as *frequency locking*.

Figure 14.20 shows the circuit, where two Van der Pol oscillators are coupled through a resistor.

Following the guidelines provided in Sect. 12.3, you can check that the normalized state equations are

$$
\begin{cases}
\dot{x}_1 = \varepsilon \left(1 - x_1^2\right) x_1 - x_2 + \sigma(x_3 - x_1), \\
\dot{x}_2 = x_1, \\
\dot{x}_3 = \varepsilon \left(1 - x_3^2\right) x_3 - (1 + \beta)x_4 + \sigma(x_1 - x_3), \\
\dot{x}_4 = x_3,
\end{cases}
\tag{14.19}
$$

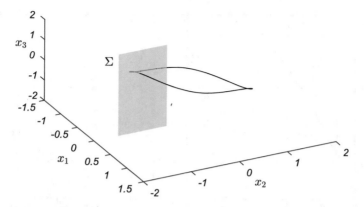

Fig. 14.21 Limit cycle in the space (x_1, x_2, x_3) asymptotically reached by a trajectory (black line) in periodic regime by setting $\sigma = 0.4$ and $\beta = 0.5$ in Eq. 14.19. The Poincaré section Σ (gray planar surface) corresponding to $\dot{x}_2 = 0$ is projected in the same space

Fig. 14.22 Time plots of $x_1(\tau)$ (black line) and $x_3(\tau)$ (gray line) over an interval of 50 time units at regime

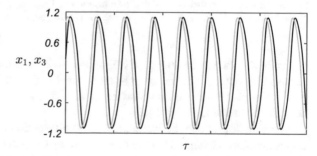

where $\sigma = g\dfrac{V_0}{I_0} > 0$ measures the coupling strength and β determines a mismatch between the two oscillators; for $\beta = 0$ they are identical, whereas for $\beta \neq 0$ they are different. We fix $\varepsilon = 1$, and we analyze the behavior of this system for different values of σ and $\beta > 0$.

For a fixed value of the mismatching parameter β and for large enough values of the coupling strength σ, the two oscillators are strongly coupled and oscillate synchronously, with the same amplitude and period but with a phase lag, thus describing a limit cycle in the 4-dimensional state space. An example (obtained for $\sigma = 0.4$ and $\beta = 0.5$) is shown in Fig. 14.21, together with the projection of the Poincaré section Σ (gray planar surface) corresponding to $\dot{x}_2 = 0$ (hyperplane $x_1 = 0$), which is intersected once per period. Henceforth, we refer to Σ as the Poincaré section, meaning that it is a planar projection of the hyperplane $x_1 = 0$.

Figure 14.22 shows $x_1(\tau)$ (black line) and $x_3(\tau)$ (gray line) over an interval of 50 time units. It is apparent that $x_1(\tau)$ and $x_3(\tau)$ are synchronized.

On the one hand, by decreasing σ, the two oscillators become weakly coupled. On the other hand, by increasing β, the mismatch between the two coupled oscillators increases. Due to both effects, the circuit undergoes a Neimark–Sacker bifurcation.

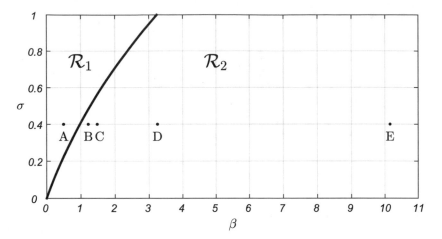

Fig. 14.23 Locus of Neimark–Sacker bifurcation points (black line) in the (β, σ)-plane. The points labeled with a letter correspond to synchronization (A, 1-cycle), quasiperiodic regime (B, torus), phase-locking (C, D, E, knotted cycles within Arnold tongues)

Figure 14.23 shows the locus of Neimark–Sacker bifurcation points (black line), numerically detected in the (β, σ)-plane. The black dots along the horizontal line $\sigma = 0.4$ mark β values corresponding to other figures of this section. For instance, point A corresponds to Fig. 14.21.

In region $\mathcal{R}_1$, above the Neimark–Sacker bifurcation curve, we find stable limit cycles qualitatively similar to the one shown in Fig. 14.21. In region $\mathcal{R}_2$, below the Neimark–Sacker bifurcation curve and above the horizontal line $\sigma = 0$, the two oscillators either synchronize (phase-lock), meaning that the whole circuit reaches a periodic steady state, or oscillate independently with incommensurate periods, thus leading the circuit to a quasiperiodic steady state. The phase-locking regions in the (β, σ)-plane are infinitely many Arnold tongues (not shown in Fig. 14.23), most of which are very thin and difficult to detect, corresponding to all possible combinations of p and q. As stated above, the boundaries of these tongues correspond to fold bifurcations of limit cycles.

We illustrate this scenario by keeping $\sigma = 0.4$ and changing β, thus moving along the horizontal line $\sigma = 0.4$ in the parameter plane of Fig. 14.23.

For $\sigma = 0.4$ and $\beta = 1.22$ (point B in Fig. 14.23), we overcome the Neimark–Sacker bifurcation curve and set the system in a parameter region where the regime is quasiperiodic. Figure 14.24 shows the corresponding torus and Poincaré section Σ in the projection (x_1, x_2, x_3) of the state space. The intersections of the system trajectory with Σ are displayed in Fig. 14.25; they densely cover a closed curve, as expected.

Finally, Fig. 14.26 shows $x_1(\tau)$ (black line) and $x_3(\tau)$ (gray line) over an interval of 100 time units. It is apparent that $x_1(\tau)$ and $x_3(\tau)$ are not synchronized.

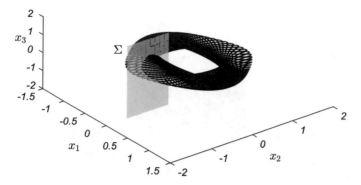

Fig. 14.24 Torus in the space (x_1, x_2, x_3) asymptotically covered by a trajectory (black line) in quasiperiodic regime by setting $\sigma = 0.4$ and $\beta = 1.22$ in Eq. 14.19; see point B in Fig. 14.23. The Poincaré section is denoted by Σ (gray planar surface)

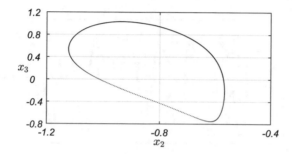

Fig. 14.25 Intersections of the asymptotic trajectory with Σ

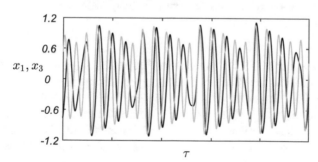

Fig. 14.26 Time plots of $x_1(\tau)$ (black line) and $x_3(\tau)$ (gray line) over an interval of 100 time units at regime

By increasing β on the horizontal line $\sigma = 0.4$ below the Neimark–Sacker bifurcation curve, the prevalent regime is quasiperiodic (corresponding to a torus in the state space), with the exception of intervals (usually very thin) corresponding to cuts of the Arnold tongues. In these intervals, the regime is periodic and corresponds to knot limit cycles with different combinations of p and q.

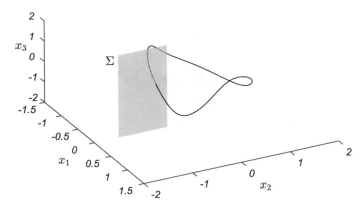

Fig. 14.27 $\sigma = 0.4$ and $\beta = 3.25$; stable knot cycle (black line) and Poincaré section Σ (gray planar surface) projected in the space (x_1, x_2, x_3)

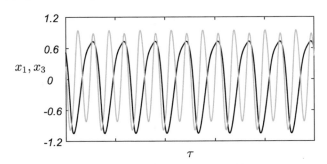

Fig. 14.28 Time plots of $x_1(\tau)$ (black line) and $x_3(\tau)$ (gray line) over an interval of 50 time units at regime

For instance, for $\beta = 3.25$ (point D in Fig. 14.23), we are in an Arnold tongue with $p/q = 1/2$, and the circuit evolves in a periodic regime also called *resonance* 1 : 2. Figure 14.27 shows the corresponding limit cycle and Poincaré section Σ in the projection (x_1, x_2, x_3) of the state space. The intersection of the system trajectory with Σ is equal to p (number of turns around the center), that is, just one point. The knot cycle makes $q = 2$ oscillations around the unstable cycle, not shown.

Figure 14.28 shows $x_1(\tau)$ (black line) and $x_3(\tau)$ (gray line) over an interval of 50 time units. It is apparent that $x_1(\tau)$ and $x_3(\tau)$ are distinct and with differing periods, but over one period x_1 makes one complete oscillation, whereas x_3 makes two oscillations.

For $\sigma = 0.4$ and $\beta = 1.4848$ (point C in Fig. 14.23), we are in an Arnold tongue with $p/q = 5/7$, also called resonance 5:7. Figure 14.29 shows the corresponding knot limit cycle and Poincaré section Σ in the projection (x_1, x_2, x_4) of the state space. As expected, there are five intersections of the system trajectory with Σ, magnified in Fig. 14.30.

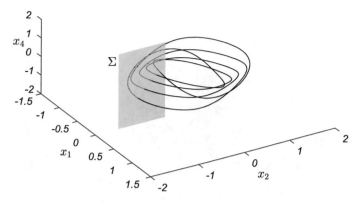

Fig. 14.29 $\sigma = 0.4$ and $\beta = 1.4848$; stable knot cycle (black line) and Poincaré section Σ (gray planar surface) projected in the space (x_1, x_2, x_4)

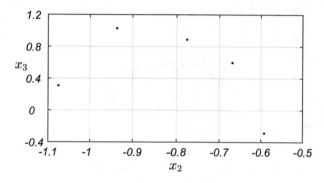

Fig. 14.30 Intersections of the stable knot limit cycle (with $p/q = 5/5$) with the plane (x_2, x_3), projection of the Poincaré section Σ

Fig. 14.31 Time plots of $x_1(\tau)$ (black line) and $x_3(\tau)$ (gray line) over an interval of 60 time units at regime

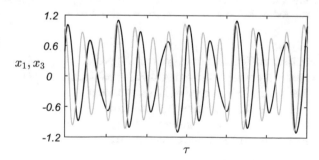

Figure 14.31 shows $x_1(\tau)$ (black line) and $x_3(\tau)$ (gray line) over an interval of 60 time units. It is apparent that $x_1(\tau)$ and $x_3(\tau)$ have the same period, but over one period $x_1(\tau)$ makes five complete oscillations, whereas $x_3(\tau)$ makes seven oscillations.

For $\sigma = 0.4$ and $\beta = 10.158$ (point E in Fig. 14.23), we are in an Arnold tongue with $p/q = 3/10$, also called resonance 3:10. Figure 14.32a [Fig. 14.32b] shows the

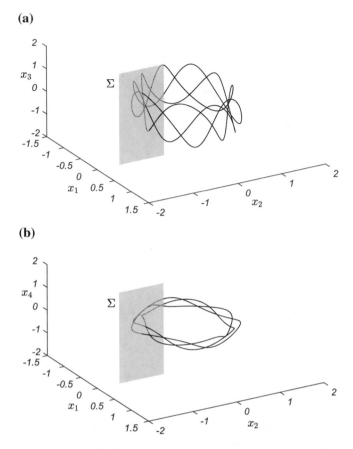

Fig. 14.32 $\sigma = 0.4$ and $\beta = 10.158$: stable knot cycle (black line) and Poincaré section Σ (gray surface) projected in the space **a** (x_1, x_2, x_3) and **b** (x_1, x_2, x_4)

corresponding knot limit cycle in the projection (x_1, x_2, x_3) $[(x_1, x_2, x_4)]$ of the state space, together with the usual Poincaré section Σ. There are three intersections of the system trajectory with Σ, as expected.

Figure 14.33 shows $x_1(\tau)$ (black line) and $x_3(\tau)$ (gray line) over an interval of 50 time units. It is apparent that $x_1(\tau)$ and $x_3(\tau)$ are different, but over one period, x_1 makes three complete oscillations, whereas x_3 makes ten oscillations.

14.6.4 Homoclinic Bifurcations

The *homoclinic bifurcation* in its most common form (also called homoclinic bifurcation to standard saddle) corresponds to the collision between the stable and unstable manifolds of the same saddle point in the control space. This corresponds also to

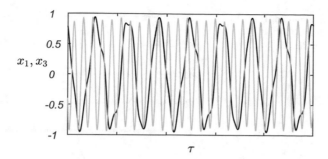

Fig. 14.33 Time plots of $x_1(\tau)$ (black line) and $x_3(\tau)$ (gray line) over an interval of 50 time units at regime

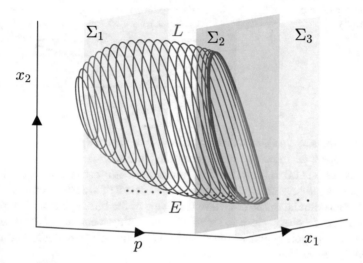

Fig. 14.34 Homoclinic bifurcation to standard saddle for a second-order circuit/system in the control space (p, x_1, x_2): by increasing the control parameter p, the (stable) limit cycle L (thin light green lines) approaches the saddle point E (red), which in this example remains fixed for all p. The sections are taken far from the bifurcation (Σ_1), at the bifurcation (Σ_2), and after the bifurcation (Σ_3)

the collision of a limit cycle L corresponding to a periodic solution (either stable or unstable) and a saddle E corresponding to an unstable equilibrium point. It is a *global* bifurcation, and as such, it cannot be characterized in terms of eigenvalue degeneracy.

As qualitatively shown in Fig. 14.34 in the case of a stable limit cycle, when the parameter p approaches the homoclinic bifurcation value, L gets closer to E, so that the period $T(p)$ of the limit cycle becomes larger: the closer the state to the equilibrium point, the slower the circuit/system dynamics. Therefore, at the bifurcation, the limit cycle degenerates to a homoclinic trajectory, that is, a trajectory starting from the unstable manifold of the saddle and asymptotically tending to the stable

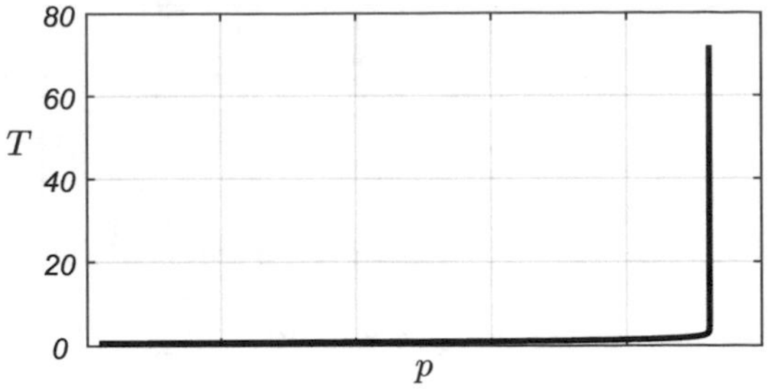

Fig. 14.35 The period of L
increases monotonically as
long as L approaches E

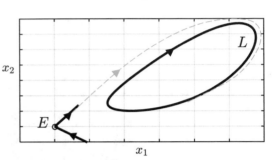

Fig. 14.36 State portrait on
section Σ_1. Solid thick line:
limit cycle L. Circle: saddle
point E. Dashed thin gray
line: trajectory

manifold of the same saddle. Correspondingly, we have $T(p) \to \infty$, meaning that
there is no longer periodicity. This is illustrated in Fig. 14.35.

This property is often used to detect homoclinic bifurcations through numerical
simulations. Another geometric property used to detect this kind of homoclinic bifurcation through simulations is related to the shape of the limit cycle: as L approaches
E, it becomes "pinched" close to the saddle, assuming a characteristic *drop shape*.
The angle of the pointy end of the drop is the angle between stable and unstable
manifolds of the saddle.

Figure 14.36 shows the state portrait on the section Σ_1: the limit cycle L is relatively far from the saddle E. A trajectory (dashed gray line) starting in the direction
tangent to the unstable manifold of E (i.e., along the eigenvector corresponding to
the positive eigenvalue of the system linearized about E) converges to L.

Figure 14.37 shows one period of $x_1(\tau)$ at steady state for a parameter value very
close to the homoclinic bifurcation. It is apparent that the state variable remains
nearly constant for a large percentage of the period: this corresponds to the fact that
L is very close to E, and when the trajectory approaches the saddle, it slows down.
The "pulse" of $x_1(\tau)$ corresponds to the part of L farther from E. This is another
fingerprint of an approaching homoclinic bifurcation.

Figure 14.38 shows the state portrait on the section Σ_2: the limit cycle L collides
with the saddle E and degenerates to a homoclinic trajectory (dashed black line)
starting on the unstable manifold of E and asymptotically tending to the stable
manifold of the same saddle. The collision between the stable and unstable manifolds
of the saddle point E is apparent. The figure shows the directions tangent to these
manifolds, that is, the eigenvectors of the system linearized about E.

Fig. 14.37 Time plot of $x_1(\tau)$ over a period at steady state very close to the homoclinic bifurcation

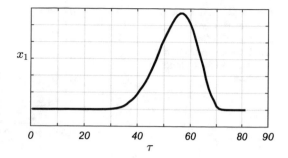

Fig. 14.38 State portrait on section Σ_2. Dashed thick line: limit cycle L degenerating to a homoclinic trajectory. Circle: saddle point E

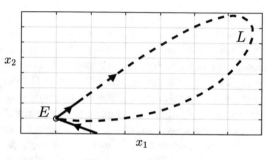

Fig. 14.39 State portrait on section Σ_3. Circle: saddle point E. Dashed thin gray line: trajectory

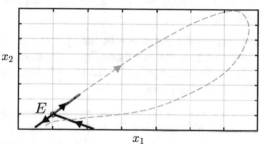

Finally, Fig. 14.39 shows the state portrait on the section Σ_3: the limit cycle L no longer exists. A trajectory (dashed gray line) starting in the direction tangent to the unstable manifold of E now turns around E and leaves the unstable manifold of E at its right.

By looking at the above state portraits in reverse order, we see that the homoclinic bifurcation is a further mechanism that explains the birth of a limit cycle. In the case of Hopf bifurcations, the emerging limit cycle is degenerate in its amplitude, which is null; in the case of homoclinic bifurcations, the cycle degeneracy is in its period, which is infinitely long.

The emerging limit cycle is stable in the above example, but by reversing the arrows of all trajectories, the same figures could be used to illustrate the case of an unstable emerging cycle. Therefore, homoclinic bifurcations in second-order systems are generically associated with a cycle emerging from the homoclinic trajectory existing at $p = p^*$ by suitably perturbing the parameter.

Fig. 14.40 Period T as a "heavily damped" wiggly function of Q (in semilogarithmic scale) for $g \approx 11.54$ in the Colpitts oscillator

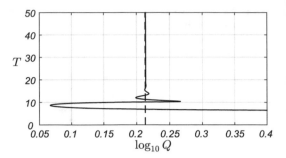

The stability of the emerging cycle can be easily predicted by looking at the sign of the so-called *saddle quantity* σ, which is the sum of the two eigenvalues of the Jacobian matrix of the system computed at the saddle, i.e., the trace of this matrix: if $\sigma < 0$, the cycle is stable, whereas if $\sigma > 0$, the cycle is unstable [2]. A remarkable extension of this theory to three-dimensional systems is due to Shil'nikov [2]. In particular, in 3D systems, in addition to the "simple" bifurcation mechanism illustrated for 2D systems, we can have a more complex form of homoclinic bifurcation, which involves not only one cycle, as in the above case, but infinitely many cycles. There are two main conditions for having this kind of homoclinic bifurcation in smooth systems:

- E is a saddle-focus, that is, the system linearized about E has one real (say, λ_1) and two complex conjugate ($\lambda_{2,3}$) eigenvalues;
- the saddle quantity, which for three-dimensional systems is defined as $\sigma = \lambda_1 + \Re\{\lambda_{2,3}\}$, is positive.

In this case, as the parameter p approaches the bifurcation value p^*, an infinite number of bifurcations (mainly, fold of limit cycles and flip) results. The dependence of the period of the involved cycles on p is no longer monotone as for the "simple" case (Fig. 14.35), but oscillating, wiggly. The cycles corresponding to different curls differ in the number of rotations near the saddle-focus E; the greater the period, the greater the number of rotations.

We illustrate this nonintuitive result through the Colpitts oscillator.

Figure 14.40 shows the oscillations of the period T with respect to the parameter Q, in semilogarithmic scale. This curve was obtained through MatCont [15], a MAT-LAB package for the bifurcation analysis of dynamical systems based on numerical continuation [2, 14, 16], which allows one to follow a given invariant set regardless of its stability, detecting its bifurcations. The numerical continuation was started from the stable cycle, say L_1, which was born through the Hopf bifurcation described in Sect. 12.7.2. By fixing Q and increasing g, the limit cycle soon becomes unstable through a flip bifurcation; see Sect. 14.6.2. By following the unstable cycle L_1 with increasing g, no further bifurcations are detected; L_1 is continued by increasing g up to ≈ 11.54; at this point, g remains fixed and L_1 is continued by decreasing Q, and its period slightly increases (lowest part of the curve shown in Fig. 14.40) until a

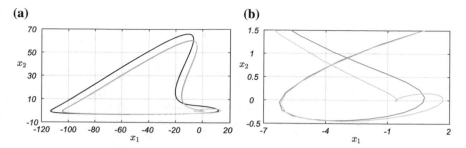

Fig. 14.41 **a** Projection of the limit cycles L_2 (black line), L_4 (gray line), and L_6 (light gray line) in the (x_1, x_2)-plane; **b** enlargement of panel a in the region where L_4 and L_6 increase the number of rotations around the saddle-focus, not shown

fold bifurcation of limit cycles occurs, which involves L_1 and another cycle L_2; this bifurcation corresponds to the leftmost oscillation in Fig. 14.40.

By following the limit cycle L_2 with increasing Q, its period increases until a second fold bifurcation of limit cycles is detected, which involves L_2 and another limit cycle L_3, corresponding to the rightmost oscillation in Fig. 14.40. This process repeats, thus originating the curls shown in Fig. 14.40. The curve $T(Q)$ oscillates around a value of $\log_{10} Q \approx 0.2137$, thus leading to a monotonic period increase.

The dashed vertical line is taken at $\log_{10} Q = 0.213$ and crosses six cycles L_i $(i = 1, \ldots, 6)$ with increasing period.

Figure 14.41a shows the projections of the limit cycles L_2 (black line), L_4 (gray line), and L_6 (light gray line) in the (x_1, x_2)-plane; the enlargement in Fig. 14.41b shows the increasing number of rotations around the saddle-focus (not shown in the figure) for the cycles L_4 and L_6.

Figure 14.42 shows one period of $x_1(\tau)$ for the same three cycles.

It is apparent that both period and number of oscillations increase from L_2 (panel a) to L_4 (panel b) and L_6 (panel c).

Finally, Fig. 14.43 shows the limit cycle L corresponding to $\log_{10} Q \approx 0.2137$ and to $T \approx 50$, quite close to a homoclinic trajectory.

This is even more evident in Fig. 14.44, showing the time plot of $x_1(\tau)$ over a period. It is apparent that x_1 displays the fingerprint of an approaching homoclinic bifurcation, since it remains nearly constant for a large percentage of the period, due to the fact that L is very close to the saddle-focus E. The "pulse" of $x_1(\tau)$ corresponds to the part of L farther from E and in this case is preceded and followed by oscillating tails.

Also in this case, the bifurcation corresponds to the collision between the stable and unstable manifolds of the saddle E.

We remark that the homoclinic bifurcation curve in the parameter plane (Q, g) organizes three families of subsidiary bifurcation curves: one of fold of cycles bifurcation curves (corresponding to the extrema of $T(Q)$), one of flip bifurcation curves, and one of subsidiary homoclinic bifurcations. The interested reader is referred to [17] for details.

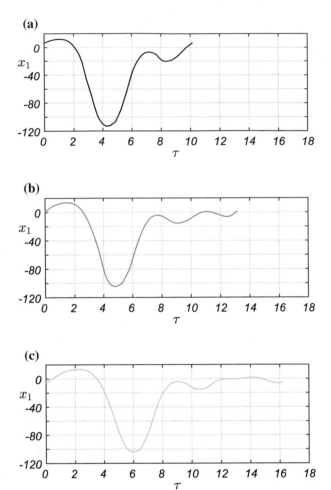

Fig. 14.42 Time plot of $x_1(\tau)$ over a period for **a** L_2, **b** L_4, and **c** L_6 at $\log_{10} Q = 0.213$

Similarly to the mechanism (period-doubling cascade) described in Sect. 14.6.2, also in this case we have a cascade of fold of cycles and flip bifurcations that accumulate around the homoclinic bifurcation and generate saddle cycles of longer and longer periods in the state space. This is another common *route to chaos*, leading to the generation of a stable chaotic attractor.

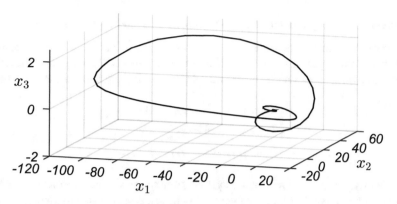

Fig. 14.43 Limit cycle L corresponding to $\log_{10} Q \approx 0.21376$ and to $T \approx 50$

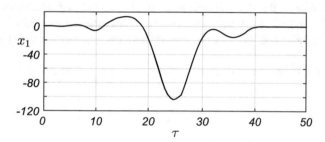

Fig. 14.44 Time plot of $x_1(\tau)$ over a period $T \approx 50$ for L at $\log_{10} Q \approx 0.2137$

14.7 Hindmarsh–Rose Neural Oscillator

In this section we analyze a biological oscillator, that is, the phenomenological neuron model proposed by Hindmarsh and Rose [18], which may be seen as a simplification of the physiologically realistic model proposed by Hodgkin and Huxley [19]. It has proven to be a good compromise between two seemingly mutually exclusive requirements: the model for a single neuron must be both computationally simple, and capable of mimicking almost all the behaviors exhibited by real biological neurons (in particular the rich firing patterns). Real neurons show a variety of dynamical behaviors, according to the values of bio-physical parameters. Among the most important ones, one may find:

- *quiescence*: the input to the neuron is below a certain threshold and the output reaches a stationary steady state;
- *spiking*: the output is made up of a regular series of equally spaced spikes (periodic steady state);
- *bursting*: the output is made up of groups of two or more spikes (called bursts) separated by periods of inactivity;
- *irregular spiking*: the output is made up of an aperiodic series of spikes;

- *irregular bursting*: the output is made up of an aperiodic series of bursts.

The model is able to reproduce all these dynamical behaviors, has been analyzed in the past with respect to one or two bifurcation parameters [20–22], and is described by the following set of ODEs, where all variables and parameters are normalized:

$$\begin{cases} \dot{x}_1 = x_2 - x_1^3 + bx_1^2 + \mathfrak{I} - x_3, \\ \dot{x}_2 = 1 - 5x_1^2 - x_2, \\ \dot{x}_3 = \mu\left(s\left(x_1 - x_r\right) - x_3\right). \end{cases} \tag{14.20}$$

Roughly, the roles played by the system parameters are the following: $\mathfrak{I}$ mimics the membrane input current for biological neurons; b allows one to switch between bursting and spiking behaviors and to control the spiking frequency; μ controls the speed of variation of the slow variable x_3 in Eq. 14.20, and in the presence of spiking behaviors, it governs the spiking frequency, whereas in the case of bursting, it affects the number of spikes per burst; s governs adaptation; x_r sets the resting potential of the system.

Henceforth, we set $\mu = 0.01$, $s = 4$, and $x_r = -1.6$ and use b and $\mathfrak{I}$ as bifurcation parameters.

14.7.1 Analysis of Equilibrium Points

First of all, we analyze the system equilibrium points by setting $b = 3.9$ e $\mathfrak{I} = 5.4$. The equilibrium condition $f(x) = 0$ gives

$$\begin{cases} x_2 - x_1^3 + bx_1^2 + \mathfrak{I} - x_3 = 0, \\ x_2 = 1 - 5x_1^2, \\ x_3 = s\left(x_1 - x_r\right), \end{cases} \tag{14.21}$$

and consequently

$$x_1^3 + (5 - b)x_1^2 + sx_1 - (sx_r + \mathfrak{I} + 1) = 0. \tag{14.22}$$

Owing to the assumptions on the parameter values, the constant term is null, and we can easily obtain the unique real solution $x^* = (0, 1, 6.4)$. The Jacobian matrix of Eq. 14.20 is

$$J(x) = \begin{pmatrix} 2bx_1 - 3x_1^2 & 1 & -1 \\ -10x_1 & -1 & 0 \\ \mu s & 0 & -\mu \end{pmatrix}. \tag{14.23}$$

By computing it at x^*, we obtain

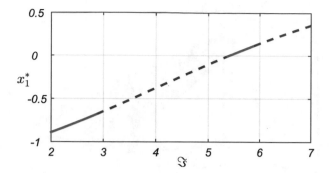

Fig. 14.45 Coordinate x_1 of the equilibrium x^* versus $\mathfrak{I}$. Solid green line denotes stability, dashed red line instability. The changes of stability are due to Hopf bifurcations, either supercritical (at $\mathfrak{I} \approx 5.38$) or subcritical (at $\mathfrak{I} \approx 2.88$ and $\mathfrak{I} \approx 5.89$)

$$J(x^*) = \begin{pmatrix} 0 & 1 & -1 \\ 0 & -1 & 0 \\ \mu s & 0 & -\mu \end{pmatrix}, \tag{14.24}$$

whose eigenvalues (one real and two complex conjugate) lie in the left complex half-plane. Therefore, x^* is a stable equilibrium point.

The analysis for different values of b and $\mathfrak{I}$ can be done numerically. For instance, keeping $b = 3.9$ fixed and changing $\mathfrak{I}$ from 2 to 7 using a numerical continuation tool such as MatCont [15] or AUTO [16], we obtain the bifurcation diagram shown in Fig. 14.45, which displays the coordinate x_1 of the equilibrium x^* versus $\mathfrak{I}$. For $\mathfrak{I} = 2$, x^* is stable (green line) and keeps its stability up to $\mathfrak{I} \approx 2.88$, where it becomes unstable (red line) through a subcritical Hopf bifurcation. Two further Hopf bifurcations, one supercritical (at $\mathfrak{I} \approx 5.38$) and one subcritical (at $\mathfrak{I} \approx 5.89$), change the stability of x^* in the considered interval.

14.7.2 Analysis of Limit Cycles

By resorting to the same continuation analysis tools, we can follow the limit cycles generated through the Hopf bifurcations. For instance, Fig. 14.46 shows the stable limit cycle L generated by the supercritical Hopf bifurcation at $\mathfrak{I} \approx 5.38$ (dark green curves). This cycle undergoes a supercritical flip bifurcation (which generates a stable 2-cycle, not shown in the figure) at $\mathfrak{I} \approx 5.24$, thus becoming unstable (red curves). It becomes stable again (light green curves) through another supercritical flip bifurcation at $\mathfrak{I} \approx 5.14$, until it collides with another unstable limit cycle (dark red curves) at $\mathfrak{I} \approx 5.13$ and disappears.

If we continue these bifurcation points numerically, we can obtain the loci of bifurcation points (that is, the bifurcation curves) in the parameter plane $(b, \mathfrak{I})$: the

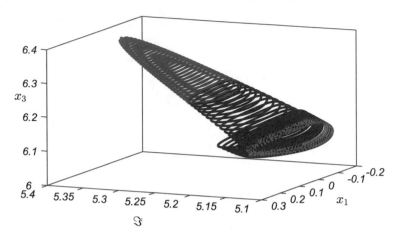

Fig. 14.46 Limit cycle generated through the supercritical Hopf bifurcation at $\Im \approx 5.38$. Green (red) curves denote stability (instability)

two supercritical flip bifurcations detected by following the limit cycle L belong to the same (cyan) bifurcation curve, labeled as f_1 in Fig. 14.47, and the fold of cycles bifurcation belongs to the blue bifurcation curve labeled as t_0.

Now we can follow the 1-cycle L, for instance starting from the curve t_0 at $b = 4$ $\Im \approx 5.1528$. If we decrease b, moving along the green dashed line in Fig. 14.47, we cross once more the flip bifurcation curve f_1. Next, many fold and flip bifurcations are detected, through the wiggly mechanism described in Sect. 14.6.4, which makes the period of the periodic solution increase its oscillating with respect to b, as shown in Fig. 14.48. In this figure, the black dots mark fold of cycles bifurcations, and the black crosses flip bifurcations. The corresponding values of b for the first bifurcations are marked in Fig. 14.47 by vertical black dashed lines (fold of cycles) and red solid lines (flip). By continuing the corresponding bifurcation points, we obtain fold bifurcation curves (blue lines) and flip bifurcation curves (cyan curves). For instance, the leftmost black dot in Fig. 14.48 corresponds to the fold curve t_1, whereas the rightmost one (at $b \approx 3.5$ and $T \approx 71$) corresponds to t_2, which is almost overlapping the flip bifurcation curve f_2. These curves accumulate around the homoclinic bifurcation curve h (black solid line in Fig. 14.47) and originate from an organizing point (red dot) located on h.

Figure 14.49 shows the three limit cycles with lowest periods along the curve T versus b, at $b = 3.47$. The cycle with lowest period is L (black line), and T increases, moving from the dark gray to the light gray limit cycle, due to the increasing number of rotations near the saddle-focus, not shown in the figure.

In order to get a more exhaustive idea of the possible periodic dynamics of our model for different values of b and $\Im$, we can exploit both Poincaré sections to analyze stable solutions and continuation methods to better understand the bifurcation scenario.

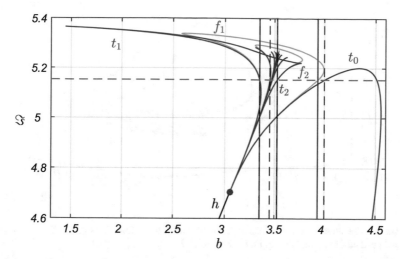

Fig. 14.47 Bifurcation curves in the parameter plane $(b, \mathfrak{I})$: fold of limit cycles (blue curves), flip (cyan curves), and homoclinic (black solid curve). Horizontal green dashed line: $\mathfrak{I} \approx 5.1528$. Vertical black dashed [red solid] lines: b values corresponding to fold of cycles [flip] bifurcations along the green dashed line, to be compared with Fig. 14.48

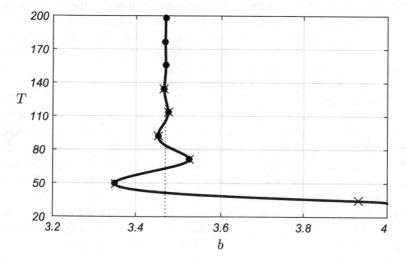

Fig. 14.48 Period T versus b for $\mathfrak{I} \approx 5.1528$ in the Hindmarsh–Rose neuron model

For instance, Fig. 14.50 shows a limit cycle corresponding to a *spiking* steady state. It is apparent that there is a unique intersection of the 1-cycle with the Poincaré section $\dot{x}_1 = 0$ (gray surface).

On the other hand, Fig. 14.51 shows a limit cycle corresponding to a *bursting* steady state, with six spikes per burst, as pointed out in panel b. It is apparent that there are six intersections of the 6-cycle with the Poincaré section $\dot{x}_1 = 0$ (gray surface).

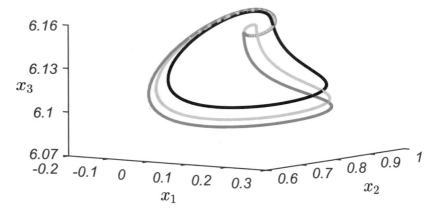

Fig. 14.49 The three lowest limit cycles corresponding to intersections of the curve T versus b with the vertical dotted line in Fig. 14.48, at $b = 3.47$

If we define a grid of values in the $(b, \Im)$-plane, we can classify the steady-state solutions based on the number of asymptotic intersections of a stable solution with the Poincaré section $\dot{x}_1 = 0$. By assigning a different color to each number, we obtain (see [20] for details) the "brute-force" two-dimensional bifurcation diagram shown in Fig. 14.52, where the following color code is used: cyan represents quiescence, green is for spiking, yellow is for bursting, and black is for chaos. Moreover, yellow changes to red as the number of spikes per burst increases, while green tends to become darker as the spiking frequency increases.

For further details about the bifurcation analysis of this oscillator, the reader is referred to [20–22].

As a general statement, valid for any oscillator, by combining "brute-force" and continuation analysis, we can determine the set of bifurcation values in the parameter space and become aware of the presence of coexisting stable states. The knowledge of the bifurcation scenario is of prime importance not only for analysis but also for synthesis. Indeed, in order to design an oscillator with the optimal operating condition, we have to select a parameter setting ensuring *structural stability*: this means that a slight variation in this parameter setting does not alter the qualitative steady-state behavior of the oscillator. In other words, this means that the parameter setting must be a point in the parameter space far enough from any bifurcation curve. Small uncertainty in the real values of the parameters or even in the model accuracy can be overcome in the presence of structural stability.

Another possible way to ensure a correct design is to obtain bifurcation diagrams directly from the physical system [23, 24].

14.7.3 Equivalent Circuit

We consider the circuit shown in Fig. 14.53, which consists of four subcircuits interacting through linear controlled sources. Each of the first three subcircuits contains

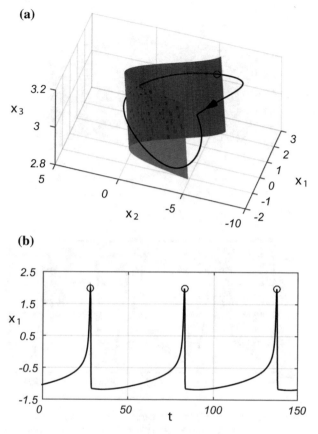

Fig. 14.50 Hindmarsh–Rose neuron model with parameters $b = 3.5$, $\mathfrak{J} = 2.5$ (spiking steady state): **a** asymptotic trajectory (black curve) and its intersections (black circles) with the Poincaré section $\dot{x}_1 = 0$ (gray surface) in the state space; **b** corresponding time evolution of $x_1(t)$

a capacitor. The last (memoryless) subcircuit includes two voltage-controlled nonlinear resistors, whose controlling voltage is v_1; their currents i_1 and i_2 and current i_r play the role of controlling currents in the other three (dynamic) subcircuits. The descriptive equations of the two nonlinear resistors are

$$
\begin{cases}
i_1 = \hat{i}_1\,(v_1) = I_0\left(\left(-\dfrac{v_1}{V_0}\right)^3 + b\left(\dfrac{v_1}{V_0}\right)^2 + \mathfrak{J}\right), \\[4mm]
i_2 = \hat{i}_2\,(v_1) = I_0\left(1 - 5\left(\dfrac{v_1}{V_0}\right)^2\right),
\end{cases}
\tag{14.25}
$$

(a)

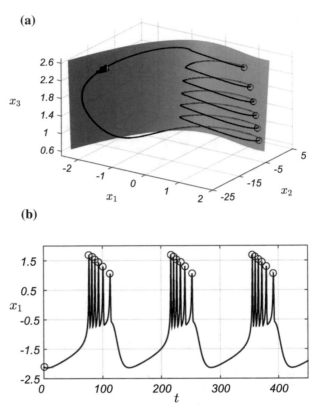

(b)

Fig. 14.51 Hindmarsh–Rose neuron model with parameters $b = 2.6$, $\mathfrak{I} = 2$ (bursting steady state with six spikes per burst): **a** asymptotic trajectory (black curve) and its intersections (black circles) with $\dot{x}_1 = 0$ (gray surface) in the state space; **b** corresponding time evolution of $x_1(t)$

where V_0 and I_0 are normalizing parameters for circuit voltages and currents, respectively. The state equations of the whole circuit are

$$\begin{cases} C\dfrac{dv_1}{dt} = \hat{i}_1(v_1) + \dfrac{v_2}{R_0} - \dfrac{v_3}{R_0}, \\[2mm] C\dfrac{dv_2}{dt} = -\dfrac{v_2}{R_0} + \hat{i}_2(v_1), \\[2mm] C\dfrac{dv_3}{dt} = -\dfrac{v_3}{R_q} + \dfrac{v_1 - V_r}{R}. \end{cases} \qquad (14.26)$$

We can now define a dimensionless time variable τ through the equation $t = R_0 C \tau$. By this definition we can write

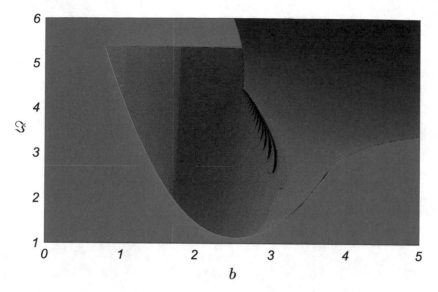

Fig. 14.52 Brute-force bifurcation diagram in the parameter plane $(b, \mathfrak{I})$, showing different kinds of asymptotic dynamics of the Hindmarsh–Rose model: quiescence (cyan), spiking (green), bursting (yellow), irregular (chaotic) spiking/bursting (black). Yellow changes to red as the number of spikes per burst increases. The darker the green, the higher the spiking frequency

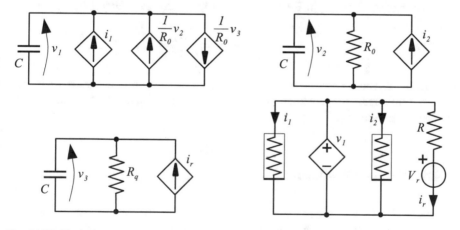

Fig. 14.53 Equivalent circuit of the Hindmarsh–Rose neural oscillator; see Eq. 14.20

$$C\frac{d}{dt} = \frac{1}{R_0}\frac{d}{d\tau}; \quad \frac{dv_k}{d\tau} = \dot{v}_k \quad (k = 1, 2, 3),$$

and the state equations can be recast in the equivalent form

$$\begin{cases} \dot{v}_1 = R_0 \hat{i}_1 (v_1) + v_2 - v_3, \\[2mm] \dot{v}_2 = -v_2 + R_0 \hat{i}_2 (v_1), \\[2mm] \dot{v}_3 = -\dfrac{R_0}{R_q} v_3 + \dfrac{R_0}{R} (v_1 - V_r). \end{cases} \tag{14.27}$$

Now we can choose $R_0 I_0 = V_0$ and define the normalized voltages

$$\frac{v_1}{V_0} = x_1; \quad \frac{v_2}{V_0} = x_2; \quad \frac{v_3}{V_0} = x_3; \quad \frac{V_r}{V_0} = x_r.$$

Therefore, the state equations become

$$\begin{cases} \dot{x}_1 = \left(-x_1^3 + bx_1^2 + \Im\right) + x_2 - x_3, \\[2mm] \dot{x}_2 = -x_2 + \left(1 - 5x_1^2\right), \\[2mm] \dot{x}_3 = -\dfrac{R_0}{R_q} x_3 + \dfrac{R_0}{R} (x_1 - x_r), \end{cases} \tag{14.28}$$

where the terms $\hat{i}_1$ and $\hat{i}_2$ of Eq. 14.25 have been written explicitly. Finally, taking

$$\frac{R_0}{R_q} = \mu; \quad \frac{R_q}{R} = s \quad \Rightarrow \quad \frac{R_0}{R} = \mu s,$$

we obtain the state equations in the form of Eq. 14.20, thus proving the equivalence.

For real circuit implementations of the Hindmarsh–Rose neural oscillator, the reader is referred to [24–26].

14.8 Forced Oscillators

In this section we consider, as an exception to the rest of this chapter, nonautonomous circuits driven by an external input. In particular, we focus as a case study on the van der Pol oscillator forced by a sinusoidal input. To this end, we add to the autonomous oscillator a sinusoidal current source $a(t) = A\cos(\omega t)$, as shown in Fig. 14.54.

Consequently, you can check (see Sect. 12.3) that Eq. 12.24 is modified as follows:

$$\begin{cases} \dot{x} = \varepsilon \left(1 - x^2\right) x - y + U \cos(\Omega \tau), \\[2mm] \dot{y} = x, \end{cases} \tag{14.29}$$

where $\Omega = \sqrt{LC}\omega$ and $U = \dfrac{A}{I_0}$.

Fig. 14.54 Van der Pol oscillator with sinusoidal forcing

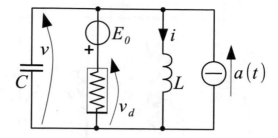

This second-order nonautonomous system can be equivalently written as a third-order autonomous system:

$$\begin{cases} \dot{x}_1 = \varepsilon \left(1 - x_1^2\right) x_1 - x_2 + U \cos(\Omega x_3), \\ \dot{x}_2 = x_1, \\ \dot{x}_3 = 1. \end{cases} \qquad (14.30)$$

The main drawback of this representation is that the trajectories evolve on a cylinder, whose axis is x_3, that is, apart from a constant shift, the time variable τ. Therefore, the trajectories cannot close, and we cannot have equilibrium solutions or limit cycles. This prevents the use of continuation methods, which are based on the presence of these invariant sets.

To overcome this problem, for carrying out continuation analysis we can recast Eq. 14.29 by adding a nonlinear oscillator with the desired periodic forcing as one of the solution components [16, 27]. Figure 14.55 shows the corresponding principle circuit scheme. In particular, for a sinusoidal forcing we can use the following auxiliary oscillator:

$$\begin{cases} \dot{x}_3 = \alpha x_3 + \Omega x_4 - x_3(x_3^2 + x_4^2), \\ \dot{x}_4 = -\Omega x_3 + \alpha x_4 - x_4(x_3^2 + x_4^2), \end{cases} \qquad (14.31)$$

which for $\alpha > 0$ has the asymptotically stable solution $x_3 = \sin(\Omega \tau), x_4 = \cos(\Omega \tau)$.
Therefore, Eq. 14.29 can be recast as

$$\begin{cases} \dot{x}_1 = \varepsilon \left(1 - x_1^2\right) x_1 - x_2 + U x_4, \\ \dot{x}_2 = x_1, \\ \dot{x}_3 = \alpha x_3 + \Omega x_4 - x_3(x_3^2 + x_4^2), \\ \dot{x}_4 = -\Omega x_3 + \alpha x_4 - x_4(x_3^2 + x_4^2). \end{cases} \qquad (14.32)$$

In so doing, we no longer have a state variable proportional to time, and we can find equilibrium points and limit cycles in the state space. Therefore, we can use this system to carry our continuation analysis of the original forced oscillator.

The steady-state circuit behavior depends on the values of the parameters ε, U, and Ω. In this section we fix $\varepsilon = 5$ and analyze the system bifurcations with respect

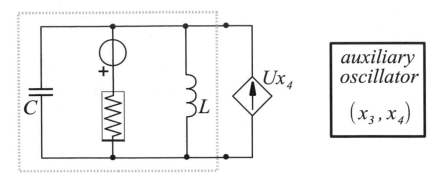

Fig. 14.55 Principle circuit scheme to reinterpret the forced oscillator as a combination of two autonomous oscillators

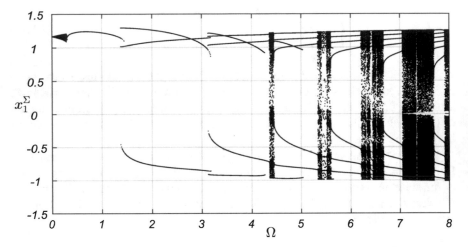

Fig. 14.56 One-dimensional bifurcation diagram for $U = 2$ showing the first coordinate x_1^Σ of the intersection of the asymptotic trajectory with the Poincaré section Σ ($\dot{x}_1 = 0$)

to the normalized amplitude U and angular frequency Ω of the input, which are assumed to be positive.

A first set of results is based on numerical integration of Eq. 14.29. For $U = 2$, by increasing Ω in the range $[0, 8]$ with small steps and taking as Poincaré section Σ the surface $\dot{x}_1 = 0$, we obtain the 1-dimensional bifurcation diagram shown in Fig. 14.56, where x_1^Σ is the first coordinate of the intersection of the asymptotic trajectory with Σ, in the direction corresponding to the maxima of x_1.

Neglecting the behaviors for too small Ω, stable periodic oscillations (*mode locking* or *frequency locking* or *entrainment*) are evident for $\Omega \in [0.5, 4]$ with odd periods $1, 3, 5$. Three examples are shown in Fig. 14.57: 1-cycle for $\Omega = 1$ (one maximum of x_1, black thick curve), 3-cycle for $\Omega = 2$ (two maxima of x_1, black thin curve), 5-cycle for $\Omega = 3.5$ (five maxima of x_1, gray curve).

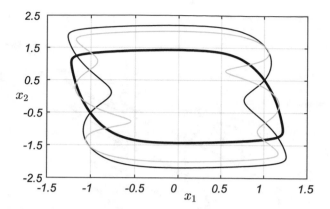

Fig. 14.57 Asymptotic trajectories in the (x_1, x_2)-plane for $U = 2$ and three values of Ω: $\Omega = 1$ (thick black curve), $\Omega = 2$ (thin black curve), $\Omega = 3.5$ (gray curve)

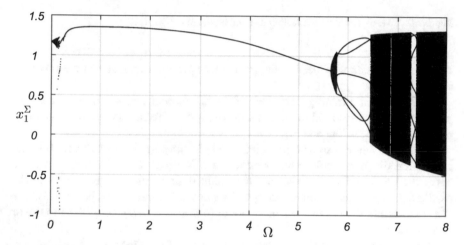

Fig. 14.58 One-dimensional bifurcation diagram for $U = 5$ showing the first coordinate x_1^Σ of the intersection of the asymptotic trajectory with the Poincaré section Σ ($\dot{x}_1 = 0$)

For higher values of Ω, limit cycles with higher odd periods alternate with densely filled regions.

For $U = 5$, by varying Ω in the same range as before and taking the same Poincaré section Σ, we obtain the 1-dimensional bifurcation diagram shown in Fig. 14.58.

This diagram points out the presence of a Neimark–Sacker bifurcation in the Ω range [5, 6], followed by an alternation of stable periodic solutions and densely filled regions. In particular, these densely filled regions at the transitions from 5-cycle to 7-cycle, from 7-cycle to 9-cycle, and so on, behave in the same way [28] and similarly to what happens for other forced oscillators, such as the nonautonomous Duffing equation [29, 30]. A complete study of the system dynamics and bifurcations within

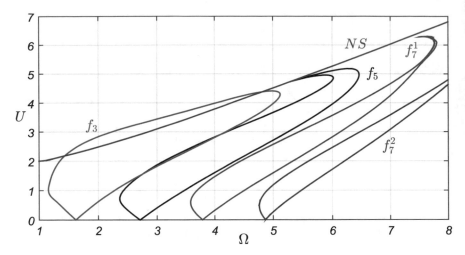

Fig. 14.59 Two-dimensional bifurcation diagram on the plane (Ω, U)

these regions is impossible, but the parameter space is organized in regions with similar behaviors.

In order to get a more general idea of the bifurcation scenario in the parameter plane (Ω, U), we resorted to continuation analysis. This second set of results was obtained using MatCont and Eq. 14.32.

Figure 14.59 shows some bifurcation curves in the parameter plane (Ω, U). The bifurcation diagram displays the presence of a Neimark–Sacker bifurcation (blue curve NS) that splits the plane into two main regions: above the blue curve the oscillator has periodic regimes corresponding to stable limit cycles; below the curve NS, the asymptotic regime can be periodic (*frequency locking*), quasiperiodic, or even chaotic.

The closed magenta curve f_3 bounds the region of existence of a stable 3-cycle, which, in the squeezed portion of plane below f_3 and above NS, coexists with the 1-cycle that undergoes the Neimark–Sacker bifurcation.

Frequency-locking phenomena occur that create and destroy stable and unstable limit cycles in the region below the blue curve. An infinite number of narrow Arnold tongues are rooted at the Neimark–Sacker bifurcation curve, which are delimited by closed fold bifurcation curves forming "islands" and corresponding to a collision between stable and saddle periodic orbits. Some examples of these fold curves are shown in the figure for a 5-cycle (black curve f_5) and for a 7-cycle (green curve f_7^1). You can check that the regions of existence of these stable limit cycles correspond to what is shown in the brute-force one-dimensional slices of Figs. 14.56 and 14.58. The saddle cycles involved in these bifurcations usually have tangencies of stable and unstable manifolds, thus giving rise to homoclinic bifurcations and to chaotic regimes. The complete picture between two islands includes also flip bifurcations; it

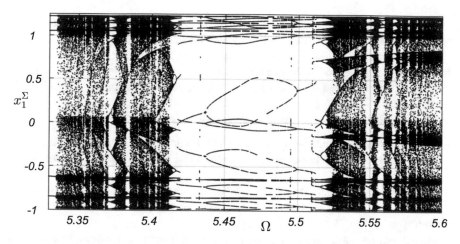

Fig. 14.60 Enlargement of the one-dimensional bifurcation diagram for $U = 2$ of Fig. 14.56 in the Ω-interval $[5.33, 5.6]$

is really complex, and the complete sequence of events is unknown [2, 30]: *hic sunt leones*!

Figure 14.60, for instance, shows an enlargement of Fig. 14.56 for $\Omega \in [5.33, 5.6]$, corresponding to the region between the islands created by the fold curves f_7^1 and f_7^2. The presence of period-doubling routes to chaos, for example in the intervals around $\Omega = 5.37$ and $\Omega = 5.52$, is apparent. Moreover, a relatively large frequency-locking interval is clearly visible in the central part of the diagram.

14.9 Networks of Coupled Oscillators

In the conclusion of this chapter and of this volume, we return to the conclusion of Vol. 1. When we deal with nonlinear dynamical circuits, the analysis usually becomes quite hard and almost always requires numerical computations. The examples of oscillators considered in this chapter are seemingly simple, but they can produce complex dynamics, including chaos. When we connect many of these oscillators in a network, the interactions can make collective behaviors emerge. Of course, the connections play a fundamental role as well as the oscillators.

Once again, graphs are a powerful tool to compactly represent complex networks: each oscillator of the network corresponds to a node of the graph—that is, it plays the role of a macro-component—whereas each connection (even nonlinear) corresponds to an edge.

When dealing with networks (of any nature), we have to deal with two basic issues: topology—how does one characterize the graph of a food web or the internet or the metabolic network of a bacterium?—and collective behavior emerging from a network of interacting dynamical systems, be they neurons, power stations, or lasers, given their individual dynamics and coupling architecture [31, 32].

These aspects are strongly related, because structure always affects function. For instance, the topology of social networks affects the spread of information and disease, and the topology of the power grid affects the robustness and stability of power transmission.

The overall picture can be made more complex, because:

- a wiring diagram could be an intricate tangle and could change over time; for instance, in the World-Wide Web, many pages and links are created and lost every minute;
- the links between nodes could have different weights, directions, and signs; for instance, synapses in a neuron network can be strong or weak, inhibitory or excitatory;
- there could be many different kinds of nodes, and each node could have its own level of complexity.

The node dynamics can lead the single (isolated) dynamical system's state to asymptotically reach an equilibrium point (stationary steady state), a limit cycle (periodic steady state), a torus (quasiperiodic steady state), or a chaotic attractor (aperiodic steady state). If we couple many such systems together, their collective behavior can provide unexpected emerging dynamics. In particular, the case of periodic oscillators is particularly common in nature, with examples ranging from rhythmically flashing fireflies [33] and chirping crickets [34], to oscillating living cells (cardiac, intestinal, neural, just to cite the best-known examples) [35]. Collective behavior in mechanical systems was discovered by Christiaan Huygens,[8] who, in the seventeenth century, observed the synchronization of two pendulum clocks hanging from (i.e., mechanically coupled through) a wooden beam. From that time, the phenomenon of synchronization has received increasing attention from the scientific community [36].

Arrays of identical oscillators coupled through linear resistors (*diffusive coupling*) often synchronize or generate patterns that depend on the symmetry properties of the underlying network [37], but under proper conditions they can produce traveling waves (in 1D arrays) or rotating spiral waves (in 2D arrays) [38, 39]. Other kinds of coupling (e.g., *pulse coupling*, as for fireflies, crickets, and neurons [40]), perhaps including delays, inhibition, and other realistic features, can be fruitfully used to reproduce and explain the emergence of other collective behaviors in networks of oscillators, identical or not.

For instance, networks of periodic oscillators have been used to model complex systems from crowd dynamics on a wobbly bridge [41, 42] to ecosystems [43], from neuron populations of various complexity [44, 45] to power grids [46].

---

[8]Christiaan Huygens (1629–1695) was a Dutch physicist, mathematician, astronomer, and inventor, who is widely regarded as one of the greatest scientists of all time. His most famous invention was the pendulum clock, in 1656.

There is a third key factor to be accounted for, since a collective behavior emerges from the interplay of dynamics of oscillators, coupling mechanism, and network topology.

The most common network topologies are regular (chains, grids, lattices, fully connected architectures) or random, as well as mixtures of regular structure and random ingredients: this is the case of *small-world networks* [47] and *scale-free networks* [48].

A rigorous analysis of this topic is beyond the scope of this book. The interested reader is referred to the references mentioned above. Here, we limit ourselves to considering a few examples of networks of increasing complexity.

14.9.1 Example 1: 3-Cell Network

Our first example is a little network of three van der Pol oscillators with nonlinear coupling, which models a *central pattern generator* (CPG), that is, a network of neurons of the peripheral nervous system able to generate rhythmic activity even in the absence of sensory input. The concerted action of their constitutive elements orchestrates the coordinated activity of muscles (activated in a precise and sequential manner), producing rhythmic movements. For instance, these networks are responsible for regular pacing of the heart muscles in crustaceans, for swimming in marine mollusks, for chewing and propulsion of food in crustaceans, for locomotion in legged insects. As we consider animals with increasing complexity, thus passing to vertebrates, the CPGs also become more complex and regulate a larger variety of activities, including breathing and, of course, locomotion. The reader is referred to [49, 50] for comprehensive surveys about this topic.

The example is taken (mutatis mutandis) from [51, 52], and the corresponding network is shown in Fig. 14.61a.

Each synapse is inhibitory and is represented as an arc with a dot-shaped head. For instance, the synapse labeled g_{31} represents the inhibitory action exerted by oscillator 1 on oscillator 3, with weight g_{31}.

The topology is regular (a ring), the van der Pol oscillators are identical, with $\varepsilon = 0.1$, and the nonlinear synaptic actions are modeled following the fast-threshold modulation (FTM) paradigm [53]. Therefore, the ith cell obeys the following equations ($i = 1, 2, 3$):

$$\begin{cases} \dot{x}_i = \varepsilon \left(1 - x_i^2\right) x_i - y_i + \sum_{j=1}^{3} g_{ij} h(x_i, x_j), \\ \dot{y}_i = x_i, \end{cases} \tag{14.33}$$

where

$$h(x_i, x_j) = \frac{\gamma - x_i}{1 + e^{\nu(x_j - \vartheta)}}. \tag{14.34}$$

(a)

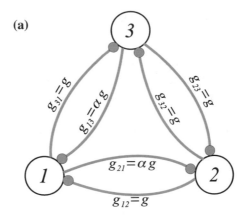

(b)

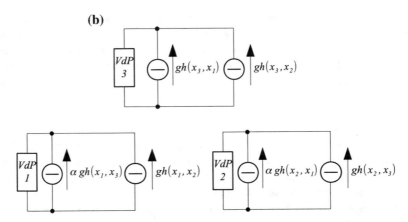

Fig. 14.61 Example 1: central pattern generator with three nodes. **a** Network topology: each node is a van der Pol oscillator and the connections are inhibitory synapses; each dot denotes the head of a synapse. **b** Circuit representation of the network

We set $g_{12} = g_{23} = g_{31} = g_{32} = g = 10^{-2}, g_{21} = g_{13} = \alpha g, g_{ii} = 0 \, (i = 1, 2, 3)$, $\gamma = -1.5, \nu = 100$, and $\vartheta = 0$. By increasing the parameter α between 1 and 2, we increase the inhibition exerted by oscillator 1 on 2 and by 3 on 1.

Figure 14.61b shows a corresponding circuit representation of this network.

The existence and stability of rhythmic patterns generated by CPGs are usually analyzed using so-called *phase lags*. All oscillators are assumed to have and maintain relatively close temporal characteristics; this means that the ith oscillator generates a stable limit cycle of period T_i in the state space. The current state position on the periodic orbit can be determined using a phase variable $\phi_i(t)$. This variable keeps values in the range [0, 1) and is associated with a double reset rule: ϕ_i is reset to 0 when it reaches 1 and is also reset to 0 at times $t_i^{(k)}$ $(k = 0, 1, \dots)$, when the variable

x_i overcomes a threshold x_{th}. The reference value $t_i^{(0)}$ is the time instant when the variable x_i overcomes the threshold x_{th} for the first time. The phase-lag representation of an N-cell network employs $N-1$ state variables (instead of the original nN in the case of oscillators with n state variables) describing the phase lags $\Delta_{1i}(t) = [\phi_i(t) - \phi_1(t)]$ mod 1 (where x mod $1 = x - \lfloor x \rfloor$) between the reference oscillator 1 and the other oscillators coupled within the network. In other words, working with phases is a way to reduce the problem's dimensionality, analogously to what is done with Poincaré sections.

The time evolutions of these state variables, being quite complex due to nonlinear interactions, can be determined through numerical simulations. For that purpose, we compute the phase lags between coupled cells in a discrete set of time instants as (for $k = 1, 2, \ldots$)

$$\Delta_{1i}^{(k)} = \frac{t_i^{(k)} - t_1^{(k)}}{t_1^{(k)} - t_1^{(k-1)}} \quad \text{mod } 1. \tag{14.35}$$

As time progresses, the phase lags $\Delta_{1i}^{(k)}$ can converge to one or several stable equilibrium points, corresponding to as many phase-locked states of the network. The presence of multistability can be established by integrating the system of equations governing the network, densely sweeping initial conditions for phases. In this way, in the considered example, the analysis of the original 6-dimensional system is carried out by studying a reduced 2-dimensional system with state variables Δ_{12} and Δ_{13}).

Figure 14.62 shows the results of these numerical simulations (performed using the software tool CEPAGE [52]) for $\alpha \in [1, 2]$, with $x_{th} = 0$. The colored curves in the plane $\alpha = 1$ are a state portrait revealing the existence of five stable equilibrium points. The basin of attraction of each equilibrium point in the plane $(\Delta_{12}, \Delta_{13})$ is shown with a different color code. By increasing α, these equilibrium points and the related basins of attraction change. The black lines mark the positions of these points. For larger values of α, four of these equilibrium points undergo fold bifurcations with as many unstable equilibria (not shown), and the CPG generates only one stable (red) pattern, which is apparent in the plane $\alpha = 2$.

Figure 14.63 shows the time evolution of five different initial conditions in the plane $(\Delta_{12}, \Delta_{13})$ for $\alpha = 1$, one per basin of attraction (the color code is the same as in Fig. 14.62). The five steady-state equilibrium points are reached after a transient. This figure reveals that two pairs of equilibrium points (blue–green and magenta–yellow) are symmetric: Δ_{12} and Δ_{13} simply swap their roles.

Figure 14.64 shows the steady-state time evolution of the first state variable of the three oscillators corresponding to the blue curves in Fig. 14.63: two oscillators, 1 and 3, are in phase (therefore $\Delta_{13} = 0$), whereas oscillator 2 has almost the opposite phase, corresponding to $\Delta_{12} \approx 0.513$. For the green basin of attraction, the roles played by Δ_{12} and Δ_{13} swap.

Figure 14.65 shows similar plots for the red basin of attraction: two oscillators, 2 and 3, are in phase, whereas oscillator 1 has almost the opposite phase, corresponding to $\Delta_{12} = \Delta_{13} \approx 0.486$.

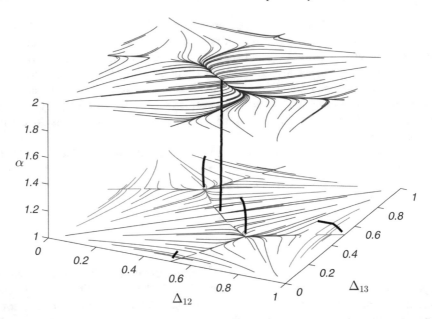

Fig. 14.62 Bifurcations in the control space $(\Delta_{12}, \Delta_{13}, \alpha)$. Colored curves: trajectories in the plane $(\Delta_{12}, \Delta_{13})$ for $\alpha = 1$ (lower plane) and $\alpha = 2$ (upper plane); in the lower plane, the different colors indicate the basins of attraction of five stable equilibrium points. Black curves: position of the stable equilibrium points

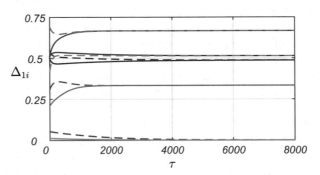

Fig. 14.63 Time plots of $\Delta_{1i}(\tau)$ $(i = 2, 3)$ for $\alpha = 1$ and for initial conditions corresponding to five different basins of attraction: $\Delta_{12}(\tau)$ (solid lines), $\Delta_{13}(\tau)$ (dashed lines). The color code is the same as in Fig. 14.62

Figure 14.66 shows similar plots for the magenta basin of attraction: the three oscillators provide the maxima of their oscillations in sequence, with $\Delta_{13} \approx 0.333$ and $\Delta_{12} \approx 0.666$. For the yellow basin of attraction, the roles played by Δ_{12} and Δ_{13} swap.

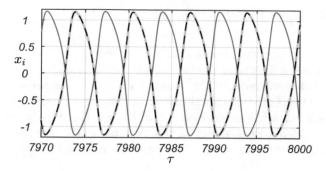

Fig. 14.64 Steady-state time plots of $x_i(\tau)$ ($i = 1, 2, 3$) for $\alpha = 1$ and initial conditions corresponding to the blue basin of attraction in Fig. 14.63: $x_1(\tau)$ (dashed black line), $x_2(\tau)$ (dark gray line), $x_3(\tau)$ (light gray thick line)

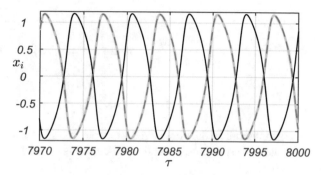

Fig. 14.65 Steady-state time plots of $x_i(\tau)$ ($i = 1, 2, 3$) for $\alpha = 1$ and initial conditions corresponding to the red basin of attraction in Fig. 14.63: $x_1(\tau)$ (black line), $x_2(\tau)$ (dashed dark gray line), $x_3(\tau)$ (light gray thick line)

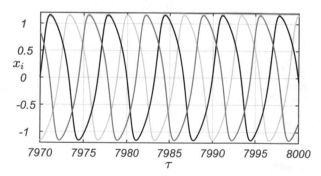

Fig. 14.66 Steady-state time plots of $x_i(\tau)$ ($i = 1, 2, 3$) for $\alpha = 1$ and initial conditions corresponding to the magenta basin of attraction in Fig. 14.63: $x_1(\tau)$ (black line), $x_2(\tau)$ (dark gray line), $x_3(\tau)$ (light gray line)

14.9.2 Example 2: 4-Cell Network

As a second example, we consider a network of four van der Pol oscillators that mimics the CPG of the marine mollusk *Melibe leonina*, also known as the hooded nudibranch, lion nudibranch, or lion's mane nudibranch. This network is a variant of the CPG studied in [54, 55], and its structure is shown in Fig. 14.67.

The van der Pol oscillators are identical, with $\varepsilon = 0.1$, and the chemical synaptic actions are modeled again following the fast-threshold modulation (FTM) paradigm. Therefore, the ith cell satisfies the following equations ($i = 1, \ldots, 4$):

$$
\begin{cases}
\dot{x}_i = \varepsilon \left(1 - x_i^2\right) x_i - y_i + \sum_{j=1}^{4} g_{ij}^{ex} h^{ex}(x_i, x_j) + \sum_{j=1}^{4} g_{ij}^{in} h^{in}(x_i, x_j) + \sum_{j=1}^{4} g_{ij}^{el} h^{el}(x_i, x_j), \\
\dot{y}_i = x_i,
\end{cases}
$$

$$(14.36)$$

where

$$
h^{xx}(x_i, x_j) = \frac{\gamma^{xx} - x_i}{1 + e^{\nu(x_j - \vartheta)}}, \quad \text{with } xx \in \{in, ex\}
$$

$$(14.37)$$

and

$$
h^{el}(x_i, x_j) = x_j - x_i.
$$

$$(14.38)$$

This is a general formulation; in the considered example, many coefficients g_{ij}^{xx} are zero, meaning that the corresponding synaptic connection is absent.

We set

$$
g^{in} = g
\begin{bmatrix}
0 & 1 & 0 & 0.25 \\
1 & 0 & 0.25 & 0 \\
0 & 0 & 0 & \alpha \\
0 & 0 & \alpha & 0
\end{bmatrix},
$$

$$(14.39)$$

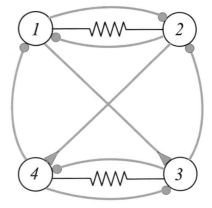

Fig. 14.67 Example 2: central pattern generator with four nodes. Network topology: each node is a van der Pol oscillator, and the connections are synapses of three kinds: chemical excitatory (represented by arcs ending with a triangle), chemical inhibitory (represented by arcs ending with a dot), and electrical (represented by resistors)

Fig. 14.68 Phase-lag evolution over the torus corresponding to the state subspace $(\Delta_{12}, \Delta_{13})$ for $\alpha = 1.68$. Phase lags converge to a stable equilibrium point (black dot), corresponding to a phase-locked state

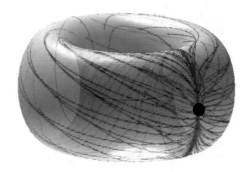

$$g^{ex} = \frac{g}{4} \begin{bmatrix} 0 & 0 & 0 & 0 \\ 0 & 0 & 0 & 0 \\ 1 & 0 & 0 & 0 \\ 0 & 1 & 0 & 0 \end{bmatrix},$$
(14.40)

$$g^{el} = \frac{g}{4} \begin{bmatrix} 0 & 1 & 0 & 0 \\ 1 & 0 & 0 & 0 \\ 0 & 0 & 0 & 2 \\ 0 & 0 & 2 & 0 \end{bmatrix},$$
(14.41)

and $g = 2.5 \cdot 10^{-3}$, $\gamma^{ex} = 1.5$, $\gamma^{in} = -1.5$, $\nu = 100$, and $\vartheta = 0$. By acting on the parameter α, we change the reciprocal inhibition exerted by oscillators 3 and 4.

Using once more the phase lags, as in Sect. 14.9.1, the analysis of the original 8-dimensional system is carried out by studying a reduced 3-dimensional system with state variables Δ_{12}, Δ_{13}, and Δ_{14}, all ranging in the interval $[0, 1)$, where 0 and 1 denote the same phase value. Owing to the circular symmetry of the phase-lag variables, we can represent a 2D projection of the 3D state space on a torus.

For a relatively low value of α ($\alpha = 1.68$), the network reaches a phase-locked state, corresponding to the black dot in Fig. 14.68, which shows the phase trajectories obtained by integrating Eq. 14.36, sweeping initial conditions for phases, using CEPAGE [52] and setting $x_{th} = 0$.

Figure 14.69 shows the time evolution of one initial condition for phases for $\alpha = 1.68$. After a transient, the phase lags $\Delta_{12}(\tau)$ (blue line), $\Delta_{13}(\tau)$ (red line), and $\Delta_{14}(\tau)$ (green line) reach a steady-state equilibrium point, which corresponds to the black dot in Fig. 14.68.

This figure points out that after a transient, nodes 1 and 3 become synchronized (blue and green curves converge to the same value 0.5), and also nodes 2 and 4 oscillate synchronously, but with opposite phase with respect to nodes 1 and 3. Indeed, the red curve $\Delta_{13}(\tau)$ converges to 0, meaning that nodes 1 and 3 oscillate synchronously with a half period of difference with respect to nodes 2 and 4. This is the mechanism that underlies alternating left–right swim motor pattern generation in *Melibe leonina* [55].

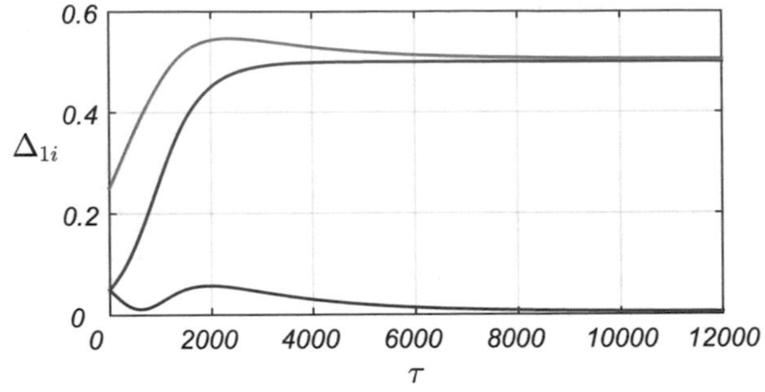

Fig. 14.69 Time plots of $\Delta_{1i}(\tau)$ ($i = 2, 3, 4$) for $\alpha = 1.68$ and for one of the initial conditions used in Fig. 14.68: $\Delta_{12}(\tau)$ (blue line), $\Delta_{13}(\tau)$ (red line), and $\Delta_{14}(\tau)$ (green line)

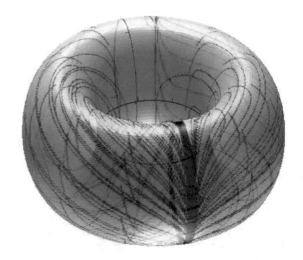

Fig. 14.70 Phase-lag evolution over the torus corresponding to the state subspace (Δ_{12}, Δ_{13}) for $\alpha = 10.68$. Phase lags converge to a limit cycle (thick black curve) corresponding to a phase-slipping state

By increasing α, the equilibrium in the phase-lag space undergoes a Hopf bifurcation, and instead of converging to an equilibrium point, the phases start slipping, thus covering a limit cycle on the torus and preventing the system from reaching a phase-locking condition. Figure 14.70 shows once more the phase trajectories obtained by integrating Eq. 14.36, sweeping initial conditions for phases, by using CEPAGE [52] and setting $x_{th} = 0$. In this case, the phase lags converge to a limit cycle (thick black curve) corresponding to a phase-slipping state of the network.

Figure 14.71 shows the time evolution of one initial condition for phases for $\alpha = 10.68$. After a transient, the phase lag $\Delta_{12}(\tau)$ (blue line) reaches a stationary steady-state, whereas $\Delta_{13}(\tau)$ (red line) and $\Delta_{14}(\tau)$ (green line) periodically oscillate. This corresponds to the thick black limit cycle in Fig. 14.70.

This figure points out that after a transient, nodes 1 and 2 become synchronized, whereas nodes 3 and 4 oscillate with no synchrony. This behavior does not correspond to any rhythmic movement.

Fig. 14.71 Time plots of $\Delta_{1i}(\tau)$ $(i = 2, 3, 4)$ for $\alpha = 10.68$ and one of the initial conditions used in Fig. 14.70: $\Delta_{12}(\tau)$ (blue line), $\Delta_{13}(\tau)$ (red line), and $\Delta_{14}(\tau)$ (green line)

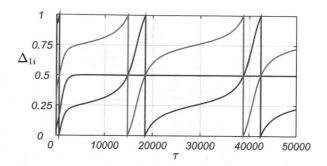

14.9.3 Example 3: 30-Cell Network

Our last example is concerned with a network of 30 van der Pol oscillators interconnected by inhibitory and electrical synapses. The topology is random (with probability of about 98% for both kinds of connection and no self-connections or autapses), the van der Pol oscillators can have a mismatch (modeled through parameter β_i, as in Sect. 14.6.3), with $\varepsilon = 1$, and the inhibitory actions are once more modeled following the fast-threshold modulation (FTM) paradigm [53]. Therefore, the ith cell satisfies the following equations $(i = 1, \ldots, 30)$:

$$
\begin{cases}
\dot{x}_i = \varepsilon \left(1 - x_i^2\right) x_i - (1 + \beta_i) y_i + \alpha^{in} \sum_{j=1}^{30} g_{ij}^{in} \frac{\gamma - x_i}{1 + e^{\nu(x_j - \vartheta)}} + \alpha^{el} \sum_{j=1}^{30} g_{ij}^{el} (x_j - x_i), \\
\dot{y}_i = x_i,
\end{cases}
$$

(14.42)

where $\gamma = -1.5$, $\nu = 100$, and $\vartheta = 0$. By increasing the positive parameters α^{in} and α^{el}, we increase the strength of the corresponding type of connection. All entries of the matrices g^{in} (nonsymmetric, in general) and g^{el} (symmetric) are set to 1, with the exception of the main diagonal (no autapses are allowed) and of inhibitory entries (1, 23), (2, 28), (3, 16), (5, 13), (6, 21), (10, 18), (11, 14), (16, 12), (20, 19), (21, 9), (21, 24), (22, 1), (23, 25), (25, 6), (26, 2), (26, 28) and electrical entries (1, 10), (1, 15), (1, 18), (1, 27), (2, 24), (3, 11), (4, 6), (5, 23), (6, 14), (7, 9), (8, 30), (9, 24), (11, 21), (14, 19), (14, 23), (16, 23), (20, 28), (21, 24) and symmetric elements, which are set to 0. Completely similar results would be obtained, mutatis mutandis, for a different random network with the same topological features.

The strongly connected structure of this network allows the onset of *partial* or *cluster synchronization*.

To start with, we set all $\beta_i = 0$ (all oscillators are identical).

- **Global synchronization**
  By setting $\alpha^{in} = 0$, we remove all inhibitory synaptic connections, and the cells are coupled only electrically. In this case, by setting $\alpha^{el} = 0.0015$, we obtain *global synchronization* of the network, that is, all cells oscillate synchronously.

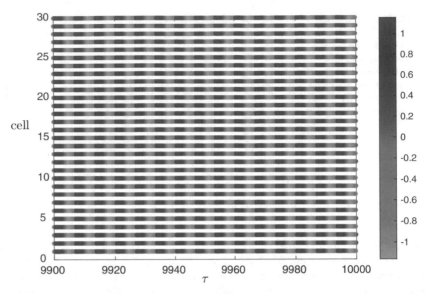

Fig. 14.72 Plots of $x_i(\tau)$ ($i = 1, \ldots, 30$) for $\alpha^{in} = 0$, $\alpha^{el} = 0.0015$, and for random initial conditions. The value of $x_i(\tau)$ is color-coded according to the color bar at the right

Figure 14.72 shows the steady-state plots of $x_i(\tau)$ for ($i = 1, \ldots, 30$) for random initial conditions. The value of $x_i(\tau)$ is color-coded according to the color bar at the right. It is apparent that all cells are synchronous.

This is confirmed by the result shown in Fig. 14.73, which shows the time plots $x_i(\tau)$ for $i = 1, \ldots, 30$, perfectly overlapping.

- **Cluster synchronization**

  Now we set $\alpha^{el} = 0.0015$ and $\alpha^{in} = 0.0035$. After quite a long transient, the network reaches a steady state in which only some groups of cells are synchronized, which corresponds to the so-called *cluster synchronization*.

  Figure 14.74 shows the steady-state plots of $x_i(\tau)$ for ($i = 1, \ldots, 30$) for random initial conditions. Again, the value of $x_i(\tau)$ is color-coded according to the color bar at the right.

  It is apparent that only some groups of cells (*clusters*) are synchronous. For instance, cells 9, 12, 13, 17, and 29 form a cluster, as pointed out in Fig. 14.75, showing the same plots as in Fig. 14.74 only for the cells of the cluster, and in Fig. 14.76, showing the time plots $x_i(\tau)$ for $i = 9, 12, 13, 17, 29$, which are perfectly overlapping.

  Other clusters are made up of cells $\{14, 15, 27\}$ (time plots in Fig. 14.77), $\{8, 19, 30\}$ (time plots in Fig. 14.78), and $\{5, 16\}$ (time plots in Fig. 14.79).

  The discussion about the basin of attraction of each stable cluster would require a deeper analysis, which goes beyond the scope of this section.

  Finally, we analyze the case of nonidentical oscillators, by setting $\alpha^{in} = 0$ and introducing a mismatch through parameters β_i. In a first simulation we introduced

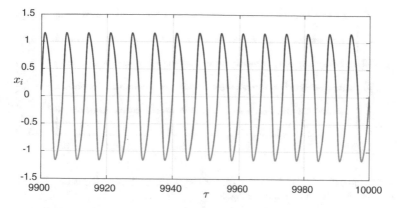

Fig. 14.73 Time plots $x_i(\tau)$ for $i = 1, \ldots, 30$ (all traces overlap)

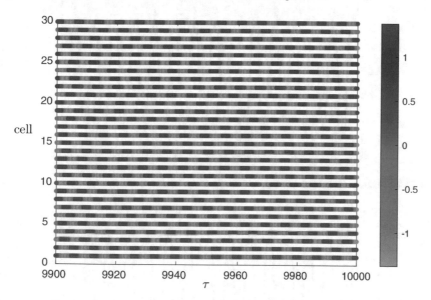

Fig. 14.74 Plots of $x_i(\tau)$ ($i = 1, \ldots, 30$) for $\alpha^{in} = 0.0035$, $\alpha^{el} = 0.0015$, and for random initial conditions. The value of $x_i(\tau)$ is color-coded according to the color bar at the right

a slight mismatch, by using random values of β_i uniformly distributed over the interval $[0, 0.01]$. In this case, as shown in Fig. 14.80, the synchronization is slightly perturbed. As usual, the value of $x_i(\tau)$ is color-coded according to the color bar at the right.

The absence of synchronization is confirmed by the result shown in Fig. 14.81, which shows the time plots $x_i(\tau)$ for $i = 1, \ldots, 30$, not overlapping.

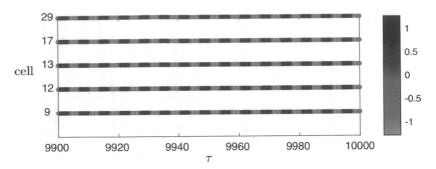

Fig. 14.75 Plots (from Fig. 14.74) of $x_i(\tau)$ for $i = 9, 12, 13, 17, 29$. The value of $x_i(\tau)$ is color-coded according to the color bar at the right

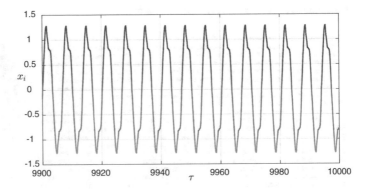

Fig. 14.76 Time plots $x_i(\tau)$ for $i = 9, 12, 13, 17, 29$ (all traces overlap)

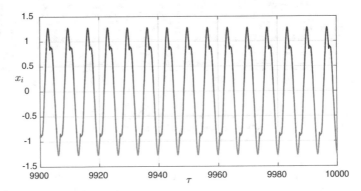

Fig. 14.77 Time plots $x_i(\tau)$ for $i = 14, 15, 27$ (all traces overlap)

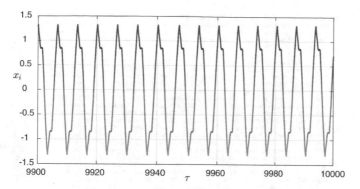

Fig. 14.78 Time plots $x_i(\tau)$ for $i = 8, 19, 30$ (all traces overlap)

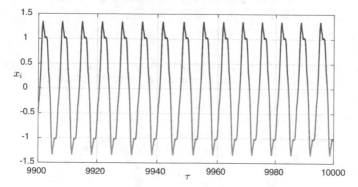

Fig. 14.79 Time plots $x_i(\tau)$ for $i = 5, 16$ (all traces overlap)

The loss of synchronization is even more evident for larger mismatches. For instance, Fig. 14.82 shows the simulation result obtained from using random values of β_i uniformly distributed over the interval $[0, 0.5]$.

14.10 Summarizing Comments

The concepts and examples proposed in this chapter not only are important from a methodological standpoint in circuit theory, but also point out the role played by nonlinear oscillators as models of phenomena of paramount importance in nature. The properties of nonlinear oscillators have been the last stage of the journey taken by the reader through linear and nonlinear dynamical circuits, with the help of the roadmap given by the index of this second volume. In this volume we have chosen to study only nonlinear behaviors induced by memoryless components, thus considering dynamical circuits whose memory components are linear. From a didactic point of view, this choice allowed us to provide the reader with a set of *tools*—according

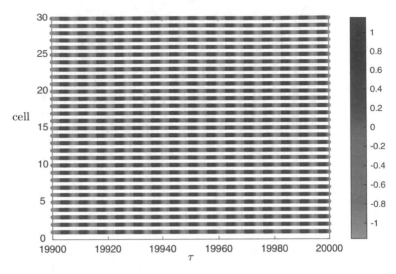

Fig. 14.80 Plots of $x_i(\tau)$ $(i = 1, \ldots, 30)$ for $\alpha^{in} = 0$, $\alpha^{el} = 0.0015$, random initial conditions, and random values β_i, uniformly distributed over the interval $[0, 0.01]$. Value of $x_i(\tau)$ color-coded according to the color bar

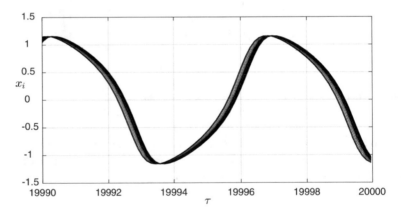

Fig. 14.81 Time plots $x_i(\tau)$ for $i = 1, \ldots, 30$

to the general philosophy of this book—useful for the **analysis** of linear and non-linear circuits. The third and last volume of our book will build on this background knowledge, with two main goals:

- to treat components and circuits from a broader perspective, generalizing their capacity of representation; the conceptual key for this generalization consists in broadening the possible choices for the descriptive variables of a given component;
- to introduce the dual aspect of **synthesis** of linear and nonlinear circuits able to realize a given model or to possess assigned features.

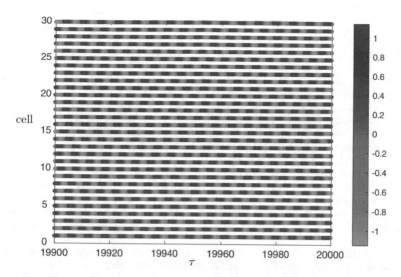

Fig. 14.82 Plots of $x_i(\tau)$ $(i = 1, \ldots, 30)$ for $\alpha^{in} = 0$, $\alpha^{el} = 0.0015$, random initial conditions, and random values β_i, uniformly distributed over the interval $[0, 0.5]$. Value of $x_i(\tau)$ color-coded according to the color bar

References

1. Strogatz S (2014) Nonlinear dynamics and chaos with applications to physics, biology, chemistry, and engineering. Westview Press, Boulder
2. Kuznetsov Y (2004) Elements of applied bifurcation theory. Springer, New York
3. Apostol T (1969) Calculus, volume II: multi-variable calculus and linear algebra, with applications to differential equations and probability. Wiley, New York
4. Jordan D, Smith P (1999) Nonlinear ordinary differential equations: an introduction to dynamical systems. Oxford University Press, New York
5. Hartman P (2002) Ordinary differential equations. Classics in applied mathematics, vol 38, Corrected reprint of the second (1982) edn. Society for Industrial and Applied Mathematics (SIAM), Philadelphia
6. Guckenheimer J, Holmes P (2013) Nonlinear oscillations, dynamical systems, and bifurcations of vector fields. Springer, New York
7. Oseledets VI (1968) A multiplicative ergodic theorem. Characteristic Lyapunov exponents for dynamical systems. Trudy Moskovskogo Matematicheskogo Obshchestva 19:179–210 (in Russian)
8. Benettin G, Galgani L, Giorgilli A, Strelcyn JM (1980) Lyapunov characteristic exponents for smooth dynamical systems and for Hamiltonian systems; a method for computing all of them. Part 1: theory. Meccanica 15:9–20
9. Benettin G, Galgani L, Giorgilli A, Strelcyn JM (1980) Lyapunov characteristic exponents for smooth dynamical systems and for Hamiltonian systems; a method for computing all of them. Part 2: numerical application. Meccanica 15:21–30
10. Wolf A, Swift JB, Swinney HL, Vastano JA (1985) Determining Lyapunov exponents from a time series. Phys D: Nonlinear Phenom 16:285–317
11. Andronov A, Leontovich E, Gordon I, Maier A (1971) Theory of bifurcations of dynamic systems on a plane. Israel Program for Scientific Translations, Jerusalem
12. Kolmogorov AN (1957) Théorie générale des systèmes dynamiques de la mécanique classique. Séminaire Janet. Mécanique analytique et mécanique céleste 1(6):1–20

13. Arnold Vl (2012) Geometrical methods in the theory of ordinary differential equations. Springer, New York
14. Allgower E, Georg K (2012) Numerical continuation methods: an introduction. Springer, New York
15. Dhooge A, Govaerts W, Kuznetsov YA (2003) MatCont: a MATLAB package for numerical bifurcation analysis of ODEs. ACM Trans Math Softw (TOMS) 29:141–164
16. Doedel EJ, Fairgrieve TF, Sandstede B, Champneys AR, Kuznetsov YA, Wang X (2007) AUTO-07P: continuation and bifurcation software for ordinary differential equations. Technical report
17. De Feo O, Maggio GM, Kennedy MP (2000) The Colpitts oscillator: families of periodic solutions and their bifurcations. Int J Bifurc Chaos 10:935–958
18. Hindmarsh JL, Rose RM (1984) A model of neuronal bursting using three coupled first order differential equations. Proc R Soc Lond Ser B Biol Sci 221:87–102
19. Hodgkin AL, Huxley AF (1952) A quantitative description of membrane current and its applications to conduction and excitation in nerve. J Physiol 117:500–544
20. Storace M, Linaro D, de Lange E (2008) The Hindmarsh–Rose neuron model: bifurcation analysis and piecewise-linear approximations. Chaos Interdiscip J Nonlinear Sci 18:033128(1–10)
21. Barrio R, Shilnikov A (2011) Parameter-sweeping techniques for temporal dynamics of neuronal systems: case study of Hindmarsh-Rose model. J Math Neurosci 1:6(1–22)
22. Linaro D, Champneys A, Desroches M, Storace M (2012) Codimension-two homoclinic bifurcations underlying spike adding in the Hindmarsh-Rose burster. SIAM J Appl Dyn Syst 11:939–962
23. De Feo O, Maggio GM (2003) Bifurcations in the Colpitts oscillator: from theory to practice. Int J Bifurc Chaos 13:2917–2934
24. Linaro D, Poggi T, Storace M (2010) Experimental bifurcation diagram of a circuit-implemented neuron model. Phys Lett A 374:4589–4593
25. Pinto R, Varona P, Volkovskii A, Szücs A, Abarbanel H, Rabinovich M (2000) Synchronous behavior of two coupled electronic neurons. Phys Rev E 62:2644–2656
26. Poggi T, Sciutto A, Storace M (2009) Piecewise linear implementation of nonlinear dynamical systems: from theory to practice. Electron Lett 45:966–967
27. Bizzarri F, Linaro D, Oldeman B, Storace M (2010) Harmonic analysis of oscillators through standard numerical continuation tools. Int J Bifurc Chaos 20:4029–4037
28. Parlitz U, Lauterborn W (1987) Period-doubling cascades and devil's staircases of the driven van der Pol oscillator. Phys Rev A 36:1428–1434
29. Kawakami H (1984) Bifurcation of periodic responses in forced dynamic nonlinear circuits: computation of bifurcation values of the system parameters. IEEE Trans Circuits Syst CAS-31:248–260
30. Mira C, Touzani-Qriouet M, Kawakami H (1999) Bifurcation structures generated by the nonautonomous Duffing equation. Int J Bifurc Chaos 9:1363–1379
31. Strogatz SH (2001) Exploring complex networks. Nature 410:268–276
32. Newman M, Barabasi A-L, Watts D (2011) The structure and dynamics of networks. Princeton University Press, Princeton
33. Buck J, Buck E (1968) Mechanism of rhythmic synchronous flashing of fireflies: fireflies of Southeast Asia may use anticipatory time-measuring in synchronizing their flashing. Science 159:1319–1327
34. Walker TJ (1969) Acoustic synchrony: two mechanisms in the snowy tree cricket. Science 166:891–894
35. Glass L, Mackey M (1988) From clocks to chaos: the rhythms of life. Princeton University Press, Princeton
36. Pikovsky A, Rosenblum M, Kurths J (2003) Synchronization: a universal concept in nonlinear sciences. Cambridge University Press, Cambridge
37. Stewart I (2004) Networking opportunity. Nature 427:601–604
38. Winfree A (2001) The geometry of biological time. Springer Science & Business Media, New York

39. Kuramoto Y (2003) Chemical oscillations, waves, and turbulence. Courier Corporation, North Chelmsford
40. Mirollo RE, Strogatz SH (1990) Synchronization of pulse-coupled biological oscillators. SIAM J Appl Math 50:1645–1662
41. Strogatz S, Abrams D, McRobie A, Eckhardt B, Ott E (2005) Theoretical mechanics: crowd synchrony on the Millennium Bridge. Nature 438:43–44
42. Belykh I, Jeter R, Belykh V (2017) Foot force models of crowd dynamics on a wobbly bridge. Sci Adv 3:e1701512(1–12)
43. May RM (2001) Stability and complexity in model ecosystems. Princeton University Press, Princeton
44. Eliasmith C, Stewart TC, Choo X, Bekolay T, DeWolf T, Tang Y, Rasmussen D (2012) A large-scale model of the functioning brain. Science 338:1202–1205
45. Danner SM, Shevtsova NA, Frigon A, Rybak IA (2017) Computational modeling of spinal circuits controlling limb coordination and gaits in quadrupeds. Elife 6:e31050(1–27)
46. Motter AE, Myers SA, Anghel M, Nishikawa T (2013) Spontaneous synchrony in power-grid networks. Nat Phys 9:191–197
47. Watts DJ, Strogatz SH (1998) Collective dynamics of 'small-world' networks. Nature 393:440–442
48. Barabasi AL, Albert R (1999) Emergence of scaling in random networks. Science 286:509–512
49. Kiehn O, Dougherty K (2016) Locomotion: circuits and physiology. In: Pfaff D, Volkow N (eds) Neuroscience in the 21st century: from basic to clinical, 1337–1365. Springer, New York
50. Ijspeert AJ (2008) Central pattern generators for locomotion control in animals and robots: a review. Neural Netw 21:642–653
51. Schwabedal JT, Knapper DE, Shilnikov AL (2016) Qualitative and quantitative stability analysis of penta-rhythmic circuits. Nonlinearity 29:3647–3676
52. Lodi M, Shilnikov A, Storace M (2017) CEPAGE: a toolbox for central pattern generator analysis. In: 2017 IEEE international symposium on circuits and systems (ISCAS), Piscataway (NJ), pp 1–4. IEEE
53. Somers D, Kopell N (1993) Rapid synchronization through fast threshold modulation. Biol Cybern 68:393–407
54. Jalil S, Allen D, Youker J, Shilnikov A (2013) Toward robust phase-locking in Melibe swim central pattern generator models. Chaos 23:046105
55. Sakurai A, Gunaratne CA, Katz PS (2014) Two interconnected kernels of reciprocally inhibitory interneurons underlie alternating left-right swim motor pattern generation in the mollusk Melibe leonina. J Neurophysiol 112:1317–1328

Appendix A
Complex Numbers

In this appendix we briefly summarize the main properties of complex numbers in the context of this book.

A.1 Imaginary Unit

The equation $x^2 = 1$ has no real solutions. The *imaginary unit i* is defined as a solution of this equation, that is, a nonreal (complex) number such that $i^2 = 1$. The sixteenth-century Italian mathematician Gerolamo Cardano[1] is credited with introducing complex numbers in his attempts to find solutions to cubic equations, around 1545.

In this book, the symbol i is commonly used to denote currents; therefore, for the sake of clarity, the imaginary unit is denoted by j.

A.2 Representations of Complex Numbers

A complex number z can be expressed in different forms.

A.2.1 Standard Form and Geometric Representation

The most common form is $z = a + jb$, where a and b are real numbers: a is called the *real part* of z, and b is called the *imaginary part* of z.

---

[1]Gerolamo Cardano (1501–1576) had actually a broad spectrum of interests, including (besides mathematics) physics, biology, medical science, chemistry, astrology, astronomy, and philosophy. He was one of the key figures in the foundation of probability theory.

© Springer Nature Switzerland AG 2020
M. Parodi and M. Storace, *Linear and Nonlinear Circuits: Basic and Advanced Concepts*,
Lecture Notes in Electrical Engineering 620,
https://doi.org/10.1007/978-3-030-35044-4

(a) **(b)**

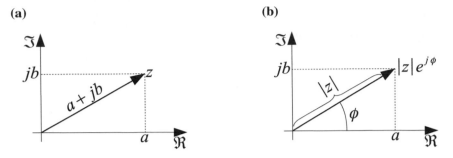

Fig. A.1 Geometric representation of a complex number $z = a + jb$: **a** standard form; **b** polar form

Geometrically, complex numbers extend the concept of the one-dimensional space of real numbers to the two-dimensional *complex plane* (or z-plane or Argand plane),[2] using the horizontal (or real) axis for the real part and the vertical (or imaginary) axis for the imaginary part. Therefore, the complex number $z = a + jb$ corresponds to the point (a, b) in the complex plane, as shown in Fig. A.1a. This geometric representation is also known as an Argand diagram.

A complex number whose real part is zero corresponds to a point lying on the imaginary axis of the complex plane and is said to be *purely imaginary*.

Conversely, a complex number whose imaginary part is zero corresponds to a point lying on the real axis of the complex plane and can be viewed as a real number.

A.2.2 Polar Form

Complex numbers can also be represented in *polar form*, which associates each complex number with its distance from the origin (its *magnitude*) and with a particular angle known as the *argument* of the complex number.

The magnitude is the absolute value, or modulus, of the complex number. By the Pythagorean theorem, it can be expressed as $|z| = \sqrt{a^2 + b^2}$.

The argument is $\phi = \arg(z) = \arctan \dfrac{b}{a}$ and is usually taken in the interval $[0, 2\pi]$. Thus we can write $z = |z|e^{j\phi}$, where e is the base of the natural logarithm or Euler's number or Napier's constant. Geometric representation and relationships with the standard notation are shown in Fig. A.1b.

Using the Euler's formula $e^{j\phi} = \cos \phi + j \sin \phi$, we obtain a third possible expression for the complex number z: $z = |z| \cos \phi + j|z| \sin \phi$, as shown in Fig. A.2a.

---

[2]It is named after the French amateur mathematician Jean-Robert Argand (1768–1822). In 1806, while managing a bookstore in Paris, he published the idea of the geometric interpretation of complex numbers known as the Argand diagram and is known for the first rigorous proof of the fundamental theorem of algebra.

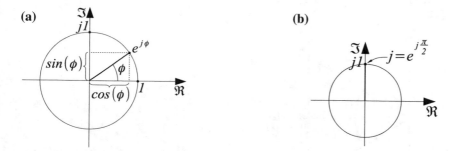

Fig. A.2 Geometric representation of **a** Euler's formula and **b** the imaginary unit

The imaginary unit j, corresponding to an argument of $\phi = \dfrac{\pi}{2}$ radians (see Fig. A.2b), can also be represented as $j = e^{j\frac{\pi}{2}}$.

A.2.3   A Particular Case

What is the result of the seemingly strange operation j^j? To find the answer, it is sufficient to recast it as $\left(e^{j\frac{\pi}{2}} \right)^j$. Therefore,

$$j^j = e^{-\frac{\pi}{2}}.$$

The result is (quite surprisingly!) a real number, expressed in terms of the two most famous irrational numbers (e and π).

A.3   Some Elementary Operations

A.3.1   Complex Conjugate

The complex conjugate of the complex number $z = a + jb = |z|e^{j\phi} = |z| \cos\phi + j|z| \sin\phi$ is defined to be $z^* = a - jb = |z|e^{-j\phi} = |z| \cos\phi - j|z| \sin\phi$.

Geometrically, z^* is the "reflection" of z about the real axis, as shown in Fig. A.3a.

(a)   **(b)**

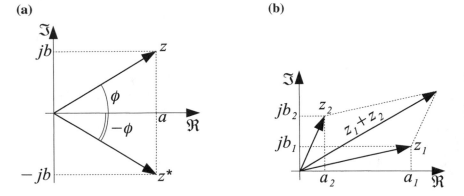

Fig. A.3 Geometric representation of **a** the complex conjugate and **b** the sum of complex numbers

A.3.2 Sum

Complex numbers are added by separately adding the real and imaginary parts of the summands. That is to say,

$$z_1 + z_2 = (a_1 + jb_1) + (a_2 + jb_2) = (a_1 + a_2) + j(b_1 + b_2).$$

Geometrically, the sum of two complex numbers z_1 and z_2, interpreted as points of the complex plane, is the point obtained by building a parallelogram, three of whose vertices are the origin, z_1, and z_2, as shown in Fig. A.3b.

A.3.3 Multiplication

The multiplication of two complex numbers is defined as follows:

$$z_1 \cdot z_2 = (a + jb)(c + jd) = (ac - bd) + j(ad + bc).$$

In particular, by multiplying a complex number z and its conjugate z^*, we obtain the square of the absolute value:

$$z \cdot z^* = (a + jb)(a - jb) = a^2 + b^2 = |z|^2.$$

Appendix B
Synoptic Tables

B.1   Metric System Prefixes

Table B.1 summarizes the most frequently used prefixes of the metric system.

Table B.1 Metric system prefixes

Prefix	Symbol	Value	Colloquial
femto	f	10^{-15}	Quadrillionth
pico	p	10^{-12}	Trillionth
nano	n	10^{-9}	Billionth
micro	μ	10^{-6}	Millionth
milli	m	10^{-3}	Thousandth
centi	c	10^{-2}	Hundredth
deci	d	10^{-1}	Tenth
/	/	10^{0}	One
deka	da	10^{1}	Ten
hecto	h	10^{2}	Hundred
kilo	k	10^{3}	Thousand
mega	M	10^{6}	Million
giga	G	10^{9}	Billion
tera	T	10^{12}	Trillion
peta	P	10^{15}	Quadrillion

© Springer Nature Switzerland AG 2020   479
M. Parodi and M. Storace, *Linear and Nonlinear Circuits: Basic and Advanced Concepts*,
Lecture Notes in Electrical Engineering 620,
https://doi.org/10.1007/978-3-030-35044-4

Table B.2 Properties of the main memoryless components

Component	Linear	Time-invariant	Bases	Energetic Behav.	Recipr.	Symmetr.
Resistor	Y	Y	v, i	Passive	Y	/
Voltage source	N	N	i	Active	N	/
Current source	N	N	v	Active	N	/
Diode (Shockley)	N	Y	v, i	Passive	N	/
CCCS	Y	Y	(i_1, v_2)	Active	N	N
CCVS	Y	Y	(i_1, i_2)	Active	N	N
VCVS	Y	Y	(v_1, i_2)	Active	N	N
VCCS	Y	Y	(v_1, v_2)	Active	N	N
Nullor	N	Y	/	Active	N	N
Ideal transformer	Y	Y	(v_1, i_2) (i_1, v_2)	Nonenergic	Y	N (if $n \neq 1$)
Gyrator	Y	Y	(i_1, i_2) (v_1, v_2)	Nonenergic	N	N

B.2 Properties of the Main Memoryless Components

Table B.2 summarizes the fundamental properties of the most relevant memoryless components.

B.3 Reciprocity and Symmetry Conditions for Linear, Time-Invariant, and Memoryless Two-Ports

Table B.3 summarizes the reciprocity and symmetry conditions for linear, time-invariant and memoryless two-ports described by at least one matrix.

Table B.3 Reciprocity and symmetry conditions for linear, time-invariant and memoryless two-ports

Matrix	Reciprocity condition	Symmetry condition	Reciprocity and symmetry condition
R	$R_{12} = R_{21}$	$R_{12} = R_{21}$ $R_{11} = R_{22}$	$R_{12} = R_{21}$ $R_{11} = R_{22}$
H	$H_{12} = -H_{21}$	$H_{11} = \dfrac{H_{11}}{det(H)}$ $H_{12} = -\dfrac{H_{21}}{det(H)}$ $H_{21} = -\dfrac{H_{12}}{det(H)}$ $H_{22} = \dfrac{H_{22}}{det(H)}$	$H_{12} = -H_{21}$ $det(H) = 1$
T	$det(T) = 1$	$T_{11} = \dfrac{T_{22}}{det(T)}$ $T_{12} = \dfrac{T_{12}}{det(T)}$ $T_{21} = \dfrac{T_{21}}{det(T)}$ $T_{22} = \dfrac{T_{11}}{det(T)}$	$det(T) = 1$ $T_{11} = T_{22}$

Solutions

Problems of Chap. 9

Solutions of problems of Chap. 9.

9.1 $i_1 = \dfrac{C}{n^2}\dfrac{dv_1}{dt}$. The two-terminal is a capacitor with scaled capacitance C/n^2.

9.2 1. $R_{22}C\dfrac{dv_2}{dt} + v_2 = R_{21}a(t)$.

2. $\lambda = -\dfrac{1}{R_{22}C}$.

3. $R_{22} > 0$ (assuming $C > 0$, as usual).

4. $v_2(t) = 0$ for $t < 0$; $v_2(0^+) = v_2(0^-) = 0$; $v_2(t) = R_{21}A(1 - e^{\lambda t})$ for $t > 0$.

5. $v_1(t) = R_{11}a(t) - R_{12}C\dfrac{dv_2}{dt} = \left(R_{11} - \dfrac{R_{12}R_{21}}{R_{22}}e^{\lambda t}\right)Au(t)$.

9.3 1. $g^2LR\dfrac{di}{dt} + (1 + Rg)i = -g(1 + Rg)e(t)$.

2. $\lambda = -\dfrac{1 + Rg}{g^2LR}$.

3. $1 + Rg > 0$ (assuming $L, R > 0$, as usual).

4. $i(t) = 0$ for $t < 0$; $i(0^-) = 0$; $i(0^+) = -\Phi\frac{1+Rg}{gLR}$; $i(t) = -\Phi\dfrac{1 + Rg}{gLR}e^{\lambda t}$ for $t > 0$.

9.4 1. $R_1C\dfrac{dv}{dt} + v = e(t)$.

2. $\lambda = -\dfrac{1}{R_1C}$.

3. Absolutely stable without conditions (assuming $R_1, C > 0$, as usual).

© Springer Nature Switzerland AG 2020
M. Parodi and M. Storace, *Linear and Nonlinear Circuits: Basic and Advanced Concepts*,
Lecture Notes in Electrical Engineering 620,
https://doi.org/10.1007/978-3-030-35044-4

4. $v(t) = E_0$ for $t < 0$; $v(0^-) = E_0$; $v(0^+) = v(0^-) = E_0$; $v(t) = \underbrace{(E_0 - E_1)e^{\lambda t}}_{v_{tr}(t)} +$

$\underbrace{E_1}_{\hat{v}}$ for $t > 0$.

9.5 1. $\dfrac{L}{R}(2R + R_0)\dfrac{di}{dt} + (3R + 2R_0)i = e(t)$.

2. $\lambda = -\dfrac{3R + 2R_0}{2R + R_0}\dfrac{R}{L}$.

3. $\dfrac{3R + 2R_0}{2R + R_0}R > 0$ (assuming $R_0, L > 0$, as usual).

4. $i(t) = 0$ for $t < 0$; $i(0^-) = 0$; $i(0^+) = i(0^-) = 0$; $i(t) = (1 - e^{\lambda t})\dfrac{E}{3R + 2R_0}$ for $t > 0$.

9.6 1. $R_1 C\dfrac{dv}{dt} + \dfrac{R_1 + R_2}{R_2}v = e(t)$.

2. $\lambda = -\dfrac{R_1 + R_2}{R_1 R_2 C} = -\dfrac{1}{R_p C}$, where $R_p = \dfrac{R_1 R_2}{R_1 + R_2}$ (parallel connection between R_1 and R_2).

3. Absolutely stable without conditions (assuming $R_1, R_2, C > 0$, as usual).

4. $v(t) = 0$ for $t < 0$; $v(0^-) = 0$; $v(0^+) = \dfrac{\Phi}{R_1 C}$; $v(t) = \underbrace{\dfrac{\Phi}{R_1 C}e^{\lambda t}}_{v_{tr}(t)}$ for $t > 0$.

9.7 1. $L\dfrac{di}{dt} + Ri = e(t) - Ra(t)$.

2. $\lambda = -\dfrac{R}{L}$.

3. $i(t) = 0$ for $t < 0$; $i(0^-) = 0$; $i(0^+) = -\dfrac{QR}{L}$; $i(t) = \underbrace{-\left(\dfrac{QR}{L} + \dfrac{E}{R}\right)e^{\lambda t}}_{i_{tr}(t)} + \underbrace{\dfrac{E}{R}}_{\hat{i}}$

for $t > 0$.

9.8 1. $RC(n + 1)^2\dfrac{dv}{dt} + (\alpha n + 1)v = n(\alpha n + 1)e(t)$.

2. $\lambda = -\dfrac{\alpha n + 1}{RC(n + 1)^2}$.

3. $\alpha n + 1 > 0$.

4. $v(t) = 0$ for $t < 0$; $v(0^-) = 0$; $v(0^+) = v(0^-) = 0$; $v(t) = \underbrace{-nEe^{\lambda t}}_{v_{tr}(t)} + \underbrace{nE}_{\hat{v}}$ for $t > 0$.

5. $p = \dfrac{\alpha}{R}\dfrac{(ne - v)(e - nv)}{(n + 1)^2}$.

9.9 1. $a_{NR} = G(e_1 + e_2) - e_2/R$; $R_{NR} = \dfrac{R}{Rg_3 + 1}$. Notice that if $g_3 = -1/R$, the Norton equivalent resistor becomes an open circuit, and the state equation is $Cdv/dt = -a_{NR}$. Therefore, in this case, the circuit's natural frequency is $\lambda = 0$, corresponding to simple stability.

2. $R_{NR}C\dfrac{dv}{dt} + v = -R_{NR}a_{NR}$.

3. $\lambda = -\dfrac{1}{R_{NR}C}$.

4. $R_{NR} > 0$, i.e., $g_3 > -1/R$.

5. $v(t) = 0$ for $t < 0$; $v(0^-) = 0$; $v(0^+) = \dfrac{1 - RG}{RC}\Phi$; $v(t) = \underbrace{[v(0^+) + R_{NR}GE]e^{\lambda t}}_{v_{tr}(t)}$

$\underbrace{-R_{NR}GE}_{\hat{v}}$ for $t > 0$.

9.10   1. 2 WSVs, 1 SSV (since there is an algebraic constraint between v and the current flowing in the inductor $i_L = -\alpha i = \alpha(v - e)/R$).

2. $RC\dfrac{dv}{dt} + (\alpha + 1)v = (\alpha + 1)e(t)$.

3. $\lambda = -\dfrac{\alpha + 1}{RC}$.

4. $\alpha > -1$.

5. $v(t) = E$.

6. $v(0^-) = E$.

7. $v(0^+) = E + \dfrac{\alpha + 1}{RC}\Phi$.

8. $v(t) = v_{tr}(t) = v(0^+)e^{\lambda t}$.

9.11   1. $(R_2 + R_3)C\dfrac{dv}{dt} + (1 - \alpha)v = (R_2 + \alpha R_3)a(t)$.

2. $\lambda = -\dfrac{1 - \alpha}{(R_2 + R_3)C}$ $(< 0$, since $\alpha < 1$ by assumption$)$.

3. $v(t) = 0$.

4. $v(0^-) = 0$.

5. $v(0^+) = v(0^-) = 0$.

6. $v(t) = \dfrac{R_2 + \alpha R_3}{1 - \alpha}A(1 - e^{\lambda t})$.

7. $v_3(t) = R_3\left(a(t) - \dfrac{v + R_3 a(t)}{R_2 + R_3}\right) = \dfrac{R_3 A}{1 - \alpha}\left(\dfrac{R_2 + \alpha R_3}{R_2 + R_3}e^{\lambda t} - \alpha\right)$.   (The last equality holds only for $t > 0$.) Notice that $v_3(t)$ (which is not a state variable) has a first-order discontinuity at $t = 0$, i.e., it has the same degree of discontinuity as the input.

9.12 The circuit has three WSVs (i.e., the voltages across the three capacitors), but just one SSV, since there are two independent loops, one containing only the capacitors C_1, C_2, C_3 and one containing only the voltage source and capacitors (either C_1 or C_2, C_3). The SSV can be the voltage across either C_2 or C_3, not the voltage across C_1 $(= e(t))$. You can easily check that the circuit's natural frequency is $\lambda = -\dfrac{R_2 + R_3}{R_2 R_3 (C_2 + C_3)}$; then the circuit is absolutely stable.

9.13 1. For a generic resistance matrix, we would find $e_{TH} = \dfrac{R_{21}}{R_{11}} e(t)$ and $R_{TH} = \dfrac{det([R])}{R_{11}}$. For the assigned resistance matrix, this reduces to $e_{TH} = \dfrac{e(t)}{2}$ and $R_{TH} = \dfrac{R}{2}$.

2. For a generic resistance matrix, we would find $\lambda = -\dfrac{det([R])}{R_{11}L}$ (which is < 0 for $det([R])$ and R_{11} with opposite signs). For the assigned resistance matrix, this reduces to $\lambda = -\dfrac{R}{2L} < 0$.

3. For a generic resistance matrix, we would find $i(\infty) = \dfrac{R_{21}}{det([R])} E$. For the assigned resistance matrix, this reduces to $i(\infty) = \dfrac{E}{R}$.

9.14 1. The circuit has two WSVs, but no SSV, due to the following two independent algebraic constraints: $v = 3R(a_1 - a_2)$ and $i = 3RC\dfrac{d(a_1 - a_2)}{dt}$. This implies that there are no natural frequencies.

2. $v(t) = 3R(A_1 - A_2)$.
3. $i(t) = 0$.

9.15 1. The circuit has three WSVs, but one SSV, due to the loop containing only $e_1(t)$ and the left capacitor C ($v_3 = e_1$), and to the algebraic constraint $v_1 + e_2 = v_2$. This implies that there is only one natural frequency.

2. $2RC\dfrac{dv_1}{dt} + 3v_1 = 2e_1 - 3e_2$.

3. The I/O relationship for $v_2(t)$ is $2RC\underbrace{\dfrac{dv_2}{dt}}_{\delta} + \underbrace{3v_2}_{step} = \underbrace{2e_1}_{\delta} + 2RC\underbrace{\dfrac{de_2}{dt}}_{\delta}$. According to the discontinuity balance, v_2 has a first-order (step) discontinuity at $t = 0$.

4. $\lambda = -\dfrac{3}{2RC}$.
5. The circuit is absolutely stable ($\lambda < 0$) without any conditions (according to the usual assumptions).
6. $v_1(t) = -E$ for $t < 0$.
7. $v_1(0^-) = -E$.
8. $v_1(0^+) = \dfrac{\Phi}{RC} + v_1(0^-)$.
9. $v_1(t) = v_1(0^+)e^{\lambda t}$ for $t > 0$.

9.16 1. $L(R_0 + 2R)\dfrac{di}{dt} + 2R_0 Ri = 2Re(t)$.

2. $\lambda = -2\dfrac{R_0}{R_0 + 2R}\dfrac{R}{L}$ (< 0 according to the problem assumptions).

3. $i(t) = 0$ for $t < 0$; $i(0^+) = i(0^-) = 0$ according to the discontinuity balance;
$i(t) = \underbrace{-\dfrac{E}{R_0}e^{\lambda t}}_{i_{tr}(t)} + \underbrace{\dfrac{E}{R_0}}_{\hat{i}} = \dfrac{E}{R_0}\left(1 - e^{\lambda t}\right)$ for $t > 0$.

4. $i(t) = 0$ for $t < 0$; $i(0^+)(\neq i(0^-)) = 2\dfrac{R}{R_0 + 2R}\dfrac{\Phi}{L}$ according to the discontinuity balance; $i(t) = i_{tr}(t) = i(0^+)e^{\lambda t}$ for $t > 0$.

9.17 1. $(\alpha + Rg)L\dfrac{di}{dt} + R(1 + \beta)i = e(t)$.

2. $\lambda = -\dfrac{R(1 + \beta)}{(\alpha + Rg)L}$.

3. We have to impose $\dfrac{1 + \beta}{\alpha + Rg} > 0$, i.e., either $\beta > -1$ and $\alpha > -Rg$ or $\beta < -1$ and $\alpha < -Rg$.

4. $i(t) = \dfrac{E}{R(1 + \beta)} = i(0^-)$.

5. $i(0^+) = \dfrac{\Phi}{(\alpha + Rg)L} + i(0^-)$.

6. $i(t) = i_{tr}(t) = i(0^+)e^{\lambda t}$.

7. It is the energy stored in the inductor at $t = 0^+$, since the circuit is absolutely stable by assumption and there is no input for $t > 0$: $w = \dfrac{1}{2}Li^2(0^+)$.

9.18 1. The circuit has four WSVs, but two SSVs, due to the presence of a cut-set containing only the inductors L_1 and L_2, and a cut-set containing only the current source $a(t)$ and the inductor L_3.

9.19 1. The circuit has two WSVs, but one SSV, due to the algebraic constraint $i = -a$. This implies that there is only one natural frequency.

2. $(R_1 + R_3)C\dfrac{dv}{dt} + v = -R_3a(t)$.

3. $\lambda = -\dfrac{1}{(R_1 + R_3)C} < 0$.

4. $v(t) = 0(= v(0^-))$ for $t < 0$; $v(0^+) = -\dfrac{R_3}{R_1 + R_3}\dfrac{Q}{C}$; $v(t) = \underbrace{(v(0^+) + R_3A)e^{\lambda t}}_{v_{tr}(t)}$

$\underbrace{-R_3A}_{\hat{v}} = \underbrace{v(0^+)e^{\lambda t}}_{ZIR} + \underbrace{R_3A\left(e^{\lambda t} - 1\right)}_{ZSR}$ for $t > 0$.

9.20 1. The circuit has three WSVs, but one SSV, due to the algebraic constraints $i_1 = 0$ and $i_2 = a - e/R$. This implies that there is only one natural frequency.

2. $rC\dfrac{dv}{dt} + v = \dfrac{R - r}{R}e(t) + ra(t) + \dfrac{L_2}{R}\dfrac{de}{dt} - L_2\dfrac{da}{dt}$.

3. $\lambda = -\dfrac{1}{rC} < 0$.

4. $v(t) = \dfrac{R - r}{R}E(= v(0^-))$ for $t < 0$; $v(0^+) = -\dfrac{L_2}{rC}\left(\dfrac{E}{R} + A\right) + v(0^-)$; $v(t) = \underbrace{\left[v(0^+) - rA\right]e^{\lambda t} + rA}_{v_{tr}(t)} = \underbrace{v(0^+)e^{\lambda t}}_{ZIR} + \underbrace{rA\left(1 - e^{\lambda t}\right)}_{ZSR}$ for $t > 0$.

9.21 1. $4R_1C\dfrac{dv}{dt} + 4\dfrac{R_1}{R_2}v = e(t)$.

2. $\lambda = -\dfrac{1}{R_2C} < 0$.

3. $v(t) = \dfrac{R_2}{4R_1}E(= v(0^-))$ for $t < 0$; $v(0^+) = \dfrac{\Phi}{4R_1C} + v(0^-)$; $v(t) = v_{tr}(t) =$ $v(0^+)e^{\lambda t}$ for $t > 0$.

9.22 1. The circuit has two WSVs, but one SSV, due to the algebraic constraint $i = a$.

2. $RC\dfrac{dv}{dt} + v = -L\dfrac{da}{dt}$.

3. $\lambda = -\dfrac{1}{RC}$.

4. $v(0^-) = 0$.

5. $v(0^+) = -\dfrac{LA}{RC} + v(0^-)$.

6. $v(t) = v(0^+)e^{\lambda t}$.

7. $i_x = \dfrac{1 + Rg}{R}v + a(t) + \dfrac{L}{R}\dfrac{da}{dt}$.

9.23 1. The circuit has two WSVs, but one SSV, due to the algebraic constraint $v_2 = v_1 + R_1a$.

2. $C_1\dfrac{dv_2}{dt} = a(t) + R_1C_1\dfrac{da}{dt}$.

3. $\lambda = 0$.

4. $v_2(0^-) = V_0$.

5. $v_2(0^+) = R_1A + v_2(0^-)$.

6. $v_2(t) = v_2(0^+) + \dfrac{A}{C_1}t$.

9.24 1. $\left[RC + (1 - \beta)\dfrac{L}{R}\right]\dfrac{dv}{dt} + v = Ra(t)$.

2. $\lambda = -\dfrac{1}{RC + (1 - \beta)L/R}$ (there is only one SSV, because $i_2 = -v/R$).

3. $v(0^-) = RA$.

4. $v(0^+) = \dfrac{R^2Q}{R^2C + (1 - \beta)L} + v(0^-)$.

5. $v(t) = v(0^+)e^{\lambda t}$.

Problems of Chap. 13

Solutions of problems of Chap. 13.

10.1 1. Since $i(0^+) = A$, the condition is $A > I_0$.

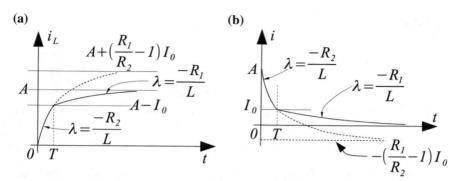

Fig. B.1 Problem 10.1

2. For $t \in (0, T)$ we obtain

$$i_L = \left(A + \left(\frac{R_1}{R_2} - 1 \right) I_0 \right) \left(1 - e^{-\frac{R_2}{L}t} \right); \qquad i = A - i_L.$$

3. $T = \dfrac{L}{R} \ln \left(1 + \dfrac{A - I_0}{\dfrac{R_1}{R_2} I_0} \right).$

4. Defining $t' = t - T$ we obtain (for $t' \geq 0$),

$$i_L(t') = A - I_0 e^{-\frac{R_1}{L}t'}; \qquad i = A - i_L = I_0 e^{-\frac{R_1}{L}t'}.$$

Therefore, for all $t' > 0$ we have $i(t') < I_0$, and the resistor works on the branch
ⓐ of its characteristic.

The curves $i_L(t)$ and $i(t)$ are shown in Fig. B.1.

10.2 1. Since $v(0^+) = \dfrac{Q}{C}$, the condition is $\dfrac{Q}{C} > V_0$.

2. For $t \in (0, T)$ we obtain

$$v = \frac{Q}{C} - \frac{V_0}{RC}t.$$

3. $T = \dfrac{RC}{V_0} \left(\dfrac{Q}{C} - V_0 \right).$

4. Defining $t' = t - T$ we obtain (for $t' \geq 0$):

$$v(t') = V_0 e^{-\frac{t'}{RC}}.$$

Fig. B.2 Problem 10.2

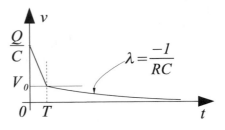

Therefore, for all $t' > 0$ we have $v(t') < V_0$, and the resistor works on the branch ⓐ of its characteristic.

The curve $v(t)$ is shown in Fig. B.2.

10.3 1. Suppose for the sake of obtaining a contradiction that the diode works initially in the branch ⓐ of its characteristic. Then $i_d = 0$, provided that $v_d < 0$. The circuit equations can be easily formulated as follows:

$$\begin{cases} e\left(1 - \dfrac{1}{n}\right) = L\dfrac{di_L}{dt} + \dfrac{R}{n^2}i_L \\[2mm] v_d = \dfrac{e}{n} + \dfrac{R}{n^2}i_L \\[2mm] i_1 = -\dfrac{i_L}{n} \end{cases}$$

Integrating the terms of the first equation between 0^- and 0^+ and recalling that $i_L(0^-) = 0$, $n > 1$ and $e(t) = \Phi\delta(t)$, we obtain

$$\left(1 - \frac{1}{n}\right)\int_{0^-}^{0^+} \Phi\delta(\tau)d\tau = Li_L(0^+) + \frac{R}{n^2}\underbrace{\int_{0^-}^{0^+} i_L(\tau)d\tau}_{0} \;\Rightarrow\; i_L(0^+) = \frac{\Phi}{L}\left(\frac{n-1}{n}\right) > 0.$$

The second equation written for $t = 0^+$ gives

$$v_d(0^+) = \underbrace{\frac{e(0^+)}{n}}_{0} + \frac{R}{n^2}i_L(0^+) \;\Rightarrow\; v_d(0^+) = \frac{R}{n^2}i_L(0^+) > 0.$$

This result contradicts the hypothesis that the diode operates in the branch ⓐ, which holds for $v_d < 0$. This proves that the diode operates initially in the branch ⓑ.

2. For the branch ⓑ we have $v_d = 0$; $i_d \geq 0$. The circuit equations with $v_d = 0$ can be easily formulated as

$$\begin{cases} e(t) = L\dfrac{di_L}{dt}, \\[2mm] i_d = \dfrac{ne(t)}{R} + i_L. \end{cases}$$

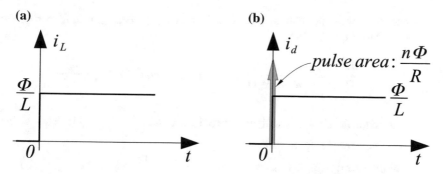

(a)   **(b)**

Fig. B.3 Problem 10.3

Therefore, we obtain

$$i_L = \frac{\Phi}{L}u(t); \qquad i_d = \frac{n\Phi}{R}\delta(t) + \frac{\Phi}{L}u(t).$$

Notice that $i_d > 0$ for all $t > 0$. Then the diode operates in the branch ⓑ for all $t > 0$. The currents $i_L(t)$ and $i_d(t)$ are shown in Fig. B.3.

10.4 In the study of the circuit we will always refer to the following general equations:

$$
\begin{cases}
e(t) = R_1 \left(i_c + C\dfrac{dv}{dt} \right) + v_{ce}, \\
v_{ce} = v + v_{be}, \\
v_{be} = R_2 \left(C\dfrac{dv}{dt} - i_b \right).
\end{cases}
\tag{C.1}
$$

1. Let us first show that in $t = 0^+$ the BJT cannot work according to pairs ①–ⓐ and ①–ⓑ.

   - For the pair ①–ⓐ, that is, assuming $i_b = 0$, $v_{be} \le V_T$, $v_{ce} = 0$, $0 \le i_c \le \beta i_b$, and with $v(0^+) = 0$, we obtain $v_{be}(0^+) = v_{ce}(0^+) = 0$; then from the third of equations C.1 we have $C\dfrac{dv}{dt}\Big|_{0^+} = 0$, which implies (first equation) $i_c(0^+) = \dfrac{E}{R_1}$. This result is incompatible with the assumption $i_c \le \beta i_b = 0$.

   - For the pair ①–ⓑ, that is, assuming $i_b = \dfrac{v_{be} - V_T}{h}$, $v_{be} > V_T$, $v_{ce} = 0$, $0 \le i_c \le \beta i_b$, and with $v(0^+) = 0$, the second of Eqs. C.1 gives $v_{be}(0^+) = 0$, which is incompatible with the assumption $v_{be} > V_T > 0$.

   Now we show that the remaining two pairs in $t = 0^+$ are compatible with the circuit under mutually exclusive conditions.

- For the pair ⓘ–ⓐ, that is, assuming $i_b = 0$, $v_{be} \leq V_T$, $v_{ce} > 0$, $i_c = \beta i_b$, we find from Eqs. C.1 that

$$\frac{dv}{dt}\bigg|_{0+} = \frac{E}{(R_1 + R_2)\,C}; \quad v_{ce}(0^+) = v_{be}(0^+) = \frac{E\,R_2}{R_1 + R_2}.$$

The requisite on v_{be} is satisfied, provided that $\dfrac{E\,R_2}{R_1 + R_2} \leq V_T$. This is the condition for the ⓘ–ⓐ pair.

- For the pair ⓘ–ⓑ, that is, assuming $i_b = \dfrac{v_{be} - V_T}{h}$, $v_{be} > V_T$, $v_{ce} > 0$, $i_c = \beta i_b$, we obtain

$$C\frac{dv}{dt}\bigg|_{0+} = \frac{E\,(R_2 + h) + V_T\,(\beta R_1 - R_2)}{R_1 R_2\,(\beta + 1) + h\,(R_1 + R_2)}; \quad v_{be}(0^+) = R_2\frac{V_T R_1\,(\beta + 1) + Eh}{R_1 R_2\,(\beta + 1) + h\,(R_1 + R_2)}.$$

Taking the second result, after some manipulations it can be shown that the requisite $v_{be}(0^+) > V_T$ implies the condition $\dfrac{E\,R_2}{R_1 + R_2} > V_T$, which is complementary to that found in the previous case, as expected.

2. Pair ⓘ–ⓐ: determination of $v_{ce}(t)$ and $v_{be}(t)$. Taking $i_b = 0, i_c = \beta i_b = 0$, from Eqs. C.1 we obtain

$$E = (R_1 + R_2)\,C\frac{dv}{dt} + v \quad \Rightarrow \quad v(t) = E\left(1 - e^{-\dfrac{t}{(R_1 + R_2)\,C}}\right).$$

Once the state variable $v(t)$ is known, $v_{ce}(t)$ and $v_{be}(t)$ can be found algebraically. Notice that from the state equation, we have

$$C\frac{dv}{dt} = \frac{E - v}{R_1 + R_2}.$$

Therefore, it follows that

$$v_{ce} = v + R_2 C\frac{dv}{dt} = v + R_2\frac{E - v}{R_1 + R_2} = E\frac{R_2}{R_1 + R_2} + v\frac{R_1}{R_1 + R_2};$$

$$v_{be} = R_2 C\frac{dv}{dt} = R_2\frac{E - v}{R_1 + R_2}.$$

The natural frequency for all these variables is $-\dfrac{1}{(R_1 + R_2)\,C}$. The resulting curves are shown in Fig. B.4. Notice that the conditions for the pair ⓘ–ⓐ hold for all $t > 0$.

The case ⓘ–ⓑ can be analyzed similarly.

Fig. B.4 Problem 10.4

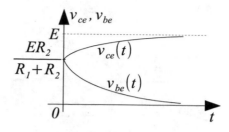

10.5 1. The general equations for the circuit of Fig. 10.52c can be formulated as follows:

$$\begin{cases} v_o + v + v_x = 0, \\ i + I + \dfrac{v_x}{R} = 0, \\ v_o = f(v_x). \end{cases} \quad (C.2)$$

Starting from these equations, the circuit can be studied in each PWL region of the device by adding the equation and the inequality defining that region.

- For the Ⓐ-region we easily obtain

$$v = R(1 + A)(i + I); \qquad |i + I| < \frac{V_s}{AR}.$$

- Assuming the Ⓑ-region, we obtain

$$v = -V_s + R(i + I); \qquad i + I \le -\frac{V_s}{AR}.$$

- Finally, for the Ⓒ-region we have

$$v = V_s + R(i + I); \qquad i + I \ge \frac{V_s}{AR}.$$

This set of results corresponds to the PWL characteristic represented in Fig. B.5a. This characteristic individuates a nonlinear resistor that admits both the current basis and the voltage basis. Therefore, both the complementary-component circuits of Fig. 10.2 are possible.

2. The general equations for the circuit of Fig. 10.52d are

$$\begin{cases} v_o + v - v_x = 0, \\ i + I - \dfrac{v_x}{R} = 0, \\ v_o = f(v_x). \end{cases} \quad (C.3)$$

Following the same procedure of the previous case, the results are

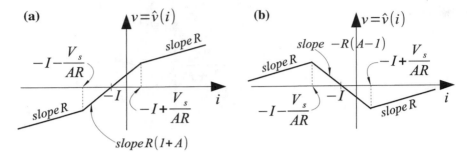

Fig. B.5 Problem 10.5. The figure has been drawn by assuming that $A > 1$

- $\textcircled{\alpha}$-region

$$v = -R(A-1)(i+I); \qquad |i+I| < \frac{V_s}{AR}.$$

- $\textcircled{\beta}$-region

$$v = -V_s + R(i+I); \qquad i+I \geq \frac{V_s}{AR}.$$

- $\textcircled{\gamma}$-region

$$v = V_s + R(i+I); \qquad i+I \leq -\frac{V_s}{AR}.$$

These results correspond to the PWL characteristic represented in Fig. B.5b.

Notice that the characteristic of Fig. B.5b individuates a nonlinear resistor that admits only the current basis. In this case, then, only the complementary-component circuit of Fig. 10.2b is possible.

Problems of Chap. 11

Solutions of problems of Chap. 11.

11.1   1. $(\alpha + Rg)L\dfrac{di}{dt} + R(1+\beta)i = e(t)$.

2. $\lambda = -\dfrac{R(1+\beta)}{(\alpha + Rg)L}$.

3. We have to ensure absolute stability due to the steady-state assumption. Therefore
$$\begin{cases} \beta > -1 \\ \alpha > -Rg \end{cases} \text{ or } \begin{cases} \beta < -1 \\ \alpha < -Rg \end{cases} \text{ (assuming } R, L > 0, \text{ as usual).}$$

4. $i(t) = \dfrac{E_0}{\rho} - \dfrac{E_1}{\rho^2 + (\omega\ell)^2}[\omega\ell\cos(\omega t) - \rho\sin(\omega t)]$, where $\rho = R(1+\beta)$ and $\ell = (\alpha + Rg)L$;

$$i(0^-) = \frac{E_0}{\rho} - \frac{\omega \ell E_1}{\rho^2 + (\omega \ell)^2}.$$

5. $i(0^+) = \dfrac{\Phi_0}{(\alpha + Rg)L} + i(0^-).$

6. $i(t) = i(0^+)e^{\lambda t}.$

7. The energy dissipated by the circuit for $t > 0$ is the energy stored in the inductor at $t = 0^+$, that is, $w = \dfrac{1}{2}Li^2(0^+).$

11.2 1. $2RC\dfrac{dv_1}{dt} + 3v_1 = 2e_1 - 3e_2.$

2. $2RC\dfrac{dv_2}{dt} + 3v_2 = 2e_1 + 2RC\dfrac{de_2}{dt}$; v_2 has a step-discontinuity at $t = 0$.

3. $\lambda = -\dfrac{3}{2RC}$ (absolutely stable). There is only one natural frequency, because due to the algebraic constraint between WSVs ($v_2 - v_1 = e_2$), there is only one SSV.

4. $v_1(t) = -E_0 = v_1(0^-).$

5. $v_1(0^+) = \dfrac{\Phi_0}{RC} + v_1(0^-).$

6. $v_1(t) = Ke^{\lambda t} + v_{AC1}(t) + v_{AC2}(t)$, with

$$v_{AC1}(t) = \frac{2E_1}{9 + (2\omega RC)^2}[3\cos(\omega t) + 2\omega RC \sin(\omega t)],$$

$$v_{AC2}(t) = \frac{3E_2}{9 + (4\omega RC)^2}[4\omega RC \cos(2\omega t) - 3\sin(2\omega t)],$$

$$K = v_1(0^+) - \frac{6E_1}{9 + (2\omega RC)^2} - \frac{12\omega RC E_2}{9 + (4\omega RC)^2}.$$

7. $\bar{v}_1 = 0$ and $v_{1eff} = \sqrt{\dfrac{1}{2}\dfrac{4E_1^2}{9 + (2\omega RC)^2} + \dfrac{1}{2}\dfrac{9E_2^2}{9 + (4\omega RC)^2}}.$

11.3 1. $\dfrac{di_2}{dt} + \dfrac{1}{\tau}i_2 = -\dfrac{e}{L}$, where $L = (1-\alpha)L_1 + L_2 + (\alpha - 2)M$, $\tau = \dfrac{L}{R(1-\alpha)}.$

It is also a state equation.

2. There are three WSVs and two algebraic constraints ($i_1 + i_2 + i_3 = 0$ and $i_3 + \alpha i_2 = 0$); thus there is only one SSV.

The natural frequency is $\lambda = -\dfrac{1}{\tau} = -\dfrac{(1-\alpha)R}{(1-\alpha)L_1 + L_2 + (\alpha - 2)M}.$

3. The absolute stability condition is

$$\begin{cases} \alpha < 1 \\ (1-\alpha)L_1 + L_2 + (\alpha - 2)M > 0 \end{cases} \text{ or } \begin{cases} \alpha > 1 \\ (1-\alpha)L_1 + L_2 + (\alpha - 2)M < 0 \end{cases}.$$

4. $i_2(t) = -\dfrac{\tau}{L}E_0 = i_2(0^-).$

5. $i_2(0^+) = -\dfrac{\Phi_0}{L} + i_2(0^-).$

6. $i_2(t) = Ke^{\lambda t} + i_{DC}(t) + i_{AC}(t)$, with

$$i_{DC}(t) = -\frac{\tau}{L}E_0,$$

$$i_{AC}(t) = -\frac{\tau}{L}\frac{E_1}{1 + (\omega\tau)^2}[\cos(\omega t) + \omega\tau\sin(\omega t)],$$

$$K = i_2(0^+) + \frac{\tau}{L}\left(E_0 + \frac{E_1}{1 + (\omega\tau)^2}\right).$$

11.4 1. $L_1\dfrac{di_2}{dt} + 3Ri_2 = Ra_1 + Ra_2 + L_2\dfrac{da_2}{dt}$, where $L_1 = 3L - 2M$, $L_2 = 2L - M$. It is also a state equation.

2. There are two WSVs and one algebraic constraint ($i_1 + i_2 = a_2$). Thus there is only one SSV.

The natural frequency is $\lambda = -\dfrac{3R}{L_1} = -\dfrac{3R}{3L - 2M}$. We remark that if we had $3L = 2M$, then the I/O relationship would be a new algebraic constraint, and the circuit would have no state.

3. The absolute stability condition is $3L > 2M$.
4. $i_2(t) = i_{DC}(t) + i_{AC}(t)$, with

$$i_{DC}(t) = \frac{A_0}{3},$$

$$i_{AC}(t) = \frac{A_1}{(3R)^2 + (\omega L_1)^2}[(3R^2 + \omega^2 L_1 L_2)\cos(\omega t) - \omega R(3L_2 - L_1)\sin(\omega t)].$$

$$i_2(0^-) = \frac{A_0}{3} + \frac{3R^2 + \omega^2 L_1 L_2}{(3R)^2 + (\omega L_1)^2}A_1.$$

5. $i_2(0^+) = \dfrac{Q_0 R}{L_1} + \dfrac{L_2}{L_1}(A_0 - A_1) + i_2(0^-).$

6. $i_2(t) = \left(i_2(0^+) - \dfrac{A_0}{3}\right)e^{\lambda t} + \dfrac{A_0}{3}.$

11.5 1. $L_1\dfrac{di_1}{dt} + Ri_1 = -e - M\dfrac{da}{dt}$. It is also a state equation.

2. There are three WSVs and two algebraic constraints ($i_2 = a$ and $v = \alpha Ra$). Thus there is only one SSV.

The natural frequency is $\lambda = -\dfrac{R}{L_1}$.

3. The circuit is absolutely stable without any conditions.
4. $i_1(t) = i_{AC1}(t) + i_{AC2}(t)$, with

$$i_{AC1}(t) = -\frac{E_1}{R^2 + (\omega L_1)^2}[R\cos(\omega t) + \omega L_1\sin(\omega t)],$$

$$i_{AC2}(t) = -\frac{2\omega M A_1}{R^2 + (2\omega L_1)^2}[R\cos(2\omega t) + 2\omega L_1\sin(2\omega t)],$$

$$i_1(0^-) = -\frac{RE_1}{R^2 + (\omega L_1)^2} - \frac{2\omega M R A_1}{R^2 + (2\omega L_1)^2}.$$

5. $\bar{p} = 2R\left(A_0^2 + \dfrac{A_1^2}{2}\right)$.

6. $i_1(0^+) = -\dfrac{\Phi_0}{L_1} + i_1(0^-)$.

7. $i_1(t) = \left(i_1(0^+) + \dfrac{E_0}{R}\right)e^{\lambda t} - \dfrac{E_0}{R}$.

11.6 1. $L\dfrac{di_2}{dt} + (1-\alpha)Ri_2 = -e + (L_1 - M)\dfrac{da}{dt}$, where $L = (1-\alpha)L_1 + L_2 + (\alpha - 2)M$. It is also a state equation.

2. There are three WSVs and two algebraic constraints ($i_1 + i_2 + i_3 = a$ and $i_3 + \alpha i_2 = 0$). Thus there is only one SSV.

The natural frequency is $\lambda = -\dfrac{(1-\alpha)R}{L}$.

3. $\alpha < 1$ and $M < 0$, which ensures $\lambda < 0$.

4. $i_2(t) = \dfrac{E_0}{(\alpha - 1)R} = i_2(0^-)$.

5. $i_2(0^+) = -\dfrac{\Phi_0}{L} + (L_1 - M)\dfrac{A_0}{L} + i_2(0^-)$.

6. $i_2(t) = Ke^{\lambda t} + i_{DC}(t) + i_{AC1}(t) + i_{AC2}(t)$, with

$$i_{DC}(t) = \frac{E_0}{(\alpha - 1)R},$$

$$i_{AC1}(t) = -\frac{E_1}{[(1-\alpha)R]^2 + (\omega L)^2}[(1-\alpha)R\cos(\omega t) + \omega L \sin(\omega t)],$$

$$i_{AC2}(t) = \frac{(L_1 - M)2\omega A_1}{[(1-\alpha)R]^2 + (2\omega L)^2}[(1-\alpha)R\cos(2\omega t) + 2\omega L \sin(2\omega t)],$$

$$K = i_2(0^+) + \frac{E_0}{(1-\alpha)R} + \frac{(1-\alpha)RE_1}{[(1-\alpha)R]^2 + (\omega L)^2} - \frac{(1-\alpha)R(L_1 - M)2\omega A_1}{[(1-\alpha)R]^2 + (2\omega L)^2}.$$

11.7 1. $L\dfrac{di_2}{dt} + Ri_2 = -Ra_2 + (L_1 - M)\dfrac{da_1}{dt}$, where $L = L_1 + L_2 - 2M$. It is also a state equation.

2. There are two WSVs and one algebraic constraint ($i_1 + i_2 = a_1$). Thus there is only one SSV.

The natural frequency is $\lambda = -\dfrac{R}{L} = -\dfrac{R}{L_1 + L_2 - 2M}$.

3. $L_1 + L_2 > 2M$.

4. $i_2(t) = -A + \dfrac{\omega(L_1 - M)A_1}{R^2 + (\omega L)^2}[R\cos(\omega t) + \omega L \sin(\omega t)]$;

$i_2(0^-) = -A + \dfrac{\omega(L_1 - M)RA_1}{R^2 + (\omega L)^2}$.

5. $\bar{i}_2 = -A$;

$$i_{2eff} = \sqrt{A^2 + \frac{1}{2}\frac{[\omega(L_1 - M)A_1]^2}{R^2 + (\omega L)^2}}.$$

6. $i_2(0^+) = -\dfrac{RQ_0}{L} + i_2(0^-).$

7. $i_2(t) = [i_2(0^+) + A_2]e^{\lambda t} - A_2.$

11.8 1. $L\dfrac{di_2}{dt} + 3Ri_2 = \gamma e$, where $L = (1 + gR)L_2 + gRM$ and $\gamma = -gR$. It is also a state equation.

2. There are three WSVs and one algebraic constraint ($i_1 = 0$). Thus there are two SSVs.

3. The two natural frequencies are $\lambda_1 = -\dfrac{3R}{L} = -\dfrac{3R}{(1 + gR)L_2 + gRM}$ and $\lambda_2 = 0$, due to the presence of the cut-set involving only a and C, which corresponds to an algebraic constraint on the derivative of v, namely $C\dfrac{dv}{dt} = -a.$

4. $(1 + gR)L_2 + gRM > 0$; this condition ensures the simple stability of the circuit, due to the presence of the null natural frequency. We remark that the state variable i_2 is sensitive only to λ_1, and thus it can reach a steady state.

5. $i_2(0^+) = -\dfrac{\ell}{L}A_0 + \dfrac{\gamma}{L}\Phi_0.$

6. $i_2(t) = Ke^{\lambda t} + i_{AC}(t)$, with

$$i_{AC}(t) = \frac{\gamma E_2}{(3R)^2 + (2\omega L)^2}[3R\cos(2\omega t) + 2\omega L\sin(2\omega t)],$$

$$K = i_2(0^+) - \frac{3R\gamma E_2}{(3R)^2 + (2\omega L)^2}.$$

11.9 1. $\tau\dfrac{dv}{dt} + v = \tau_a R\dfrac{da}{dt} + e + \tau_e\dfrac{de}{dt}$, where $\tau = 2RC_1$, $\tau_a = 2rC_1$, and $\tau_e = (2R - r)C_1.$

2. There are three WSVs and one algebraic constraint ($v_2 = e$). Thus there are two SSVs.

3. The two natural frequencies are $\lambda_1 = -\dfrac{1}{2RC_1}$ and $\lambda_2 = 0$, due to the presence of an algebraic constraint on the derivatives of SSVs, namely $L\dfrac{di}{dt} + rC_1\dfrac{dv_1}{dt} = 0.$

4. $\lambda_1 < 0, \lambda_2 = 0$, and thus the circuit is simply stable with no conditions. We remark that v is sensitive only to λ_1; thus it can reach a steady state.

5. $v(t) = v_{DC}(t) + v_{AC}(t)$, with

$$v_{DC}(t) = E_0,$$

$$v_{AC}(t) = \frac{E_1}{1 + (\omega\tau)^2}[\omega(\tau_e - \tau)\cos(\omega t) + (1 + \omega^2\tau\tau_e)\sin(\omega t)].$$

The I/O relationship is of the first order. Therefore, we need only one condition at $t = 0^-$: $v(0^-) = E_0 + \dfrac{\omega E_1(\tau_e - \tau)}{1 + (\omega\tau)^2}.$

6. $v_{eff} = \sqrt{E_0^2 + \dfrac{E_1^2}{2}\dfrac{1+(\omega\tau_e)^2}{1+(\omega\tau)^2}}.$

7. $v(0^+) = -RA\dfrac{\tau_a}{\tau} + v(0^-) = -rA + v(0^-).$

8. $v(t) = [v(0^+) - E_0]e^{\lambda_1 t} + E_0.$

11.10 1. $(2-\alpha)L\dfrac{di}{dt} + Ri = e(t)$. It is also a state equation.

2. There are two WSVs and no algebraic constraints. Thus both WSVs are SSVs.

3. The two natural frequencies are $\lambda_1 = -\dfrac{R}{(2-\alpha)L}$ and $\lambda_2 = 0$, due to the presence

   of an algebraic constraint on the derivatives of SSVs, namely $L\dfrac{di}{dt} = RC\dfrac{dv}{dt}.$

4. $\lambda_1 < 0$ for $\alpha < 2$, $\lambda_2 = 0$; thus the circuit is simply stable, provided that $\alpha < 2$; we remark that the state variable i is sensitive only to λ_1, and thus it can reach a steady state.

5. $i(0^+) = \dfrac{\Phi_0}{(2-\alpha)L}.$

6. $i(t) = \left[i(0^+) - \dfrac{E}{R}\right]e^{\lambda_1 t} + \dfrac{E}{R}.$

7. Given that (by assumption) there are no active independent sources for $t > 0$, the energy dissipated by the circuit for $t > 0$ is the energy stored in the conservative components at $t = 0^+$, that is, $w = \dfrac{1}{2}Li^2(0^+) + \dfrac{1}{2}Cv^2(0^+)$. We already know $i(0^+)$, but we have to compute $v(0^+)$, assuming that $v(0^-)$ is known. By integrating the equation $L\dfrac{di}{dt} = RC\dfrac{dv}{dt}$ between 0^- and 0^+, we obtain $v(0^+) = \dfrac{\Phi_0}{(2-\alpha)RC} + v(0^-).$

11.11 1. $L\dfrac{di_2}{dt} + Ri_2 = e - M\dfrac{da_1}{dt} + L_3\dfrac{da_2}{dt}$, where $L = L_2 + L_3 + \alpha M$. It is also a state equation.

2. There are four WSVs and two algebraic constraints ($i + i_2 = a_2$ and $a_1 + \alpha i_2 = i_1$); thus there are two SSVs.

3. The two natural frequencies are $\lambda_1 = -\dfrac{R}{L} == -\dfrac{R}{L_2 + L_3 + \alpha M}$ and $\lambda_2 = 0$, due to the presence of an algebraic constraint on the derivatives of the SSV v, namely $C\dfrac{dv}{dt} = 0.$

4. $\lambda_1 < 0$, provided that $L_2 + L_3 + \alpha M > 0$, $\lambda_2 = 0$. Thus the circuit is simply stable, provided that $\alpha > -\dfrac{L_2 + L_3}{M}$; we remark that the state variable i_2 is sensitive only to λ_1, and thus it can reach a steady state.

5. $i_2(t) = i_{DC}(t) + i_{AC}(t)$, with

$$i_{DC}(t) = \dfrac{E}{R},$$

$$i_{AC}(t) = -\frac{\omega M A_1}{R^2 + (\omega L)^2}[R\cos(\omega t) + \omega L \sin(\omega t)].$$

The I/O relationship is of the first order, and therefore we need only one condition
at $t = 0^-$: $i_2(0^-) = \dfrac{E}{R} - \dfrac{\omega M R A_1}{R^2 + (\omega L)^2}$.

6. $\bar{i}_2 = \dfrac{E}{R}$;

$$i_{2eff} = \sqrt{\left(\frac{E}{R}\right)^2 + \frac{1}{2}\frac{(\omega M A_1)^2}{R^2 + (\omega L)^2}}.$$

7. $i_2(0^+) = \dfrac{\Phi_0 - M A_0}{L} + i_2(0^-)$.

8. $i_2(t) = \left[i_2(0^+) - \dfrac{E}{R}\right]e^{\lambda_1 t} + \dfrac{E}{R}$ (see Fig. B.6a).

9. $i_2(t) = K e^{\lambda t} + K^* e^{\lambda^* t} + \dfrac{E}{R}$, with K the solution of the linear system

$$\begin{cases} i_2(0^+) = K + K^* + \dfrac{E}{R} \\ \dfrac{di_2}{dt}\bigg|_{0^+} = K\lambda + K^*\lambda^*. \end{cases}$$

The missing information is the value of the initial condition $\dfrac{di_2}{dt}\bigg|_{0^+}$.

The qualitative solution is shown in Fig. B.6b.

11.12 1. $L\dfrac{di_1}{dt} + r i_1 = e$,   where $L = \dfrac{L_1(L_2 + L_3) - M^2}{L_2 + L_3}$ and
$r = \dfrac{R(L_2 + L_3) - MR}{L_2 + L_3}$. It is also a state equation.

2. There are three WSVs and one algebraic constraint (cut-set involving only L_2 and L_3); thus there are two SSVs.

3. The two natural frequencies are $\lambda_1 = -\dfrac{r}{L} = -R\dfrac{L_2 + L_3 - M}{L_1(L_2 + L_3) - M^2}$ and $\lambda_2 = 0$, due to the presence of an algebraic constraint on the derivatives of the SSVs.

4. $\lambda_2 = 0$; thus the circuit is simply stable, provided that $\lambda_1 < 0$, that is, provided that $L_2 + L_3 - M$ and $L_1(L_2 + L_3) - M^2$ have the same sign.

5. $i_1(0^+) = \dfrac{\Phi_0}{L}$.

6. $i_1(t) = K e^{\lambda_1 t} + i_{DC}(t) + i_{AC}(t)$, where

$$i_{DC}(t) = -\frac{E_0}{r},$$

$$i_{AC}(t) = -\frac{E_1}{r^2 + (\omega L)^2}[r\cos(\omega t) + \omega L \sin(\omega t)],$$

$$K = i_1(0^+) + \frac{E_0}{r} + \frac{r E_1}{r^2 + (\omega L)^2}.$$

Fig. B.6 Qualitative
graphical solutions to
questions 8 (**a**) and 9 (**b**) of
problem 11.11

(a)

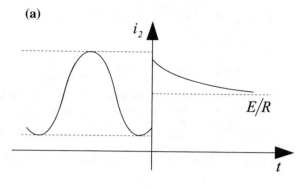

(b)

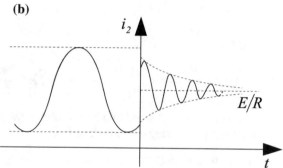

11.13 1. $L\dfrac{di}{dt} + 2Ri = e - 2Ra_2 + 2Ra_1$. It is also a state equation.

2. $LC\dfrac{d^2v}{dt^2} + \left(\dfrac{L}{R} + 2RC\right)\dfrac{dv}{dt} + 2v = e - 2Ra_2 - L\dfrac{da_1}{dt}$.

3. There are two WSVs, which are also SSVs, because there are no algebraic constraints.

4. The two natural frequencies are $\lambda_1 = -\dfrac{2R}{L}$ and $\lambda_2 = -\dfrac{1}{RC}$.

5. $\lambda_{1,2} < 0$; thus the circuit is absolutely stable.

6. $i(t) = i_{DC}(t) + i_{AC1}(t) + i_{AC2}(t)$, with

$$i_{DC}(t) = A_0,$$

$$i_{AC1}(t) = \frac{2RA_1}{(2R)^2 + (\omega L)^2}[2R\cos(\omega t) + \omega L\sin(\omega t)],$$

$$i_{AC2}(t) = \frac{RA_2}{R^2 + (\omega L)^2}[\omega L\cos(2\omega t) - R\sin(2\omega t)],$$

$$i(0^-) = A_0 + \frac{4R^2 A_1}{(2R)^2 + (\omega L)^2} + \frac{\omega L R A_2}{R^2 + (\omega L)^2}.$$

7. $\bar{i} = A_0$;

$$i_{eff} = \sqrt{A_0^2 + \frac{1}{2}\frac{(2RA_1)^2}{(2R)^2+(\omega L)^2} + \frac{1}{2}\frac{(RA_2)^2}{R^2+(\omega L)^2}}.$$

8. $i(0^+) = \dfrac{\Phi_0}{L} + i(0^-).$

9. $i(t) = \left[i(0^+) - A_0 - \dfrac{E}{2R}\right]e^{\lambda_1 t} + A_0 + \dfrac{E}{2R}.$

10. From the discontinuity balance applied to the I/O relationship for $v(t)$, we obtain
    $v(0^+) = v(0^-)$. By integrating the same equation between 0^- and 0^+, we also

    obtain $\left.\dfrac{dv}{dt}\right|_{0^+} = \dfrac{\Phi_0 + LA_1}{LC} + \left.\dfrac{dv}{dt}\right|_{0^-}$. For $t > 0$, $v(t) = K_1 e^{\lambda_1 t} + K_2 e^{\lambda_2 t} + \dfrac{E}{2}$,

    where K_1 and K_2 are solutions of the linear system

    $$\begin{cases} v(0^+) = K_1 + K_2 + \dfrac{E}{2} \\ \left.\dfrac{dv}{dt}\right|_{0^+} = K_1 \lambda_1 + K_2 \lambda_2 \end{cases}.$$

11.14 1. $LC\dfrac{d^2 i}{dt^2} + 2RC\dfrac{di}{dt} + i = -a_2 + 2RC\dfrac{da_1}{dt} + C\dfrac{de}{dt}$. It is also a state equa-
    tion.

2. There are three WSVs and two SSVs, due to the presence of one algebraic con-
    straint (loop involving only $e(t)$ and C).

3. The two natural frequencies are $\lambda_\pm = \dfrac{-2RC \pm \sqrt{(2RC)^2 - 4LC}}{2LC}$.

4. $R^2 C < L$.

5. The circuit is absolutely stable. Indeed, if the natural frequencies are complex con-
    jugate, then $\Re\{\lambda_\pm\} = -\dfrac{R}{L} < 0$. If they are real, they are also negative, because
    $\sqrt{(2RC)^2 - 4LC} < 2RC$.

6. $i(t) = i_{DC}(t) + i_{AC1}(t) + i_{AC2}(t)$, with

    $$i_{DC}(t) = -A_2,$$

    $$i_{AC1}(t) = \frac{2\omega RCA}{(1-\omega^2 LC)^2 + (2\omega RC)^2}[(1 - \omega^2 LC)\cos(\omega t) + 2\omega RC \sin(\omega t)],$$

    $$i_{AC2}(t) = \frac{2\omega C E_1}{(1-4\omega^2 LC)^2 + (4\omega RC)^2}[4\omega RC \cos(2\omega t) - (1 - 4\omega^2 LC)\sin(2\omega t)],$$

    $$i(0^-) = -A_2 + \frac{2\omega RCA(1-\omega^2 LC)}{(1-\omega^2 LC)^2 + (2\omega RC)^2} + \frac{8(\omega C)^2 RE_1}{(1-4\omega^2 LC)^2 + (4\omega RC)^2},$$

    $$\left.\frac{di}{dt}\right|_{0^-} = \frac{(2\omega RC)^2 \omega A}{(1-\omega^2 LC)^2 + (2\omega RC)^2} - \frac{4\omega^2 C(1-4\omega^2 LC)E_1}{(1-4\omega^2 LC)^2 + (4\omega RC)^2}.$$

7. From the discontinuity balance applied to the I/O relationship for $i(t)$, we obtain
    $i(0^+) = i(0^-)$. By integrating the same equation between 0^- and 0^+, we also

    obtain $\left.\dfrac{di}{dt}\right|_{0^+} = \dfrac{2RA_1}{L} + \dfrac{E_0 - E_1}{L} + \left.\dfrac{di}{dt}\right|_{0^-}.$

8. There is only a transient response. By assuming $\lambda_+ = \lambda$ and $\lambda_- = \lambda^*$, we have $i(t) = Ke^{\lambda t} + K^* e^{\lambda^* t}$, where K is a solution of the linear system

$$\begin{cases} i(0^+) = K + K^* \\ \dfrac{di}{dt}\bigg|_{0^+} = K\lambda + K^*\lambda^* \end{cases}.$$

11.15 1. $LC\dfrac{d^2v}{dt^2} + \alpha v = -Ra + 2L\dfrac{da}{dt}$.

2. The two natural frequencies are $\lambda_\pm = \pm\sqrt{-\dfrac{\alpha}{LC}}$.

3. $\alpha > 0$.

4. If $\alpha > 0$, the circuit is simply stable (two purely imaginary natural frequencies); if $\alpha = 0$, the circuit is weakly unstable ($\lambda = 0$ with multiplicity 2); if $\alpha < 0$, the circuit is strongly unstable (two real natural frequencies, one of which is positive).

5. From the discontinuity balance applied to the I/O relationship for $v(t)$, we obtain $v(0^+) = v(0^-)$. By integrating the same equation between 0^- and 0^+, we also obtain $\dfrac{dv}{dt}\bigg|_{0^+} = \dfrac{2A_0}{C} + \dfrac{dv}{dt}\bigg|_{0^-}$.

11.16 1. $L\dfrac{di}{dt} + R_2 i = e(t)$. It is also a state equation.

2. $R_1 LC\dfrac{d^2v}{dt^2} + (L + R_1 R_2 C)\dfrac{dv}{dt} + R_2 v = R_1 e$.

3. The two natural frequencies are $\lambda_1 = -\dfrac{R_2}{L}$ and $\lambda_2 = -\dfrac{1}{R_1 C}$.

4. Circuit absolutely stable (two real negative natural frequencies).

5. $i(t) = \dfrac{E_0}{R_2}$; $i(0^-) = \dfrac{E_0}{R_2}$.

6. $v(t) = \dfrac{R_1}{R_2} E_0$.

7. From the discontinuity balance applied to the I/O relationship for $i(t)$, we obtain $i(0^+) \neq i(0^-)$. By integrating the same equation between 0^- and 0^+, we also obtain $i(0^+) = \dfrac{\Phi_0}{L} + i(0^-)$.

8. $i(t) = Ke^{\lambda_1 t} + i_{AC1}(t) + i_{AC2}(t)$, where

$$i_{AC1}(t) = \dfrac{E_1}{R_2^2 + (\omega L)^2}[R_2 \cos(\omega t) + \omega L \sin(\omega t)],$$

$$i_{AC2}(t) = -\dfrac{E_2}{R_2^2 + (2\omega L)^2}[2\omega L \cos(2\omega t) - R_2 \sin(2\omega t)],$$

$$K = i(0^+) - \dfrac{R_2 E_1}{R_2^2 + (\omega L)^2} + \dfrac{2\omega L E_2}{R_2^2 + (2\omega L)^2}.$$

9. $\bar{p} = -\dfrac{1}{2}\dfrac{R_1 E^2}{[1 + (\omega R_1 C)^2][R_2^2 + (\omega L)^2]}$.

11.17 1. $RLC\dfrac{d^2 i}{dt^2} - RrC\dfrac{di}{dt} = -ra - RrC\dfrac{da}{dt} + (R+r)C\dfrac{de}{dt}$. It is also a state

equation.

2. The two natural frequencies are $\lambda_1 = 0$ and $\lambda_2 = \dfrac{r}{L}$.

3. Circuit simply stable if $r < 0$; weakly unstable if $r = 0$; strongly unstable if $r > 0$. Therefore, the stability condition is $r < 0$.

4. $i(t) = \dfrac{1}{\omega RC[(\omega L)^2 + r^2]}[-(r^2 A + \omega Lb)\cos(\omega t) + r(\omega LA - b)\sin(\omega t)]$,

where $b = \omega C\,[(R+r)E - RrA]$.

11.18 1. $2RC\dfrac{dv}{dt} + v = Ra(t)$. It is also a state equation.

2. $2RLC\dfrac{d^2 i}{dt^2} + L\dfrac{di}{dt} = -R^2 C\dfrac{da}{dt}$.

3. The two natural frequencies are $\lambda_1 = -\dfrac{1}{2RC}$ and $\lambda_2 = 0$.

4. Circuit simply stable.

5. $v(t) = -\dfrac{RA_1}{1 + (2\omega RC)^2}[2\omega RC\cos(\omega t) - \sin(\omega t)]$; $v(0^-) = -\dfrac{2\omega R^2 CA_1}{1 + (2\omega RC)^2}$.

6. $i(t) = \dfrac{R^2 CA_1}{L[1 + (2\omega RC)^2]}[2\omega RC\cos(\omega t) - \sin(\omega t)]$; $i(0^-) = \dfrac{2\omega RA_1(RC)^2}{L[1 + (2\omega RC)^2]}$.

7. From the discontinuity balance applied to the I/O relationship for $v(t)$, we obtain $v(0^+) = v(0^-)$.

From the discontinuity balance applied to the I/O relationship for $i(t)$, we obtain $i(0^+) = i(0^-)$ and $\left.\dfrac{di}{dt}\right|_{0^+} \neq \left.\dfrac{di}{dt}\right|_{0^-}$. Since we do not have $\left.\dfrac{di}{dt}\right|_{0^-}$, instead of integrating the I/O relationship for $i(t)$ between 0^- and 0^+, we can use the state equation $2L\dfrac{di}{dt} = v - Ra$, thus obtaining $\left.\dfrac{di}{dt}\right|_{0^+} = -\dfrac{RA_0}{2L} + \dfrac{v(0^+)}{2L}$.

8. $v(t) = Ke^{\lambda_1 t} + RA_0$, where $K = v(0^+) - RA_0$.

9. $i(t)$ is sensitive to both natural frequencies, one of which is null, and therefore we cannot separate free response and forced response. The solution is $i(t) = K_1 e^{\lambda_1 t} + K_2$, where K_1 and K_2 are solutions of the linear system

$$\begin{cases} i(0^+) = K_1 + K_2 \\ \left.\dfrac{di}{dt}\right|_{0^+} = K_1 \lambda_1 \end{cases}.$$

11.19 1. The two natural frequencies are $\lambda_\pm = \dfrac{-RC \pm \sqrt{(RC)^2 - 3LC}}{LC}$. Under the usual assumptions $(R, L, C > 0)$, the circuit is absolutely stable for $R^2 C \neq 3L$.

2. $\bar{i}_\infty = \dfrac{E_0 + 2RA_0}{3R}$.

3. $\bar{p}_1 = Ri_{\infty eff}^2$, with $i_{\infty eff}^2 = \left(\dfrac{E_0 + 2RA_0}{3R}\right)^2 + \dfrac{1}{2}\left[\left(\dfrac{E_1}{3R}\right)^2 + \left(\dfrac{2A_1}{3}\right)^2\right]$.

11.20 1. $L_1 C \dfrac{d^2 v}{dt^2} + \tau \dfrac{dv}{dt} + v = \alpha\tau\dfrac{de}{dt} - M\dfrac{da}{dt}$, where $\tau = \dfrac{L_1}{R}$. It is also a state equation.

2. The two natural frequencies are $\lambda_\pm = \dfrac{-L_1 \pm \sqrt{L_1^2 - 4L_1 R^2 C}}{2L_1 RC}$.

3. Circuit absolutely stable (under the usual assumptions: $R, L, C > 0$).

4. $v(t) = \dfrac{\alpha\omega\tau E_1}{\beta^2 + (\omega\tau)^2}[\beta\cos(\omega t) + \omega\tau\sin(\omega t)]$, where $\beta = 1 - \omega^2 L_1 C$;

$v(0^-) = \dfrac{\alpha\beta\omega\tau E_1}{\beta^2 + (\omega\tau)^2}$;

$\left.\dfrac{dv}{dt}\right|_{0^-} = \dfrac{\alpha\omega^3\tau^2 E_1}{\beta^2 + (\omega\tau)^2}$.

5. From the discontinuity balance applied to the I/O relationship for $v(t)$, we obtain $v(0^+) = v(0^-)$ and $\left.\dfrac{dv}{dt}\right|_{0^+} \neq \left.\dfrac{dv}{dt}\right|_{0^-}$. By integrating the same equation between 0^- and 0^+, we also obtain $\left.\dfrac{dv}{dt}\right|_{0^+} = -\dfrac{\alpha\tau E_0}{L_1 C} - \dfrac{M A_0}{L_1 C} + \left.\dfrac{dv}{dt}\right|_{0^-}$.

6. $v(t) = K_+ e^{\lambda_+ t} + K_- e^{\lambda_- t}$, where K_+ and K_- are solutions of the linear system

$$\begin{cases} v(0^+) = K_+ + K_- \\ \left.\dfrac{dv}{dt}\right|_{0^+} = K_+\lambda_+ + K_-\lambda_- \end{cases}.$$

11.21 1. $T^2 \dfrac{d^2 v_2}{dt^2} + \tau\dfrac{dv_2}{dt} + \beta v_2 = ra(t)$, where $T^2 = n^2 LC_2$, $\tau = RC_2$, $\beta = n(\alpha + n)$, $r = (1 - n)R$.

2. The three natural frequencies are $\lambda_\pm = \dfrac{-\tau \pm \sqrt{\tau^2 - 4\beta T^2}}{2T^2}$ (from the I/O relationship at point 1) and $\lambda_3 = 0$ (due to the presence of a cut-set involving only the independent current source and C_1).

3. The natural frequencies $\lambda_\pm$ are complex conjugates, provided that $\tau^2 < 4\beta T^2$, that is (in terms of the original circuit parameters), $R^2 C_2 < 4n^3(\alpha + n)L$. If this condition is satisfied, the circuit is simply stable (under the usual assumptions $R, L, C > 0$), since $\Re\{\lambda_\pm\} = \dfrac{-\tau}{2T^2} < 0$.

4. $v_2(t) = \dfrac{rA}{\beta} = v_2(0^-)$; $\left.\dfrac{dv_2}{dt}\right|_{0^-} = 0$.

5. From the discontinuity balance applied to the I/O relationship for $v(t)$, we obtain $v_2(0^+) = v_2(0^-)$ and $\left.\dfrac{dv_2}{dt}\right|_{0^+} \neq \left.\dfrac{dv_2}{dt}\right|_{0^-}$. By integrating the same equation between 0^- and 0^+, we also obtain $\left.\dfrac{dv_2}{dt}\right|_{0^+} = \dfrac{rQ_0}{T^2}$.

6. $v_2(t) = K_+ e^{\lambda_+ t} + K_- e^{\lambda_- t}$, where K_+ and K_- are solutions of the linear system

$$\begin{cases} v_2(0^+) = K_+ + K_-, \\ \dfrac{dv_2}{dt}\bigg|_{0^+} = K_+\lambda_+ + K_-\lambda_-. \end{cases}$$

11.22 1. $\bar{i} = -\dfrac{A}{4}, \bar{v} = \dfrac{5}{8}RA.$

2. Notice that under the prescribed assumptions, the circuit contains (in addition to the independent source) only passive elements, and some of them are dissipative. You can check that it is absolutely stable. The required mean power is

$$\bar{p} = R i_{Reff}^2 = R\left(I_0^2 + \frac{|\dot{I}_1|^2}{2}\right), \text{ with } I_0 = 0 \text{ and}$$

$$|\dot{I}_1|^2 = \frac{(\omega L A_1)^2}{R^2(1 - 2\omega^2 LC)^2 + \omega^2(L + R^2C)^2}.$$

11.23 Notice that the circuit contains (in addition to the independent current sources) only passive elements, and some of them are dissipative. You can check that it is absolutely stable.

1. $\bar{v}_a = \dfrac{15}{8}RA, \bar{i}_e = \dfrac{E}{8R}.$

2. $i_{1eff} = \sqrt{\bar{i}_1^2 + \dfrac{|\dot{I}_1|^2}{2}}$, with $\bar{i}_1 = A$ and $|\dot{I}_1|^2 = \dfrac{(\omega C M E_1)^2}{L_2^2}$. Recall that the closely coupled inductors assumption implies $L_1 L_2 = M^2$.

11.24 1. $\bar{i} = \dfrac{1}{3 + 2Rg}\left[A_0 + \dfrac{E_0}{R}(1 + Rg)\right], \bar{v} = \dfrac{E_0 - 2RA_0}{3 + 2Rg}.$

2. $\bar{p} = 2R i_{eff}^2 = 2R\left(\bar{i}^2 + \dfrac{|\dot{I}_1|^2}{2} + \dfrac{|\dot{I}_2|^2}{2}\right)$, with

$\bar{i}$ computed at point 1, but with $g = 0$ (i.e., $\bar{i} = \dfrac{A_0}{3} + \dfrac{E_0}{3R}$)

$$|\dot{I}_1|^2 = \left(\frac{E_1}{R}\right)^2 \frac{1}{9 + 4(\omega RC)^2} \text{ and}$$

$$|\dot{I}_2|^2 = \frac{A_1^2}{9 + 16(\omega RC)^2}.$$

11.25 1. $\bar{v}_a = \dfrac{3}{8}\dfrac{R}{R + R_2}E.$

2. $\bar{p} = R_2 i_{eff}^2 = R_2\left(\bar{i}^2 + \dfrac{|\dot{I}_1|^2}{2} + \dfrac{|\dot{I}_2|^2}{2}\right), \quad \text{with} \quad \bar{i} = -\dfrac{E_0}{R + R_2}, \quad |\dot{I}_1|^2 =$

$\dfrac{E_1^2}{(R + R_2)^2 + (\omega L_2)^2}$ and $|\dot{I}_2|^2 = 0.$

11.26 1. We have to find the absolute stability condition. The circuit's natural frequency is $\lambda = \dfrac{R - R_1}{R_1 R_2 C}$; therefore $\lambda < 0$ if and only if $R < R_1$. Notice that R is the gain of a CCVS, thus in general, there are no restrictions on its sign.

2. $\bar{v} = \dfrac{E_1}{6} - \dfrac{E_2}{3}$.

11.27 The circuits of Fig. 11.77a, c are absolutely stable. For instance, the state equations for the circuit of Fig. 11.77a are

$$
\begin{cases}
\dfrac{di}{dt} = \dfrac{v}{L} \\
\dfrac{dv}{dt} = -\dfrac{v}{RC} - \dfrac{i}{C} + \dfrac{a}{C}
\end{cases}
\Rightarrow
\frac{d}{dt}\begin{pmatrix} i \\ v \end{pmatrix} =
\begin{pmatrix} 0 & \dfrac{1}{L} \\ -\dfrac{1}{C} & -\dfrac{1}{RC} \end{pmatrix}
\begin{pmatrix} i \\ v \end{pmatrix} + \begin{pmatrix} 0 \\ a \end{pmatrix}.
$$

The eigenvalues λ_1, λ_2 of the state matrix are the solutions of the characteristic equation

$$
\lambda^2 + \frac{\lambda}{RC} + \frac{1}{LC} = 0,
$$

and we have $\Re\{\lambda_j\} < 0$ $(j = 1, 2)$.

A completely analogous result holds for the circuit of Fig. 11.77c. Therefore, for instance, a sinusoidal input causes, for $t \to \infty$, a sinusoidal steady state in both circuits.

Consider now the circuit of Fig. 11.77b. In this case, since $v = e$, the only SSV is the inductor current i. The state equation and the corresponding eigenvalue λ are

$$
\frac{di}{dt} = \frac{e}{L}; \quad \lambda = 0.
$$

Therefore, this circuit is simply stable. For instance, when $e(t) = E \cos(\omega t) \cdot u(t)$ and for $i(0) = I_0 \neq 0$, we immediately have, for all $t \geq 0$,

$$
i(t) = I_0 + \int_0^t E \cos(\omega \tau)\, d\tau = I_0 + \frac{E}{\omega} \sin(\omega t).
$$

Inasmuch as $i(t)$ contains the constant term I_0, this is not a sinusoidal steady state. Analogous considerations hold for the circuit of Fig. 11.77d, where the only SSV is the capacitor voltage v, and we still have a single eigenvalue $\lambda = 0$ (simply stable circuit). The remaining part of the discussion is identical, mutatis mutandis, to that developed in the previous case.

Problems of Chap. 13

Solutions of problems of Chap. 13.

13.1 1. $Z(j\omega) = \dfrac{\dot{V}}{\dot{I}} = R + \dfrac{10\alpha R_0}{1 + j\omega 10 R_0 C}$.

2. By setting $R_0 = R/10$ we obtain

$$Z(j\omega) = R(\omega) + jX(\omega); \quad R(\omega) = R + \frac{\alpha R}{1 + (\omega RC)^2}; \quad X(\omega) = \frac{-\alpha R^2 \omega C}{1 + (\omega RC)^2}.$$

For $\alpha > -1$, we have $R(\omega) > 0$ for all ω. In particular:

- for $-1 < \alpha < 0$, $X(\omega) > 0$ and $Z(j\omega)$ is resistive–inductive;
- for $\alpha > 0$, $Z(j\omega)$ is resistive–capacitive;
- for $\alpha = 0$, we have $X(\omega) = 0$ (purely resistive impedance).

13.2 1. $Y(j\omega) = \dfrac{1}{R_1 + R_2} + j\omega \left(C_1 + \dfrac{R_2 C_2}{R_1 + R_2} \right)$.

2. $P = \dfrac{1}{2} \dfrac{R_1 E^2}{(R_1 + R_2)^2}; \quad Q = -\dfrac{E^2}{2} \dfrac{\omega R_1 R_2 C_2}{(R_1 + R_2)^2}$.

3. Inasmuch as the composite two-terminal is resistive–capacitive, the element to be connected in series is an inductor L with impedance $j\omega L$ such that

$$\Im \left\{ j\omega L + \frac{1}{Y(j\omega)} \right\} = 0.$$

The result is $L = (R_1 + R_2) \dfrac{(R_1 + R_2) C_1 + R_2 C_2}{1 + \omega^2 \left((R_1 + R_2) C_1 + R_2 C_2\right)^2}$.

13.3 1. $Z(j\omega) = \dfrac{j2\omega RL\,(1 - \alpha)}{(3 - 2\alpha) R + j2\omega L}$.

2. We have

$$\begin{cases} \Re\{Z(j\omega)\} > 0 \\[2mm] \Im\{Z(j\omega)\} > 0 \end{cases}$$

for $\alpha < 1$. Therefore, $Z(j\omega)$ is resistive–inductive for $\alpha < 1$.

3. For $\alpha = 0$, the admittance of the composite two-terminal is $Y = \dfrac{1}{R} - j\dfrac{3}{2\omega L}$.

Therefore, the condition $\Im\{j\omega C + Y\} = 0$ is satisfied for $C = \dfrac{3}{2\omega^2 L}$.

13.4 1. $Z(j\omega) = R\dfrac{R\left(1 - \omega^2 LC\right) + j\omega L}{R\left(2 - \omega^2 LC\right) + j\omega\left(L + R^2 C\right)} \doteq R\dfrac{a + jb}{c + jd}$

2. Inasmuch as $Z(j\omega) = R\dfrac{(a + jb)(c - jd)}{c^2 + d^2} = R\dfrac{(ac + bd) + j\,(bc - ad)}{c^2 + d^2}$, the

condition $\Im\{Z(j\omega)\} = 0$ is met for $(bc - ad) = 0$, which implies $L = \dfrac{R^2 C}{1 + (\omega RC)^2}$.

13.5 1. The impedance matrix is

$$[Z(j\omega)] = \begin{bmatrix} R & Rg\dfrac{R+j\omega L}{1+g\,(R+j\omega L)} \\[4mm] 0 & \dfrac{R+j\omega L}{1+g\,(R+j\omega L)} \end{bmatrix}.$$

2. $Z_{TH} = R$; $\dot{E}_{TH} = Rg\dot{E}$.

13.6 $\dot{E}_{TH} = \dfrac{j\omega\,(M-L_2)\,(R+j\omega L)}{R+j\omega\,(L+L_2)}\dot{A}$; $\quad Z_{TH}(j\omega) = \dfrac{j\omega L_2\,(R+j\omega L)}{R+j\omega\,(L+L_2)}$.

13.7 $\dot{A}_{NR} = -\dfrac{j\omega Cn\dot{E}}{(n\omega)^2\,LC - 1 + j\omega nLg}$; $\quad Z_{NR}(j\omega) = \dfrac{(n\omega)^2\,LC - 1 + j\omega nLg}{n\,(g - j\omega nC)}$.

13.8 $\dot{A}_{NR} = \dfrac{j\dot{E}}{\omega L}$; $\quad Z_{NR}(j\omega) = \dfrac{j\omega L}{1 - \omega^2 LC}$.

13.9 $\dot{A}_{NR} = \dot{A}\dfrac{n + Rg}{n^2 + j\omega RC}$; $\quad Z_{NR}(j\omega) = -\dfrac{n^2 + j\omega RC}{ng + R\,(\omega C)^2 - j\omega C\,(n^2 + 1)}$.

13.10 1. $[T(j\omega)] = \begin{bmatrix} 1 & R \\[3mm] j\omega C & 1 + j\omega RC \end{bmatrix}$.

2. $\dot{E}_{TH} = -j\dfrac{\dot{A}}{\omega C}$; $\quad Z_{TH}(j\omega) = R + \dfrac{1}{j\omega C}$.

3. $P = \dfrac{R}{2}\dfrac{A^2 + (\omega CE)^2}{1 + (\omega RC)^2}$; $\quad Q = -\dfrac{\omega C}{2}\dfrac{(RA - E)^2}{1 + (\omega RC)^2}$.

Notice that P is the active power absorbed by the resistor R inside the two-port, $P = \dfrac{1}{2}R\,|\dot{i}_2|^2$, whereas Q is the reactive power absorbed by the capacitor: $Q = -\dfrac{1}{2}\omega C\,|\dot{V}_1|^2$.

13.11 1. $[T(j\omega)] = \begin{bmatrix} 2 + \dfrac{j\omega L}{R} & R + j\omega L \\[4mm] \dfrac{1}{R} & 1 \end{bmatrix}$.

2. $\dot{E}_{TH} = -RA$; $\quad Z_{TH}(j\omega) = 2R + j\omega L$; $\quad Z_{NR} = Z_{TH}$; $\quad \dot{A}_{NR} = -\dfrac{RA}{2R + j\omega L}$.

3. $[Z(j\omega)] = \begin{bmatrix} n^2\,(2R + j\omega L) & nR \\[3mm] nR & R \end{bmatrix}$. Notice that $[Z(j\omega)]$ is symmetric, and therefore the two-port is reciprocal.

13.12 1. $[T_A(j\omega)] = \begin{bmatrix} -\dfrac{1}{n} & 0 \\ 0 & -n \end{bmatrix}$.

2. $[T_B(j\omega)] = \begin{bmatrix} -\dfrac{L_1}{M} & 0 \\ -\dfrac{1}{j\omega M} & -\dfrac{L_2}{M} \end{bmatrix}$.

3. $[T(j\omega)] = T_A \cdot T_B = \begin{bmatrix} \dfrac{L_1}{nM} & 0 \\ \dfrac{n}{j\omega M} & \dfrac{nL_2}{M} \end{bmatrix}$.

4. $P + jQ = \dfrac{\dot{V_1}\dot{I_1^*}}{2} + \dfrac{\dot{V_2}\dot{I_2^*}}{2} = \dfrac{\dot{V_1}jA}{2} + \dfrac{E\dot{I_2^*}}{2}$. Therefore:

$$\begin{cases} P = 0 \\ Q = \dfrac{L_1 E A}{2nM} + \dfrac{E^2}{2\omega L_2} - \dfrac{MAE}{2nL_2} \end{cases}$$

13.13 $[T(j\omega)] = \begin{bmatrix} \dfrac{1 + j\omega RC}{1 - j\omega rC} & R \\ \dfrac{j\omega C}{1 - j\omega rC} & 1 \end{bmatrix}$.

13.14 1. Denoting by $\dot{E}$ the phasor representing the sinusoidal source $e(t)$, we have

$$[Y(j\omega)] = \begin{bmatrix} \dfrac{1}{R_1} + \dfrac{1-\beta}{R_2} - j\omega C\alpha\beta & j\omega C\alpha \\ -j\omega C\beta & j\omega C \end{bmatrix} ; \quad \dot{A_1} = \dfrac{\dot{E}}{R_2}; \quad \dot{A_2} = 0.$$

2. The reciprocity condition $Y_{12} = Y_{21}$ holds, provided that $\alpha = -\beta$.

13.15

$$[Z(j\omega)] = \begin{bmatrix} 0 & 0 \\ -2R & 0 \end{bmatrix}; \quad \dot{E_1} = \dot{E}\left(\dfrac{1}{\omega^2 LC} - 1\right); \quad \dot{E_2} = \dot{E}\left(1 - \dfrac{1}{\omega^2 LC} + \dfrac{2R}{j\omega L}\right).$$

13.16

$$[Z(j\omega)] = \begin{bmatrix} \dfrac{1}{j\omega C} & r + \dfrac{1}{j\omega C} \\[3mm] \dfrac{1}{j\omega C} & r + j\omega L + R + \dfrac{1}{j\omega C} \end{bmatrix}; \quad \dot{E}_1 = \dfrac{\dot{A}}{j\omega C}; \quad \dot{E}_2 = j\dot{A}\dfrac{1 - \omega^2 LC}{\omega C}$$

13.17 The complex power delivered by the two-port is

$$P + jQ = -\left(\frac{\dot{V}\dot{A}^*}{2} + \frac{\dot{E}\dot{I}^*}{2}\right) = -\frac{5}{6}\dot{E}\dot{A}^* - \frac{1}{6}gE^2$$

Now, setting $\dot{E} = E$ and $\dot{A} = -jA$, we obtain

$$\begin{cases} P = -\dfrac{gE^2}{6}, \\[3mm] Q = -\dfrac{5}{6}EA. \end{cases}$$

13.18 We first obtain $\dot{I}_\infty = \dfrac{E - j2RA}{3R}$ and $\dot{V}_\infty = -j\omega L\dot{I}_\infty$. Therefore, the complex power absorbed by the nullor is $P + jQ = -\dfrac{j\omega L |\dot{I}_\infty|^2}{2}$, and

$$\begin{cases} P = 0, \\[3mm] Q = -\dfrac{\omega L}{2}\left(\left(\dfrac{E}{3R}\right)^2 + \left(\dfrac{2A}{3}\right)^2\right). \end{cases}$$

13.19 1. $Z(j\omega) = \alpha R + j\omega L + \dfrac{R}{1 + j\omega RC}$.

2. $L = \dfrac{R^2 C}{1 + (\omega RC)^2}$.

3. $Z_h = Z^* = R\left(\alpha + \dfrac{1}{1 + (\omega RC)^2}\right) - j\omega\left(L - \dfrac{R^2 C}{1 + (\omega RC)^2}\right)$.

13.20 As a first step, we find $Z_2 = R\left(1 - j\dfrac{2R}{\omega L}\right)$. Therefore, we have $Z_1 = Z_2^* = R\left(1 + j\dfrac{2R}{\omega L}\right)$.

13.21 The circuit equations are

$$\begin{cases} \dfrac{\dot{E} - \dot{V_1}}{R_1} = j\omega C_2 \left(\dot{V_1} - \dot{V} \right) + j\omega C_1 \dot{V_1} \\[4mm] j\omega C_1 \dot{V_1} = -\dfrac{\dot{V}}{R_2} \end{cases}$$

From these equations we obtain, after few manipulations,

$$H_2(j\omega) = \frac{\dot{V}}{\dot{E}} = -\frac{1}{\dfrac{R_1 (C_1 + C_2)}{R_2 C_1} + j\omega R_1 C_2 - j\dfrac{1}{\omega R_2 C_1}}.$$

$H_2(j\omega)$ is a purely real value at the angular frequency ω_0 such that

$$\omega_0 R_1 C_2 = \frac{1}{\omega_0 R_2 C_1}, \quad \text{that is,} \quad \omega_0 = \frac{1}{\sqrt{R_1 R_2 C_1 C_2}}.$$

$H_2(j\omega)$ can be recast as follows:

$$H_2(j\omega) = -\frac{\dfrac{R_2 C_1}{R_1 (C_1 + C_2)}}{1 + j \left(\dfrac{\omega R_1 C_2 R_2 C_1}{R_1 (C_1 + C_2)} - \dfrac{1}{\omega R_2 C_1} \cdot \dfrac{R_2 C_1}{R_1 (C_1 + C_2)} \right)} =$$

$$= -\frac{\dfrac{R_2 C_1}{R_1 (C_1 + C_2)}}{1 + j\dfrac{1}{(C_1 + C_2)} \left(\underbrace{\dfrac{\omega}{R_1 \omega_0^2} - \dfrac{\omega_0^2 R_2 C_1 C_2}{\omega}}_{= \dfrac{1}{\omega R_1}} \right)} =$$

$$= -\frac{\dfrac{R_2 C_1}{R_1 (C_1 + C_2)}}{1 + j\underbrace{\dfrac{1}{\omega_0 R_1 (C_1 + C_2)}}_{Q} \left(\dfrac{\omega}{\omega_0} - \dfrac{\omega_0}{\omega} \right)} = -\frac{\dfrac{R_2 C_1}{R_1 (C_1 + C_2)}}{1 + jQ \left(\dfrac{\omega}{\omega_0} - \dfrac{\omega_0}{\omega} \right)}.$$

Summing up, by defining

$$\omega_0 = \frac{1}{\sqrt{R_1 R_2 C_1 C_2}}; \quad Q = \frac{1}{\omega_0 R_1 (C_1 + C_2)} = \frac{\omega_0 R_2 C_1 C_2}{(C_1 + C_2)},$$

$H_2(j\omega)$ assumes the same structure as $H_1(j\omega)$, apart from the constant coefficient $-\dfrac{R_2 C_1}{R_1 (C_1 + C_2)} \dfrac{1}{R}$, as expected.

Index

A
Active power, 348
Admittance, 328
    resistive-capacitive, 330
    resistive-inductive, 330
Apparent power, 351
Arnold tongue, 428
Attractor, 414
Autonomous, 93

B
Band-pass filter, 387
Basin of attraction, 98, 414
Bifurcation, 121, 283
    flip, 423
    fold, 283
    fold of cycles, 421
    fold of equilibria, 283
    global, 421
    homoclinic, 434
    Hopf, 284
    local, 420
    Neimark–Sacker, 427
Boucherot's theorem, 356

C
Canonical form
    first order, 71
    first-order, linear, 20
    first-order, nonlinear, 93
    higher-order, linear, 163
    higher-order, nonlinear, 274

Capacitance, 4
Complementary component, 17, 161
Complex power, 350
Conductance, 329
Control space, 124
Convolution integral, 52
Coupling coefficient
    coupled inductors, 149
Cycle, 402

D
Descriptive equations
    capacitor, 4
    coupled inductors, 143
    inductor, 6
Dirac δ-function, 42
Discontinuity balance, 44
Dissipative system, 302

E
Energetic behavior
    conservative, 7, 150
Equilibrium
    center, 270
    condition, 94, 276
    point, 94, 275
    saddle, 268
    stable focus, 269
    stable node, 267
    unstable focus, 270
    unstable node, 267

© Springer Nature Switzerland AG 2020
M. Parodi and M. Storace, *Linear and Nonlinear Circuits: Basic and Advanced Concepts*,
Lecture Notes in Electrical Engineering 620,
https://doi.org/10.1007/978-3-030-35044-4